Fundamentals of Limnology and Aquaculture Biotechnology

Fundamentals of Limnology and Aquaculture Biotechnology

Devashish Kar

2024

Daya Publications House®

A Division of

Astral International Pvt. Ltd.

New Delhi – 110 002

First Published,2007

Reprinted, 2024

ISBN: 978-81-7035-455-0 (HB)

Published by : **Daya Publishing House®**
A Division of
Astral International Pvt. Ltd.
– ISO 9001:2015 Certified Company –
4736/23, Ansari Road, Darya Ganj
New Delhi-110 002
Ph. 011-43549197, 23278134
E-mail: info@astralint.com
Website: www.astralint.com

Foreword

It has been a pleasure for me to go through the book entitled, '*Fundamentals of Limnology and Aquaculture Biotechnology*' authored by Dr. Devashish Kar, Professor and Head of the Department of Life Science, Assam (Central) University, Silchar. Ecology, which essentially includes Limnology, is still in its infancy in India. Before independence, Limnology or Freshwater Biology, as a discipline of study and research, was not much appreciated. But for the last quarter century or so, Limnology has emerged as an important discourse in many educational institutions. Limnological research is also going on in many universities and laboratories.

However, in recent years, many books on Limnology are being published from the Western countries which unfold various aspects of the said discipline. But in most cases, Limnology of tropical countries did not get due attention. In India, few books on Limnology are written following the shadow of already published books. But I am indeed very glad to find that Professor Kar has utilized his own findings in his book which is indeed a unique feature. Professor Kar has got considerable research experience in the field of Limnology and Fishery of North-East India, specially Assam, Tripura and Mizoram. Besides Limnological information, he has also included some findings on fishery, aquaculture, fish disease, fish genetics, fish biotechnology, etc.

I am confident that this book, both in content and concept, will satisfy the needs of the students, researchers, environmentalists, planners and policy makers, fishery departments and others who are interested in fishery in general and Limnology in particular.

Date: 16.1.2006

N.C. Datta
Former Professor & Head of the Department of Zoology
Former Coordinator, M. Phil in Environmental Science
University of Calcutta

Preface

'Water' is life and must be protected and conserved. Having originated in water, life has evolved itself into an enchanting world of rich and vastly diverse flora and fauna. The dependence of man on the biological wealth of rivers, lakes and seas and oceans can not be overemphasized. As such, there is need to depend our understanding of the aquatic environment with regard to its physical, chemical and biological characteristics and interactions. The aquatic ecosystems perform many important environmental functions, *e.g.*, re-cycling of nutrients, re-charge of ground water, augment and maintain stream flow and recreation of people to name a few. Unplanned developmental works coupled with rapid increase of population have led to the decline of water bodies, both lentic and lotic. Increased demand for drainage of wetlands have been accommodated by channelisation, resulting in further loss of stream habitat, which has been leading to aquatic biota becoming extinct or imperiled in increasing number and to the impairment of many beneficial uses of water, including drinking, swimming and fishing. Various anthropogenic activities have altered the physical, chemical and biological processes within aquatic ecosystems. Failure to restore these ecosystems will result in sharply increased environmental costs later; in the extinction of species or ecosystem types and in permanent ecological damage.

Fishes constitute almost half the total number of vertebrates. Of the *c* 39,900 vertebrate species known to exist so far, 21,723 are living species of fishes; of which, *c* 841 are of freshwater and the rest, 11,650 are marine species. In the Indian region, alone, of the *c* 2500 species, 930 are freshwater inhabitants and 1570 are marine.

Most of the present-day fish fauna is said to have descended from the Pleistocene migrants. The subject assumes further importance and interest in view of fact that, some of the major river systems in North-East (NE) India flow between terrains of India, Myanmar and Bangladesh; indicating, thereby, that, fishes of NE India contains both Myanmarese and Bangladeshi elements. And, further, it is quite probable that the fish fauna of N.E.India serves as an important link between those of Far-East Asia and the Indian Sub-continent.

The N.E. Region of India is considered as one of the Hotspots of Freshwater Fish Biodiversity in the World. However, there have been drastic reduction in the abundance of freshwater fishes in the region due to many reasons; notably, destruction of habitats, unsustained over-exploitation, dreadful fish diseases like the Epizootic Ulcerative Syndrome (EUS) and other anthropogenic causes. Still further, 'fish' is one of the principal staple food items and rich source of animal protein for the people of fish-eating regions.

Lack of a comprehensive treatise on the Limnology, Fishery Science and Aquaculture Biotechnology has prompted this humble piece of work and the author will consider himself to be amply rewarded if this humble piece of work proves to be useful for those it is meant.

Devashish Kar

Professor and Head, Department of Life Science,
School of Life Sciences, Assam (Central) University
Silchar – 788 011, Assam (India)

Contents

Chapter 1

Basic Concepts of Limnology

'Limnology', basically, is the study of different aspects of water which indeed is part of the broad domain of Ecology and Environment. As such, the concept of limnology would be instilled into the young mind through a series of steps.

Ecology and Environment

An organism is best suited to its favourable environment. Understanding of aquatic life requires knowledge of the external influences which directly or indirectly affect the organisms. Hence, that sub-division of Biology which has, as its special discourse, the study of interrelationships between the organism and their environment, is called 'Ecology', which continues unabatedly to enjoy universal adoption of accepted designation (Welch, 1935). Although the terms Hexiology, Natural history, etc., are sometimes considered synonymous to ecology, Shelford (1929) considered 'ecology' as the science of communities having, as its central principle, the interrelationships which exist between organisms and their environment.

Concomitant to above, the term 'Environment' denotes 'space' on, below or above the surface of the earth. These are classified as: (a) Lithosphere (the land mass) and (b) the Hydrosphere (the water mass), both standing and running, as well as, the subterranean waters. Upon the latter is based the discipline called 'Hydrography', which, in turn, includes 'Oceanography' (concerned with salt water) and 'Limnology' (deals with freshwater).

Although there are many important differences between salt and freshwaters in addition to mere extent and mass, oceanography and limnology have many things in common; and, advances in either field often figure prominently in the progress of the other.

Limnology

From the dawn of human civilization, its establishment and development have generally been in close proximity to water, *viz.*, the lakes/wetlands and the rivers. In addition to using freshwater for drinking, agriculture and navigation, man has been using freshwater for various other purposes like energy production, industrial growth and waste and effluent disposal. Further, certain other factors like contaminated run-off from expanding urban and agricultural areas, air-borne pollutants, etc., have been degrading the surface water. As such, knowledge of the interaction of the water body with its watershed is essential in determining which of these factors has the maximum influence on the quality of the water bodies with regard to physical, chemical and biological characteristics. Limnology is one of such critical sciences which plays a key role in understanding the various kinds of interactions in aquatic organisms.

Definitions of Limnology

'Limnology' has been defined by various authors in different ways, although the term 'limnology' was used historically as that branch of science which deals exclusively with 'Lakes'. However, since the days of Forel, there have grown two different usages, both of which are extant today, namely,

(a) Restricted Usage

Here, limnology refers to the study of various units of lentic or standing water series, *i.e.*, lake/wetland → pond → swamp.

(b) Extended Usage

In this context, limnology is sometimes used to include all forms of inland waters, *viz.*, lentic and lotic (running water series).

In other words, limnology attempts to embody the study of all phenomena in freshwater. Nevertheless, it would seem better to permit limnology to cover all forms of inland waters; thereby, including the inland salt waters as well and any other inland water whose chemical content might seem to exclude them from the usual concept of freshwater.

Welch (1935) conceived the problem of 'Biological productivity' as the central theme of Limnology. He had defined limnology as that branch of science which deals with all causal influences which determine it. This definition, it may be noted here that, does not preclude any of the necessary stress on the different animal and plant communities because they are the direct results of the 'Biological productivity' of waters which they occupy.

According to Schwoerbel (1987), limnology is the science of inland waters viewed as ecosystems together with their structure, materials and energy balances. Kiihnelt (1960) considered limnology as a sub-set of ecology along with 'Oceanography' (which is concerned with marine ecosystem) and 'Epheirology' (which deals with terrestrial habitats).

In short, limnology is the study of all aquatic systems, both fresh and saline (including lakes,wetlands, marshes, bogs, ponds, reservoirs, streams, rivers, oceans, etc.) with regard to their physical, chemical and biological characteristics.

Other Terminologies

In addition to the above, certain other terms, like Hydrobiology, Freshwater biology, Aquatic biology, Aquatic ecology, etc., are sometimes used as synonymous to the word 'Limnology'. But, most of these terms are names under which a diverse variety of subject matter is included; and, only a part of it is strictly limnological in nature. Hence, such terms, if they are to be used at all, be used exclusively in connection with faunistic/floristic studies. Further, the term'Rheology' is sometimes used to denote the running waters.

Essential Nature and Objectives of Limnological Study

Limnology, primarily, is ecological in its bearing. It is essentially a synthetic science composed of elements, some of which extend well beyond the limits of biology. Limnology relies on proper application and integration of certain principles and methodologies of physics, chemistry, geology, geography, meteorology, hydrography, etc. for the solution of problems, which are, ultimately, biological in nature. As such, in order to qualify it as a science, limnology must possess certain central unifying features which tie the whole subject into a coherent orderly organized field.

Further, limnology is directed to the analysis of the *modus operandi* of the complex web of climatic, physical, chemical and biological factors in all natural aquatic systems in collaboration with all disciplines of natural science which are relevant to the overall picture. For example, geology and meteorology are related to the historical and climatic; physics and physical chemistry are relevant to the mechanics and composition of water; biochemistry and molecular biology are linked with the material transformation occurring in the water body. These gave rise to various research pathways within limnology, like Limnochemistry, Limnophysics, Limnobotany, Limnozoology, Limnobacteriology, Hydrobiology, etc. In short, the task of limnology is to investigate and explain the system properties of natural waters. This provides with the basis of more down-to-earth discipline of Applied Limnology which bears the responsibility of furnishing with answers to those problems of freshwater management which are created by overpopulation and urbanization along with environmental protection. Aspects of sewage treatment, pollution of natural waters, water quality conservation and restriction are some of the concerns of applied limnology.

Further areas of application are the Fishery biology including efforts designed to increase rate of organic production in natural and artificial waters and hence to remedy the protein differences affecting human nutrition, especially in the developing countries. Many nations have introduced intensively-managed algal growth projects, particularly, fertilized by the sewage effluent and have achieved a high level of protein production.

Nevertheless, the main aim of Modern Limnology is to understand the natural circumstances responsible for the tremendous differences in natural waters and to identify and evaluate the influences which govern a particular form of productivity, in view of the fact that, inland waters display great diversity in the quality and quantity of life which they contain. Modern limnology also deals with its relationship with other branches of science.

Historical Development and Scope of Limnology

Initiation of works in the field of limnology could be traced back to Aristotle (384-322 BC) which consisted mostly of strange mixtures of facts and fancy with little scientific value followed by observance and simple accurate recordings of certain conspicuous freshwater phenomena. However, significant contributions of strictly limnological nature began to appear about one thousand nine hundred years after Aristotle with the description related to habitat, habit and life history of certain fishes, insects and aquatic macrophytes; but, most of these were isolated accumulation of correlated observations, few of which could be used by modern limnologists mostly for historical purposes.

Since the initiation of optics with Euclid (*c* 2000 BC); and, later, with the invention of microscope, there has been significant development in the field of aquatic biology and limnology because it had not only opened the door to the whole world of microscopic organisms, but also provided with a new and effective means of studying the various higher types of life in water. This was followed by description of minute aquatic organisms by Anton Van Leewenhoek (1632-1723), the pioneering classification of microscopic organisms by the Danish biologist, Otto Friedrich Müller (1786), publication of the Treatise, '*Infus Animalcules*' by Ehrenburg (1836) which marks the beginning of those advances in knowledge which occurred in the 20th century.

Peter Erasmus Müller is credited with laying the foundation stone of limnological study. Anton Fritsch could be considered as the pioneer in lacustrine limnology for his works on lakes in the Bohemian forest; and, F. Simony (1850) is regarded, sometimes, as the Founder of Limnology for his discovery of thermal stratification. However, it had remained practically everything for F.A. Forel (1841-1912), a Professor in the University of Lausanne, Switzerland, to recognize the real biological opportunity of lake investigations; and, the science of Limnology is indebted to him for his comprehensive vision and complete anticipation about the future of this subject. And, for his 110 publications, he is regarded as the Founder of Modern Limnology (Chumley, 1910). It was he who took the decisive step forward from hydrobiology to limnology through his investigations in Lake Geneva, not only from the biological point of view but also from physical and chemical stand points; thereby, formulating the concept of lake types. In addition to his pioneering works on the profundal fauna of freshwater lakes, the design of his first programme for limnological investigations in freshwater and its subsequent execution turned out to be a model for future researches. Forel has many significant publications in limnology, notably, *Instruction a'l'e'tude de la fauna profunde du la fauna profunde du Le' man* (1874-79) and *Le'man Monographie Limnologieque'* (1892-1901) in 3 volumes, which is not only the first comprehensive limnological treatise, but also opened up a new vista in biological research. His epoch-making publication entitled *'Handbook der Seenkunde, Allgemeine Limnologie'* is regarded as the first text book of Limnology.

Forel's work paved the way for establishment of 'Limnological Society in 1887 as a component of Swiss Natural History Society (in order to promote limnological works) and later the International Commission of Limnology was established in 1890. Impetus on freshwater investigations began to be felt as upsurge in the study of lakes in Europe and America manifesting itself in the establishment of Freshwater Biological Stations, notably, at Plön (Germany) by Orozaeharias (1901), at Wisconsin (USA) by Birge and Juday, and so on. For a period of time, much of the works clung to the traditional approaches to freshwater biology; which, nevertheless, had yielded a volume of information that later added materially to the early groundwork of modern limnology.

In brief, the History of Limnology could be dated back to approx. 100 years. Although certain preliminary studies had been done on the habits, nutrition, movement, behaviour, etc., on certain aquatic organisms by different workers during the 17th and 18th centuries (Otto Friedrich, Müller, Johann Christian Schaffer, etc.), these were mostly hydrobiological works and not limnological. True studies on the relationships of biota to freshwaters could be treated as initiated from Junge (1885) and Forbes (1897) who were the first to treat the native waters as microcosm.

Limnology slowly emerged out of this hydrobiological domain following the discovery of marine plankton by Johannes Müller in 1845. At this juncture, it is pertinent to emphasise on the development of the system of lake types as one of the pointers for future advancement of limnology. Apropos to the known relationships between organic production, lake depth and morphology of the water body with its oxygen balance; the observations of Hoppe-Seyler on Lake Constance in 1896; and, oxygen balance and lake type by Birge and Juday in 1911 led to the formulation of the system of lake classification. It was Swede Eince Naumann (1917) who had first emphasized the biological relationships in the lake water column. Thus, the lakes were considered for the first time as functional ecosystems from this point of view. In other words, lakes were treated as limnological entities and were classified according to their oxygen balance. Subsequently, oxygen balance of the inland waters continued to be one of the central features of limnological research.

In recent times, Gaarder and Gran (1927) made pioneering attempts at measuring the photoautotrophic production (primary production) by quantitative determination of oxygen produced by photosynthesis. Later, the direct measurement of carbon assimilation in the waterbodies was achieved in 1952 using radio-carbon method (Steemann Nielsen, 1952). The estimation of trophic dynamics concept having regard to the biomass, material turn over and energy transport along the food chain by Lindeman (1942) not only revolutionized the field of general ecology but also gave a new direction to limnology (Cook, 1977).

Without some amount of knowledge of the micro-organisms and their effects, investigations and implications concerning the oxygen budget of natural waters is incomplete. As such, Bacteria and fungi were included within the scope of investigations even at inception of limnological research (Winogradski, 1897). Both these are constituents of ecosystem which are responsible for decomposing organic matter followed by the process of mineralisation and re-mobilisation. Further, Bacteriology, in close association with Biochemistry, is concerned with the degradation of organic matter and the microbial transformation of substances in natural waters; as well as, the trophic significance of microbes for consumers. Bacteriology, as such, has progressed further in the direction of Limno-microbiology during the last 40 years. The aquatic fungi, on the other hand, have received too little attention so far in this context. In addition to the study of lakes, there had initiated the subject of research on flowing watercourses (the lotic ecosystem) as a consequence of three very considerable stimuli, *viz.*,

1. The aspect of existence of river plankton.
2. The search for glacial relics in the fauna and flora of mountain torrents
3. The effect of sewage effluents on the colonization of lotic systems with particular emphasis on indicator organisms.

Thus, the lotic systems were included within the purview of limnology in the First Congress of International Association of Theoretical and Applied Limnology held at Kiel in 1922. As such, limnology ceased to be merely the science of 'lakes'; rather, with the inclusion of rivers, it has been broadly named as the science of 'Inland waters' (Rodhe, 1974).

Henceforth, the study of lotic systems had been progressing steadily with aspects like water movement and its biological significance, structure of material budget of an open ecosystem in dynamic equilibrium and so on.

A more decisive contribution to our understanding of materials balance of natural waters is provided by applied limnology which is also concerned with the pathology and remedial treatment of polluted waters.

The progression from freshwater biology to limnology is characterized by more detailed penetration of the material balance of the natural water bodies. However, the physical, chemical and biological structure of natural water bodies have become the targets of novel methods of investigation, both *in situ* and in the laboratory.

Present Status of Limnology

Studies in limnology today comprehend three major areas, *viz.*, physics, chemistry and biology. The physical aspects include studies related to physiography, morphology, bathymetry, fluvial dynamics, transparency and thermal profile, etc. Chemical area encompasses determinations and estimations of parameters like pH, dissolved oxygen (DO), free carbon di-oxide (FCO_2), total alkalinity (TA), total hardness (TH), and the nutrients like phosphate and nitrate and trace elements like Mn, Mg, Pb, Hg, As, Cd, B, Cu, etc. The biological components in limnological study deals with in-depth discourses on the living beings notably the plankton, the nekton and the benthos covering both plants and animals ranging from micro- to macro-level.

Discovery of microscope paved the way for great stimulus to aquatic investigations. Although the knowledge of microscopic biota in water could be traced back to Leewenhoek, the discovery of 'plankton' (the heterogeneous group of suspended drifting microscopic organisms which inhabit both fresh and salt waters) was done much later by Liljeborg and Sars (Needham and Lloyd, 1930). Researches on plankton can subsequently be tracked to Johannes Müller (1845) in the North Sea using very fine net followed by Peter Erasmus Müller who discovered the micro-crustaceans in certain Swiss lakes dispelling the previously-held idea that the clear lakes were devoid of microscopic life. Hensen (1887) proposed the definite term 'plankton' to this mass of drifting microscopic life in order to include all the minute plants, animals and debris which are suspended in natural waters; and, study of plankton had been a significant component of limnological research during the last part of 19th and first part of 20th century.

The relationship between environment and organisms is well known. Temperature and oxygen profiles determine mainly the rate of biochemical reactions in the living body. Arhenius's law is applicable equally to both living and non-living beings. However, studies on experimental physiology revealed that aquatic flatworms move faster at higher temperatures, the required energy being derived from higher oxygen consumption; there by, establishing a direct correlation between oxygen consumption and temperature. Similar relationships also exist between phytoplankton and optimum temperature as well as between zooplankton and pH, DO.

Notwithstanding the above, interesting assemblage like biocoenosis as well as phyto and zoo-sociology are often evident in the lentic systems as lentic life is always established in communities which are interdependent or influence each other mutually.

Meanwhile, there have been profound impact of Marine Biology on the study of limnological science. The attention of biologists at large being attracted to the rich diversity of marine life. Although overshadowed by the overwhelming progress of marine biology during most part of the 19th century, study of freshwater biology and limnology did not suffer a setback because the development of the methods of study and an increase in biological knowledge, in general, indirectly but materially, helped the future advances of limnology. Similarly, oceanography began to take on more definite form and its gain also contributed later to the progress of limnology indirectly because there are many features common to both freshwater and marine biota.

Limnological Organisations

The International Association for Theoretical and Applied Limnology was formed in 1922, which has been instrumental in organizing annual general meetings of the International Congress of Limnology. Many countries today have committees on Hydrobiology and Aquaculture in their National Research Councils.

Literature

With an ever increasing number of research papers, an immense literature has already been built-up. In USA, limnological papers are published in general biological journals and in special publications. In Europe, several journals are devoted exclusively to the field of limnology. In Asia, almost every nation has a journal devoted to the field of limnology. Especially, Japan, China, Malaysia, Philippines and Thailand have various limnological journals. In India, today, limnological research has been progressing so fast that almost every province has a journal devoted to the field of limnology and freshwater biology. Further, various universities and institutions issue special series of publications on limnology and freshwater biology. Also, publications originating from fisheries interests frequently carry limnological informations.

Limnological Instructions

Only a few universities and institutions used to offer brief courses in limnology during the last part of 19th century which are mostly taxonomic-faunistic/floristic in nature. During the first half of 20th century, various American and European institutions initiated courses in Limnology (Needham, 1930). Today, many institutions throughout the globe run rigid courses in limnology.

Unexplored Fields

Of the vast array of freshwater lentic systems throughout the globe, very few have received adequate attention from the limnologists. Also, those which have been studied are, for the most part, confined to restricted geographical regions. The inland lentic systems, distributed along the length and breadth of Asia, Africa,

America, Canada, Europe and Australia present wide open opportunity for doing in-depth limnological research. The diversity of these lentic systems, physically, chemically and biologically, are infinitely greater than previously supposed.

Space may not permit a detailed discussion of the very interesting growth during the last century and a quarter. Phenomenal progress of the general subject of ecology inevitably had a constructive influence on limnology; and, because of its many ramifications, limnology has profited from simultaneous advance in other sciences.

Chapter 2

Limnology of Wetlands and Rivers

'Life' is said to have originated in water. Water makes up *c* 80 per cent of the living body. Electrolytes dissolved in water readily ionize because water molecules are polar solvents of high dielectric constant. Ionisation and ionic reactions are of great importance to life.

There are three major types of natural water systems in the world, *viz.*, freshwater (FW), marine and estuarine. The FW environment remained the site of most of our evolutionary history and human activities. And, their importance to man is far greater than other areas, because:

(*a*) They are the most convenient and cheapest source of water for domestic as well as industrial uses.

(*b*) They provide the convenient and cheapest waste disposal systems.

The study of FW in all their aspects, *viz.*, physical, chemical, geological and biological, is called Limnology. FW biology emphasizes the ecology of organisms in relation to their FW habitats. The FW habitats could be broadly classified into two categories, *viz.*,

(a) The Lentic system or the standing water series

(b) The Lotic system or the running water series.

The former includes the lakes, wetlands, tanks, ponds, marshes, fens, bogs in a sequential manner from big to small size. The system also includes the artificially-made reservoirs. The latter system contains the rivers, streams, tributaries and the brooks in successive series of big to small size.

THE LENTIC SYSTEM
(The Standing Water Series)

Although the name is 'standing water', actually, the water in such bodies is in motion in different ways. Hence, 'standing water' does not necessarily mean 'static'. It simply means that water does not flow.

A 'lake ' is a large body of water surrounded by land on all sides and occupies a basin. Lakes are geologically transitory in nature.

Until recently, many naturalists including biologists, thought the lakes to be much the same wherever they occur because similarities are often found among different lakes with regard to water colour, taste, hardness and aquatic biota. Although a partial survival of this idea is still prevalent among many laymen, with the advent of limnology, it has been established that, lakes as a class, manifest a most amazing physical, chemical and biological diversity. Further, as a partial indication of lake diversity, it could well be stated that lakesare large, medium or small; deep of shallow; protected or unprotected; with or without tributaries; and, outlets; fresh, brackish or salt; acidic, neutral or alkaline; hard, medium or soft; turbid or clear; surrounded by bogs, swamps, forest or open shore; high or low in dissolved content; with or without stagnant zones; with marl, muck, sand or clay bottoms; with or without vegetation beds; with high, medium or low biological productivity; young, mature or senescent; and, so on. However, there could exist many imaginable intergrades within the various groups of characters mentioned above. The remarkable lake diversity is the effect of multitudinous combinations of many of these characteristics mentioned above.

Classification of Lakes

The lakes could be classified from various points of views. The classification pattern after Forel (1901) and as modified by Whipple (1947), on the basis of temperature of surface and bottom waters as well as the annual circulation pattern is given below:

(A) In addition to the above criteria, also, on the basis of position of the lake in the globe, lakes could be of the following types:

(a) Polar lakes

Lakes in which the surface temperature is never more than 4°C

The Polar lakes could further be classified into the following orders:

(*i*) Order 1: Lakes in which the bottom water is usually at 4° C throughout the year. Such lakes generally have only one circulation period during summer.

(*ii*) Order 2: Lakes in which the bottom water temperature varies around 4 C and does not deviate much from it. Such lake usually have only one circulation period during summer.

(*iii*) Order 3: Lakes in which the temperature of bottom water is very much identical to that of surface water. In these lakes, circulation is more or less continuous except when frozen.

(b) Temperate Lakes

In these lakes, the surface temperature varies above and below 4°C. These lakes could also be of the following orders:

(*i*) Order 1: In these lakes, temperature of bottom water is at 4°C throughout the year. There could be two circulation periods, *viz.*, one in spring and the other in autumn. Sometimes, there could be none.

(*ii*) Order 2: In these lakes, temperature of bottom water fluctuates around 4°C. Without wide deviations, there could be two circulation periods, *viz.*, one in Spring and the other in Autumn.

(*iii*) Order 3: In these lakes, temperature of bottom water is generally very similar to that of surface waters. Also, in such lakes, circulation is continuous except when frozen.

(c) Tropical Lakes

In these lakes, the surface water temperature is > 4°C

These lakes could further be classified into the following orders:

(*i*) Order 1: In these lakes, the temperature of bottom water is near 4°C throughout the year. There could be only one circulation period during winter.

(*ii*) Order 2: In these lakes, temperature of bottom water could vary but not much deviating from 4°C. There could be only one circulation period during winter.

(*iii*) Order 3: In these lakes, temperature of bottom water is very similar to that of surface water. Circulation could practically be continuous throughout the year.

(B) On the basis of water circulation and mixing, the lakes could be of the following kinds (Hutchinson, 1957).

(a) Monomictic

Lakes in which there is only one mixing, are called monomitic lakes

Monomictic lakes could again be of two types, *viz.*,

(*i*) Warm monomictic: This kind of lakes occur in the warm sub-tropical regions. Here, the water temperature is never < 4°C. There is only one circulation period during the winter.

(*ii*) Cold monomictic: This kind of lake occurs in the polar regions. Here, the water temperature never goes >4°C. Seasonal overturns occur during the summer months.

(b) Dimictic

In this type of lakes, there are generally two periods of mixing or overturns.

(c) Oligomictic

These are thermally stable tropical lakes, in which, mixing occurs very rarely and slowly.

(d) Polymictic

These are high altitude equatorial lakes, in which, circulation occurs more or less continuously. Stagnation period, if any, is very brief.

(e) Amictic

Lakes, in which, there is practically no mixing being sealed off by ice from most of the changes at the earth's surface, are called amictic lakes. These lakes are sometimes found in the Antarctica and in high mountains.

(f) Meromictic

These are chemically stratified lakes which become permanently stratified by the intrusion of saline water or salts liberated from sediments; thereby, creating a permanent density difference between the surface and bottom waters and leading to the formation of a 'chemocline' between the circulating and non-circulating layers. 'Meromictic' denotes partly mixed. Free oxygen and aerobic biota could be absent in the bottom waters of such lakes.

(g) Holomictic

Lakes, in which, surface and bottom water mix wholly and periodically are called holomictic lakes.

(C) On the basis of nutrient status and productivity, lakes could be classified into the following types (Thienemann, 1925):

(a) Oligotrophic

These are deep, steep-banked lakes with the following features:

Volume of epilimnion relatively small as compared to hypolimnion; water is quite transparent with poor nutrient status; permanent abundance of oxygen at all levels; sediment low in organic matter; scanty littoral vegetation and phytoplankton generally poor with rare blooming of blue green algae (BGA).

(b) Eutrophic

These are shallow lakes with broad littoral zone and following other features:

Greater epilimnion-hypolimnion ratio; water less transparent with abundance of nutrients and calcium; hypolimnial oxygen depletion occurs during summer; sediments rich in organic matter; abundance of littoral vegetation; phytoplankton generally quantitatively rich with frequent blooms of BGA

As such, conditions in deep lakes differ from those of shallow lakes although they could be in the same climatic and edaphic zones. The hypolimnial oxygen supply in a deep oligotrophic lake during summer, in contrast to shallow eutrophic lake, is always richer to support a diversity of benthic organisms. To the contrary, only a few species could be found in the hypolimnion of shallow eutrophic lakes. Rodhe (1969) suggested different rates of inorganic carbon fixation by photosynthesis in these kinds of lakes.

(c) Mesotrophic

Lakes, in which, the nutrient status is neither too rich nor too poor, are designated as mesotrophic lakes. Mesotrophy lies somewhere annual primary production of 25 -75 g carbon fixation.

(d) Dystrophic

These are said to be special kinds of lakes which generally have high concentration of humic acid in the water. Bog lakes, which are dystrophic in nature, have low pH and gradually develop into peat bogs which culminate into marshy wetlands.

Origin and Formation of Lakes and Classification of Lakes on the Basis of Origin

Lakes are geologically transitory. They are usually born out of catastrophes occurring in nature and die quietly (Hutchinson, 1957). Thus, a lake lives through youthful stage of maturity, senescence; and, death occurs when its concave basin is finally filled-up.

Classification of lakes on the basis of their origin is given below:

(a) Glacial Lakes

Glaciers deepen and widen the valleys when they glide down. When the huge mass of ice melts, it leaves piles of moraine, which blocks valleys; and, there by, dam streams and rivers. Thus, the scraping action of glaciers lead to the formation of lakes. Such lakes are generally found in the North temperate regions of the world.

(b) Ice-scour Lakes

When huge sheets of ice move over relatively flat surfaces of hard, jointed or fractured rocks, scours or hollow basins are formed. Later, these hollow basins are filled with water resulting in lakes.

(c) Tectonic Lakes

Lakes formed by movements of earth's crust (tectonic plates) are called tectonic lakes. When a portion of the earth's surface subsides in relation to its surroundings; or, conversely, when a portion of the earth's surface is lifted, a lake is formed. The area and depth of water in the newly-formed lake will depend on the drainage pattern from the surrounding catchment; which, in turn, depends on the amount of rainfall and the size of the area from where water drains into the lake.

(d) Volcanic Lakes

Lakes which are formed due to volcanic eruptions are called 'crater lakes'. They are formed when a series of minor eruptions take place fairly close to each other. In instances, where the volcano was big and had erupted long ago, the floor of the huge crater of such volcanoes is relatively flat; and, water accumulates in some of the deeper zones resulting in the formation of lakes. Further, volcanic activity could also form a lake when the ash or lava from an eruption blocks natural drainage.

(e) Riverine Lakes

River-formed lakes, *i.e.*, the lakes formed by the action of rivers are commonly called riverine or 'oxbow' lakes. Bank erosion is quite fast in fast flowing tortuous river. At one point of time, the river breaks through the continually eroded narrow isthmus between two successive curves. This may leave a loop of the river to one side of the new water course. This isolated portion may retain enough of water to form an oxbow lake.

WETLANDS

Standing waters occur in depressions or in basins, which are shallow from the moment they are formed. Eroded sediments and plant remains from the surrounding land are washed-in and settle at the bottom along with debris from the resident aquatic plants and animals. Gradually, the lake or pond gets shallower until it becomes a 'wetland'.

'Wetlands' are, thus, basically 'wet- lands' where the soil is saturated with water for sometime during the year. According to IUCN (1970), wetlands are areas of marsh, fen, etc., temporary of permanent; natural or artificial mass of water, the depth of which generally does not exceed 6 m. Wetlands are areas which contain substantial amount of standing water and little flow.

Classification of Wetlands

Wetlands occur throughout the world in all climatic zones and are estimated to cover *c* 6 per cent of the earth's surface. The simplest classification of wetlands has been provided by IUCN's Ramsar Convention, which is briefly listed below:

1. Freshwater lakes/wetlands
2. Oxbow lakes/wetlands
3. FW ponds
4. Marshes, swamps, bogs
5. Reservoirs.

A simple key for different types of lentic systems as fish habitats is given below:

Permanent standing water with heavy vegetation and little or no open water.............Wetland

Permanent standing water with rock or completely vegetated shoreline.............Pond and small lake

Permanent standing water with atleast some rocky or sandy wave-swept shoreline.............Large lake

As stated earlier, a FW lake is a closed water body of FW which is surrounded by land on all sides.

Oxbow lakes/wetlands is created when a river changes its course leading to isolation of a segment of the said river.

A pond is a body of standing water which is too small to have a wave-swept beach.

Swamps are wetlands with trees. Bogs and fens are vegetated low-lying areas. Bogs have outlets where water from rain run off and springs flow away. Fens are basins fed by upwelling ground water. Marshes are usually wetlands dominated by a few species of plants, *e.g.*, *Cyperus*, *Scirpus*, etc.

Origin and Evolution of Wetlands

In the cold belt, the process involves receding of the glaciers; and, concomitantly, the debris which are washed out of the melting ice, fill the crevices near the edge of the ice sheet and cover isolated blocks of melting ice. When these ice blocks melt, the resulting depressions become isolated 'Kettle lakes', etc.

In some regions, ponds are formed when subterranean chambers collapse. Also, many small ponds may be formed where beavers build dams. Many lakes and big ponds are deliberately constructed for reservoirs, meant for generation of hydroelectric power; as well as, for recreational and aesthetic purposes. Further, many ponds are formed along the Highways when dirt is removed for use in constructing the bed of the Highway.

Some wetlands are small kettle holes; others are in the last stages of 'succession' which is triggered when aquatic plants produce organic matter faster than decomposing organisms can re-cycle it. The remains of the plant bodies gradually accumulate until no open water remains.

In the tropics, notably in India and adjoining places, like Bangladesh, etc., wetlands are generally shallow depressions which could normally be in the form of a basin at the centre of hillocks on all sides; or, could be abandoned segment of a river (oxbow wetland); or, a shallow portion of a river course which is detached from the main river course during the dry season. Sometimes, wetlands in NE India, are formed due to tectonic activities.

In Assam, and in adjoining Tripura and Bangladesh, 3 kinds of wetlands are generally found. They are locally called as follows:

(a) 'Beel'

Perennial wetlands which contain water throughout the year.

(b) 'Haor'

Seasonal wetlands which contain water for some period of the year only, particularly, during the rainy season. As such, they are also called 'floodplain wetlands'.

(c) 'Anua'

These are peculiar river-formed perennial oxbow-type wetlands which are generally formed due to change in river course and which may or may not retain connection with the original river.

Wetlands in North-East India

The North-Eastern (NE) region of India, a typically difficult topography with undulating terrains, however, provides enough potential for fish production which can supplement food requirement for the region and could provide answer to diminishing protein supply. Development of agriculture has, sometimes, been a constraint due to not much availability of flat land with facilities of irrigation. Nevertheless, the region provides enough scope for development of inland fisheries all over the region.

Situated between 89-97°N Longitude and 20-30°E Latitude, the region encompasses a vast area of 2,55,083 sq km out of total Indian area of 3.3. million sq km. Comprising of eight provinces, *viz.*, Arunachal Pradesh, Assam, Manipur, Meghalaya, Mizoram, Nagaland, Tripura and Sikkim; and, bordered by Tibet, Bhutan and China on the North; Myanmar in the East and Bangladesh on the West, the region occupies a strategic position both nationally and globally. The lofty Himalayas proudly stands as sentinels to guard the northern and eastern frontiers. The Chittagong Hill tracts form a crescent on the south. The region's western mountains suddenly rise from the valleys in Bangladesh, leaving only a narrow strip which opens out to the Indian mainland.

The NE, with *c* 8 per cent of the country's geographical area and *c* 4 per cent of the total population, provides a picture of mini-India. Having diversities with regard to topography, water resources, biota, climate, race, language and culture, the region harbours more than a hundred tribes differing ethnologically, yet sharing a common destiny.

The population of NE India is *c* 33.7 million out of 843.9 million in India (Govt. of India, 1996) accounting for *c* 4 per cent of the total population The multiplicity of tribes has been the unique demographic characteristic of the region. *c* 13 per cent (projected) of the population of Assam, 28.44 per cent of Tripura, 27.36 per cent of Manipur and almost the entire inherent population of Arunachal

Pradesh (69.82 per cent), Meghalaya (80.58 per cent), Nagaland (83.99 per cent), and Mizoram (93.55 per cent) belong to Scheduled tribe (ST).

Besides lotic territories, the lentic water bodies having 0.72×10^6 ha lake coverage in India, constitute great potential of fishery resources. The NE region is blessed with a number of lentic systems, locally called 'Beel, Haor, Anua, Hola, Doloni, Jalah, etc., which alone constitute *c* 81 per cent of the total lentic area (0.12×10^6 ha) in Assam. These lentic systems are generally shallow and open, ranging in size from 35 to 3458.12 ha and with depth ranging from 0.25 to 3.0 m (in some, however, the maximum depth may exceed 6.0 m) at LSL. Further, in Assam, there are *c* 1392 number of wetlands having a total of *c* 22,896 number of fisheries of different categories; out of which, the number of registered wetlands is only 394 (30.38 per cent) covering an area of *c* 70,000 ha; of which, *c* 19,000 ha is in good condition; *c* 15,000 ha is in semi-derelict condition and *c* 35,000 ha is in derelict condition (Govt. of Assam, 2001)

A brief account of the present status of the lentic (and also lotic) systems, as potential fishery beds, in the different provinces (States) of NE India is tabulated below:

Table 1

Province	*Lentic/Lotic System*	*Area/Length (ha/km)*	*Area/Length Developed So far (ha/km)*
Arunachal Pradesh	Fish ponds/Farms	1000 ha	218 ha
	Beel, lake, swamp	2500 ha	94 ha
	Paddy-cum-fish culture	2800 ha	575 ha
	Others	100 ha	1.0 ha
	Water area under RF	600 ha	
	Lotic system	2000 km	
Assam	Beels	1,00,000 ha	
	Reservoirs	10,730 ha	
	Swamps	10,000 ha	
	Ponds and Tanks	19,860 ha	
	Lotic system	5,000 km	

Contd...

Table 1–Contd...

Province	*Lentic/Lotic System*	*Area/Length (ha/km)*	*Area/Length Developed So far (ha/km)*
Meghalaya	Beels and Lakes	375 ha	
	Reservoirs	8,430 ha	
	Ponds and Tanks	1,905 ha	
	Paddy fields	5,000 ha	
	Lotic system	5,600 km	
Manipur	Lakes, Beels, Swamps	21,000 ha	
	Logtak Lake	19,000 ha	
	Ponds and Tanks	5,000 ha	
	Lotic system	5,000 km	
Mizoram	Lakes/Reservoirs	320 ha	
	Ponds and Tanks	1,595 ha	
	Paddy-cum-fish Culture	122 ha	
	Lotic system	1,748 km	
Nagaland	Beels	215 ha	
	Lakes and Reservoirs	27,100 ha	
	Ponds and Tanks	2,000 ha	
	Terrace paddy land	10,000 ha	
	Lotic system	1,600 km	
Tripura	Reservoirs	4,500 ha	
	Lakes	4,300 ha	
	Mini barrages	3,500 ha	
	Ponds	2,652 ha	
	Private ponds	3,700 ha	
	Govt. Fish Farms	120 ha	
	Co-op. Soc. Fish Farms	292 ha	
	Lotic System	1,200 km	

Some of the significant Lentic systems (Wetlands, lakes, tanks, etc.) of Southern Assam, Mizoram and Tripura are tabulated below:

Table 2

Province	Sl. No.	Wetland	Area	Reference
Assam (Southern part)	1.	Sone Beel (Biggest)	3458.12 ha	Kar (1984,1990, 1996)
	2.	Rata Beel		Kar (1984, 1998)
	3.	Sagar Beel		Kar (2003)
	4.	Rani-Meghna Beel		Kar (2003)
	5.	Angang Beel		Kar (2003)
	6.	Medha Beel		Kar (2003)
	7.	Duberi Beel		Kar (2003)
	8.	Jabhda Beel		Kar (2004)
	9.	Karkari Beel		Kar (2004)
	10.	Junamara Beel		Kar (2004)
	11.	Deochhara Beel		Kar (2003)
	12.	Asinia Beel		Kar (2003)
	13.	Dhalchhara Beel		Kar (1987, 1988)
	14.	Hatichhara Beel		Kar (1987,1889)
	15.	Doloo Beel		Kar (1987, 1990)
	16.	Chatla Haor		Kar (1998, 2000)
	17.	Bakri Haor		Kar (1999)
	18.	Puneer Haor		Kar (2000)
	19.	Baskandi Anua		Kar (1988), Dhar (2004)
	20.	Rupairbala Anua		Kar (2003)
	21.	Dungripar-Kaptanpur Anua		Kar (2002, 2003)
	22.	Satkarakandi Anua		Kar (2001, 2003)
	23.	Ram Nagar Anua		Kar (1988, 2003)
	24.	Baraknadi-Salchapra Anua		Kar (1998, 2003)
	25.	Fulbari Anua		Kar (2000, 2003)
	26.	Sibnarayanpur Anua		Kar (2000, 2003)
Mizoram	1.	Tumdil Lake		Kar (1998)
	2.	Rungdil Lake		Kar (2000)
	3.	Polokdil Lake		Kar (2002)
Tripura	1.	Rudra Sagar		Kar (2002, 2003)
	2.	Fatik Sagar		Kar (2002,2003)
	3.	Amar Sagar		Kar (2002, 2003)
	4.	Kurti Beel		Kar (1998)

THE LOTIC SYSTEM
(Running Water Series)

Freshwater ultimately starts as rain. Some portion of it evaporates as soon as it lands, some is absorbed by plants and animals; some flows over the land surface directly into streams; and, some soaks into the ground, where it makes its way through the soil and rocks into the groundwater system. Many permanent streams are fed by springs, *i.e.*, groundwater, as well as, by run- off.

A lotic system, (*e.g.*, streams / rivers) is a flowing stretch of water which generally begins from an elevation (if it is an original stream) or could arise as a branch of the original (main) river. Streams could be classified in a variety of ways.. However broadly, streams could be classified into the following types:

(a) Headwater stream

(b) Mid-reach stream

(c) Base-level stream.

The following is a simple key furnished for different types of lotic systems as fish habitats:

Small streams originating in springs or wetlands, generally with an average depth of <10 feet......................Headwater stream

Rocky bottom streams in V-shaped valleys...........Rocky creeks

Slow meandering streams with swampy or marshy banks.........................Marshy creeks

Larger streams generally >10 feet in depth.........Natural streams

Streams with conspicuous pools and riffles.....Mid-reach streams

Sluggish meandering streams; may have occasional sand bars but few or no conspicuous pools-riffles.........Base-level streams

Large rivers used for navigation

Freshwater.........................Large rivers

Flowing into oceans: Lower reaches brackish or salt water.........................Estuaries

In short, Headwater streams are those, in which, erosion occurs upstream and little sediment is deposited. Mid-reach streams are those where erosion occurs in both the banks and the bottom; and, where pools alternate with riffles. Base-level streams have low gradients, so that, most erosion is lateral and considerable deposition of sediments causes the stream to meander.

Topography and the underlying geological framework can give a stream several cycles of mid-reach and base-level sections. Therefore, it is not unusual to find a base-level stretch upstream from a mid-reach section.

Base-level streams grade into large rivers which are used for navigation and industry; and, which flourishes with the availability of water transportation.

Then, there are estuaries, water passages, where rivers flow into the sea. Salt water pushes up under the freshwater as the river meets the sea, and the water becomes increasingly saline towards the coast. The estuaries are affected by tides because they are at sea level.

Streams, broadly, could be of the following Orders:

1st Order: The Headwater stream itself is called the 1st Order stream

2nd Order: Two 1st Order streams unite to form a 2nd Order stream

3rd Oder: Two 2nd Order streams unite to form a 3rd Order stream; and, so on.

However, a 1st Order stream joining a 2nd Order stream does not give rise to a 3rd Order stream.

Investigations into the running water series, particularly from the limnological standpoint, has lagged far behind that of lakes; and, our knowledge in this regard is distinctly fragmentary. Any general limnological account, at the moment, could merely be an attempt to pack together, some of the results of the widely-scattered studies in the rivers.

The lotic environment differs from the lentic in the following fundamental aspects:

Depth

Depth of all lotic systems are generally lesser than that of lentic systems.

Width of Basin

Water of a lotic system is generally confined to a relatively narrow channel.

Current Velocity

The whole volume of water in a lotic system flows in one direction with fluctuating velocity; while, the lentic system is particularly devoid of such water currents.

Gradient

It refers to the physical inclination of the river banks which does vary in different stretches of the river from the source to the end.

Extension of Channel with Age

The length, breadth and depth of streams, generally, increase (as compared to base level) with increase in age. This is a distinct contrast to the reduction process, characteristic of all standing water units.

Erosion and Removal of Materials

At any position along the course of a running water unit, materials eroded at that point and all materials momentarily suspended or dissolved at that level are transported downstream with no opportunity to return. In lakes, such materials commonly remain within the same basin.

Absence of Prolonged Stagnation

Constant flow with accompanying mixing of water usually eliminates prolonged summer stagnation of the bottom waters, which generally do not occur in lakes.

Basic Food Materials

Most streams manufacture, within themselves, little or no basic food materials but depend much more on the contribution from the surrounding land than do most lakes.

Characteristics of Lotic Systems

Some of the characteristics of lotic systems are given below:

Reach

A 'reach' is a channel segment with repetitions of more or less homogeneous sequence of physical processes and habitat types.

Reach No. is given to individual reaches. Althouth a stream is composed of a sequence of different reaches, extending from the headwater to the end, there could also be a single reach in a stream.

Classification of Valley Reach

The valley reaches could be of the following kinds:

(a) Colluvial

Landslides from adjacent hill-slopes deliver organic matter into the river flow.

(b) Alluvial

When the sediment is transported by the stream-flow.

Factors Determining Stream-reach

The following factors could serve as determinants of stream-reach:

(a) Channel pattern

(b) Channel confinement

(c) Gradient

(d) Stream bed and bank materials

A brief explanation of identification of reach, based on the above criteria, is given below:

Reach boundaries are usually identified at the points where there is:

1. Significant changes in stream channel form or confinement, *e.g.*, change from a single channel to a multiple braided channel.
2. Significant changes in stream bed and bank materials,*e.g.*, change from erodable to non-erodable materials.
3. Changes in sediment supply, *e.g.*, an actively eroding landslide.
4. Changes in riparian vegetation, *e.g.*, from grassland to forest.
5. Changes in stream-channel confinement, *e.g.*, from a wide floodplain to a confined canyon.

6. Changes in the relative coupling among the hill slopes, valley flat and stream channel.
7. Occurrence of a tributary confluence (atleast of 2nd Order).
8. Changes in channel gradient (at least 2 per cent).
9. Changes in land use pattern.

Classification of Channel Reach

The channel reach may be classified as follows:

(a) Fall reach

In this category, the river, just after origin, sometimes, falls from an elevation

(b) Cascade Reach

Just after a river falls down from an elevation, it generally flows down the gradient in the form of staircase through the rocky substratum. Such a staircase-like stretch of the river flow is called 'cascade' type of reach. Cascades are characteristics of steepest alluvial channels. Sometimes, a small pool may be formed, but maximum volume of the flowing water cascades over and around boulders and large woody debris.

(c) Pool-riffle Reach

This kind of reach is most commonly associated with small to medium-sized streams and is a widely-prevalent type of reach in alluvial valley of low to moderate gradient.

(d) Plain-bed Reach

This kind of reach is generally characterized by relatively long and straight channel having almost uniform depth. This type of reach having laminar flow is also called 'sheet.'

(e) Step-pool Reach

Here, there is accumulation of boulders and logs which form a series of steps alternating with pools containing finer substrata.

(f) Braided Reach

This kind of reach is characterized in having many gravel and sand bars scattered throughout the channel.

Substrate

This refers to the bottom material on the substratum (base) of the water body.

Classification of substrates is tabulated below:

Table 3

Class	*Size Group*	*Description*	*Estimation*
Fines	< 2.0 mm	Lady bug size	
Gravels	2.0–64.0 mm	Lady bug to tennis ball size	
Cobbles	64.0–256.0 mm	Tennis ball to basket ball size	All visual
Boulders	> 256.0 mm	Bigger than basket ball	
Bedrock	> 4000 mm	Boulders and blocks >4.0 m	

The substrates determine the roughness of the stream, which, in turn, influences channel hydraulics. Further, substrate provides micro-habitats needed by many fish species for spawning. Also, substrates provide clue to local and watershed influences on stream habitat quality.

Cover and Shelter for Fish

A 'cover' may be defined as a structural material (boulder, log, etc.), channel features (hedges, vegetation, etc.) or water features (turbulence, depth, etc.) in the wetted portion of the lotic system or within one metre above the water surface which provides hiding, resting or feeding places for a fish.

Cover is often described in terms of compostion and abundance in order to ascertain and estimate the amount of shelter available for a fish.

Cover Types

The various types of covers are detailed below:

(a) Turbulence

It is considered as a cover when the velocity of water in a river at a given point varies erratically in magnitude and direction and disrupts reaches with laminar flow.

(b) Woody Log

Woody log debris is considered as efficient refuge for a fish.

Logs, which are 8-10 cm in length and 1-2 cm in width, are generally called small woody logs which are considered as refuge for small fishes. Conversely, woody logs, which are >10 cm in length and >2 cm in diameter, are generally called large woody logs which could be potential shelters for large-growing fishes.

(c) Depth

It is an effective parameter for concealment and shelter for fishes.

(d) Turbidity

In the case of shallow rivers, where depth is less, turbidity is an efficient feature for the fishes to hide.

(e) Boulder

These large components of the river substrate often provide niche' for a fish for its breeding, feeding as well as hiding. Sometimes, they may form scoured-out pool and overhang the stream.

(f) Undercut Bank

These are peculiar kinds of micro-habitats created by erosion of the base of a stream bank by fast flowing water current resulting in a overhanging of the remaining part of the bank above the stream water surface.

(g) Vegatation

These are the macrophytes which could be found on the surface of the streams as well as those found overhanging. The vegetation not only gives shelter to a fish, but also has a profound impact on its physiology.

Riparian Vegetation

The vegetation on the land surface or on land adjacent to the normal high water level of the stream is called riparian vegetation.

Riparian vegetation could be classified into the following types:

1. Hydroriparian: The soil/substrate is rarely/briefly dry. Wet riparian plants dominate the vegetation.
2. Mesoriparian:The soil/substrate is dry seasonally
3. Xeroriparian:The soil/substrate is wet less than one month a year.

The Rivers of South Assam, Mizoram and Tripura in North-East India

The meandering rivers, which criss-cross the North-Eastern Provinces, have, through the ages, played a dominant role in shaping its history. The most striking feature of the Region's drainage is the magnificent flow of the mighty Brahmaputra and the Barak reckoned amongst the world's most majestic rivers. Originating in a Himalayan glacier, and after traversing *c* 1600 km in Tibet, the Tsangpo, as the Tibetans call the river, cuts the Himalayas and enter Arunachal Pradesh as Siang. Two other rivers, *viz.*, Dibong and Lohit, join the river Brahmaputra; and, the main river, thus replenished, flows down the Assam plains as the River Brahmaputra.

With an antecedent drainage system, the Brahmaputra flows for *c* 730 km in Assam before entering Bangladesh at Golokganj in Dhubri in Assam. 47 major tributaries, (notably, Bhogdoi, Dhansiri, Kopili, Kolong, Barni, Puthimari, Manas, Beki, Dikrong, etc.) originating from both North and South directions, meet the Brahmaputra in Assam valley.

The river Barak originates from Japvo peak in Nagaland (*c* 3353.65m MSL), and flows through Karang village along the Manipur-Nagaland border. It drains almost the entire Manipur valley before entering Assam. It flows through the southern part of Assam (thus, forming the Barak valley); and, after traversing through a stretch of *c* 532 km from its origin, the River Barak bifurcates into two branches, *viz..*, the Surma and the Kushiara at village Harinagar (Haritikar) at the Indo-Bangladesh border. After flowing a short spell along the Indo-Bangla border, both these rivers flow into Bangladesh to join the Meghna basin before entering the Bay of Bengal. Barak has a number of tributaries joining it on both north and south banks, *viz.*, Jiri, Chiri, Madhura, Jatinga, (north bank tributaries) and Sonai, Rukni, Dhaleswari, Ghagra (south bank tributaries). In addition to these, the River Shingla and the River Longai, both originating from the Mizo Hills, join the River Kushiara in Barak valley, the former flowing via Sone Beel, the biggest wetland in Assam. Also, the River Baleshwar joins River Surma in Barak valley.

The province of Mizoram is drained by eight major rivers belonging to three principal drainage systems as mentioned below:

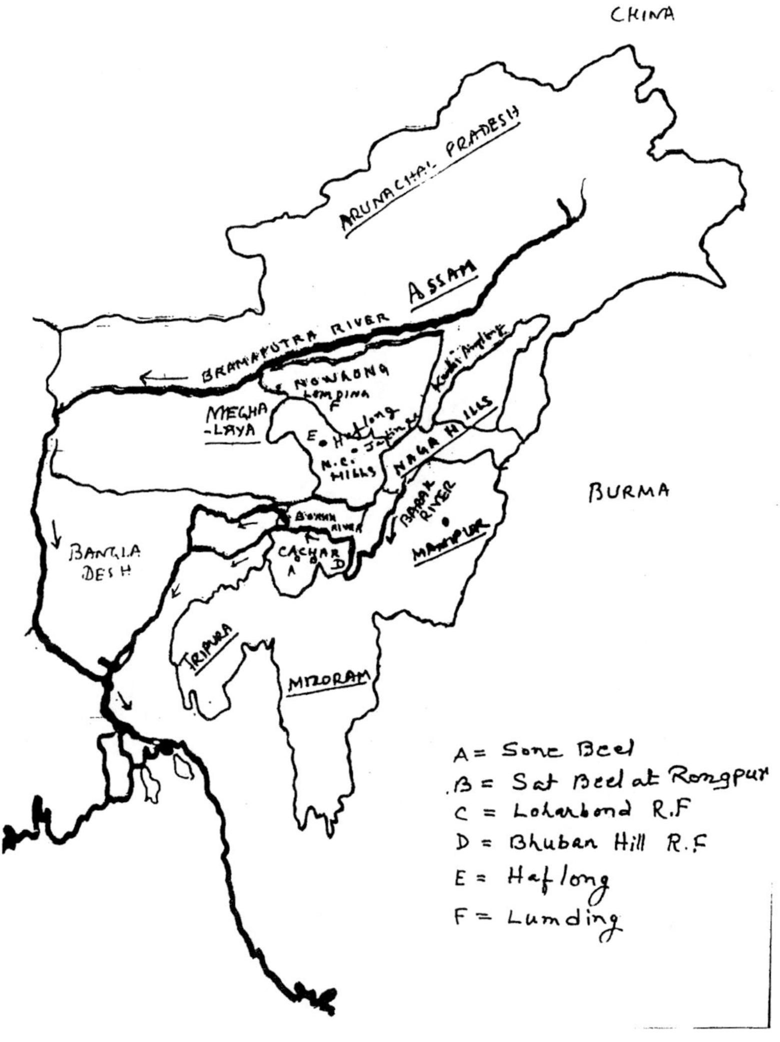

Figure 1: Map of North-East India Showing Principal Drainages

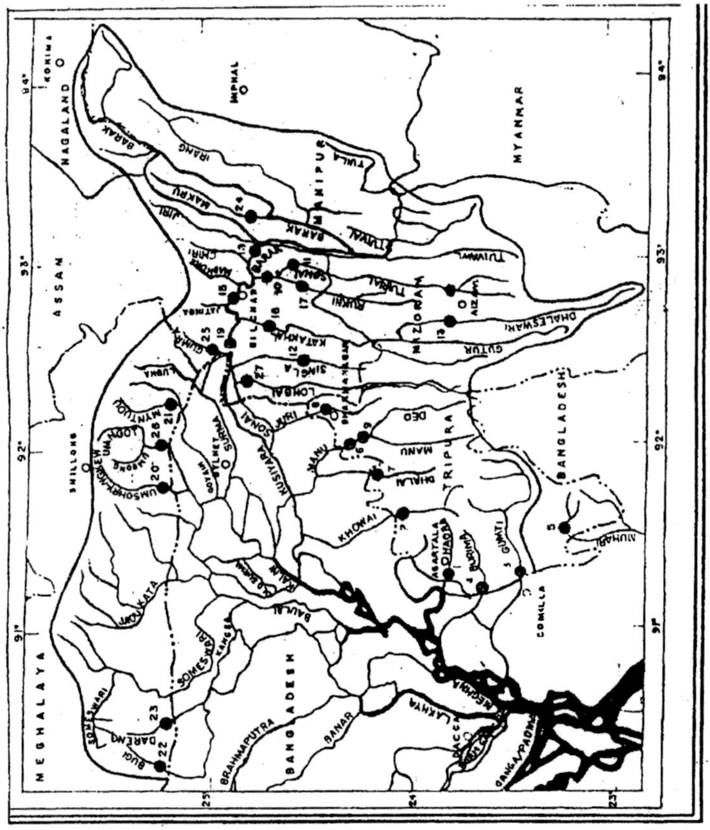

Figure 2: Map of N.E. India Showing a Number of Drainages in Assam, Mizoram and Tripura

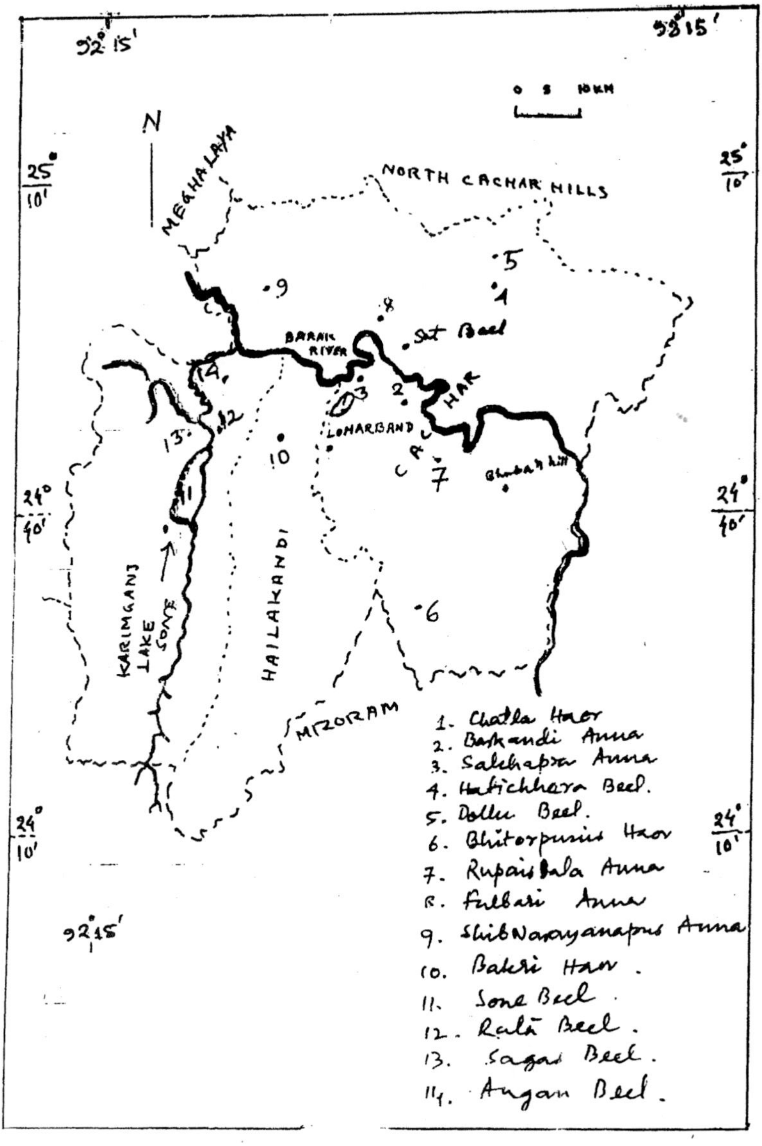

Figure 3: Map of Barak Valley Region of Assam Showing River Barak, Its Principal Tributaries and the Principal Wetlands

1. River Tuirial
2. River Tlawng
3. River Serlui Barak-Meghna drainage system: Indo-Bangladesh
4. River Tuivai
5. River Mat
6. River Kolodyne Kolodyne-Irrawaddy drainage system: Indo-Myanmar
7. River Tuichong
8. River Karnafuli Tuichong-Karnafuli drainage system: Indo-Bangladesh

The province of Tripura is drained by four major rivers forming four independent basins as given below:

1. River Manu: Manu basin
2. River Khowai: Khowai basin
3. River Gomati: Gomati basin
4. River Feni: Feni basin

All these rivers flow into Bangladesh independently.

HYDROTOPOGRAPHY

A General Description of Catchment

India is a land of varied water resources. It is a land endowed with varied aquatic habitats. The mighty rivers, *viz.*, the Ganges, the Brahmaputra and the Barak, along with their tributaries, provide precious supply of water and merge into the bigger basin called the Meghna, before going into the sea. Further, North-East region of India sustains nearly full range of microcosms, from tiny bogs followed by marshes, fens, floodplains and very big perennial wetlands, locally called 'Beels'. Similarly, the flowing water system ranges from creeks, canals, rivulets to the mighty Barak and Brahmaputra rivers, which, along with their tributaries, drain about 1/3 rd of the Sub-continent.

The diverse geological environment of North-East (NE) India practically contains the entire history of geological evolution of the region compressed in its lithography, tectonic history, palaeoenvirons and so on. Rocks range from 3600 million years upto the most recent. They also bear distinct impressions of the various changes based on relatively modern theory of Plate tectonics. Further, NE India appears to be at the destructive junction of all the characteristics of plate tectonics.

The NE Region of India could be divided into five physiographic units, *viz.*, the Assam Himalayas, the Brahmaputra valley, the Shillong plateau, the Barak valley and the South-East Hill region. However, except the Barak and the Brahmaputra valleys in Assam, and the Imphal valley in Manipur, the rest of the region constituting > 65 per cent of the area, is covered with hills and mountains. Of these, Arunachal Pradesh occupies the maximum area (32.83 per cent) closely followed by Assam (30.75 per cent), Tripura (4.11 per cent), and so on.

Further study of geological history of the region reveals that the Assam Himalayas is said to have emerged only during the Tertiary age from below the ancient Sea of Tethys. This phenomenon was said to have been aided by volcanic eruptions and crustal movements of the earth. It is believed that the rising bed had gradually folded itself into parallel ranges. The Meghalaya plateau is said to be an extension of the ancient Gondwana rocks. The alluvial Brahmaputra valley is formed mostly by the River Brahmaputra and its tributaries. The River Brahmaputra, *c* 700 km in length, forms a flat plain nested between two ranges. The Barak valley, on the other hand, is formed mainly by the River Barak and its tributaries. This valley is unusually diversified by low hills intervened by swampy flats. Further, the South-Eastern hill region is highly packed with mountain ranges separated by narrow parallel valleys. Moreover, the diverse catchment includes place(s) in Tripura which is/are < 150 m MSL; while, in Mizoram, the Blue Mountain is *c* 2164 m MSL. However, the Loktak lake forms the lowest part of Imphal valley in Manipur. Interestingly, further, the Barail range creeps into Nagaland via North Cachar Hills, the highest peak in Nagaland being the Japvo peak (*c* 3353 m MSL), from where, the River Barak is said to have originated.

A Brief Account of the General Origin of North-Eastern Region

It is assumed that, from the beginning of Pleistocene when the dismemberment of the westward-flowing drainage at the foot of the

Himalayas had occurred, the migration southwards towards the peninsula had taken place entirely during the Pleistocene. It is, further, believed that the sea level had undergone repeated fluctuations during the Pleistocene, caused due to the formation of large ice-caps on the higher latitudes during the glacial period and their melting during the inter-glacial periods. The former process led to the lowering of sea level and the latter to its rise. Hora (1951) opined that, during the glacial periods, the Garo-Rajmahal gap could have become few hundred metres higher than the sea level and the colder climate had favoured increased precipitation vis-à-vis less evaporation, and greater run-off in streams/rivers like the Narmada, Tapti along the Satpura-Vindhya. Such a condition might have helped the spread of rheophilic and torrential fishes from the region of AssamHimalayas to the Peninsular India across the Garo-Rajmahal bridge.

But the rise in the sea level during the Arid periods, naturally checked fish migration temporarily isolating certain forms and leading to rapid speciation. The peninsular isolates, *e.g.*, *Bavania*, *Silurus*, etc., showing Malayan affinities, are said to be products of rapid speciation. Their dispersal took place in different waves of migration corresponding to the five Glacial periods of the Pleistocene. The Pluvial periods had facilitated migration while the Arid periods isolated them; there by, induced rapid speciation in the Peninsula (Silas, 1952; Jhingran, 1991). However, there could be different other views in this regard.

Catchment and Hydrotopography of Barak Drainage

The Barak Sub-basin,situated between the geographical limits of 22 32′ N and 25 35′ N latitude and 92° 15′E and 94° 18′ E longitude, is one of the biggest in this region. It is endowed with enormous water resources and holds a great potential for all-round aquacultural development of this region. The topography, combined with quite uneven distribution of rainfall, account for the uniqueness.

Estimation revealed that the catchment area of Barak from its origin upto Village Harinagar in India is *c* 25086 km^2; while, including Myanmar, the total catchment area is 41723 km^2. In fact, the catchment of Barak sub-basin starts from a place situated at an elevation of > 2331 m MSL in the southern slopes of the lofty Patkai ranges around the village Karong along the Manipur-Nagaland border.

The position of original Barak valley region is like an amphitheatre formed by hills on three sides and opening into the plains of Sylhet to the west. *The hill ranges surround the valley and the sediments of this region are among the recent formations in the world.* They show different stages of morphometric development as well as diversity of geological composition. In fact, their constitution, structure, history and dynamic behaviour are all inextricably interwoven with the evolution of the Barak system.

Geomorphology

The Barak river system involves five distinct geomorphological domains. These are as follows:

1. The structurally-controlled ridges of Nagaland forming the headwaters of the Barak system. These hills are composed of strongly-folded strata of hard compact sand stone and shale (Barail and Dishing groups) which are the oldest rocks in the catchment.
2. A belt of robust coalescing ridges and valleys in western Nagaland and Manipur. These rocks, sandstones and shales are younger, less deformed and safer than those of the first unit (Surma group).
3 .The zone of 1st order topography,*i.e.*,The anticlinal hills and synclinal valleys of Mizoram and West Manipur which is the zone of 1st order topography. These hills have broad flanks and sharp clefts. The valleys carry very little amount of sediment mainly because they are narrow and V-shaped; although, the whole system is strongly linear and trends N-S. The fold patterns are simple and the rocks are mostly shales and sandstones.
4. Continuation of the previous area northwards which is distinguished by lower and less sharp hills and lesser intensity of folding. However, more open valleys are found floored with some amount of sediment.
5. The typical hills of the Barail ranges situated in the North Cachar Hills district of Assam which seem to be quite mature geomorphologically and composed structurally of highly deformed sandstones and shales (Barail group).

It is significant to note here that, the Barak valley in Assam is the youngest manifestation of the major units of Barail range contributing to the Barak system.

Rocks and Lithographic Groups

The rocks of the Barak sub-basin pertain to the stratigraphic unit of the mobile belt of NE India and are mostly sedimentary in nature belonging to the Tertiary age. The list of the major lithographic groups, beginning from the oldest, is given below:

1. The Dishang group consists of a variety of shales along with some sandstones which increase progressively upward in a sequence; and, thus, forms a transition for the overlying unit. Though very compact, these shales give rise to thick mud when wet, and lose all strength. Thus, slopes consisting of Dishang shales are very prone to landslides. This formation is > 10,000 m thick and is found only in the upper reaches of some of the NE tributaries of the Barak system.
2. The Dishang group continues upward, through a zone of transition, into the dominant arenaceous Barail Group which consists of 3 major sub-divisions, *viz.*, the Laisond, the Jenam and the Renji formations. The Laisond and Renji are both composed of strongly deformed coloured sandstones while the intervening Jenam consists of a variety of shales which may crumble easily on exposure; and, thus, form very vulnerable slopes. An almost same pattern continues from the Barail rocks upwards in the stratigraphic sequence, *i.e.*, an alternation of dominant sandy and shaly units. However, the younger rocks are, generally, softer and less consolidated than the older. *The hills between the Lubha gorge and the Cachar valley are typical examples of the Barail outcrops.*
3. The Barail rocks, in turn, are succeeded by a very similar assemblage known as the 'Surma Group', which is differentiated by a distinctly lower degree of induration. The Surma group has two sub-divisions, *viz.*,
   (*a*) The Bhuban sub-group, which, in turn, comprises of arenaceous lower and upper sections separated by an argillaceous section in between. The typical example

of this is the type area called Bhuban Hill where the upper Bhuban sandstones form the precipitous crown.

(*b*) The Bokabil formation which consists mainly of shale with sub-ordinate sand stones.

4. The rocks lying just above the Surma group are mainly soft sand stones with some amount of clay. They pass upwards into mottled clay of the Tipam sandstone and Girujan clay units which are well-developed in the area. Fossil wood, petrified or silicified things are quite common in these rocks.
5. A mixture of laterised sandstones and pebble beds constitute the youngest lithified sediments in the area. They are said to contain pockets and layers of clay. These rocks have been categorised under the Dupitila formation and the Dihing formation. The former form the low hillocks (locally called 'tila') in the Cachar plains; while, the latter, by and large, form the lower slopes of the hills around the Tea gardens.

Geology

This region is unique geologically because the processes of morphogeny, lithogeny and tectonics are said to be active simultaneously. The catchment of the River Barak is formed by two major geotectonic features, *viz.*, (a)the Nagaland-Haflong-Diyung thrust to the North and (b) the ophiolite belt of Nagaland and Manipur to the East (across the Manipur valley). This strongly-folded mobile belt is separated, by the northern thrust boundary, from the relatively stable platform of Assam, which is said to carry sedimentary cover of the same age. In fact, the ophiolite belt separates the zone from the Myanmarese platform. It is, however, believed that the ophiolite represents the margin of the Indian plate. It continues southward into the Andaman and Nicobar Islands through the Arakan Yoma (around Myanmar) and then into the Indonesian chain of Islands. However, in the folded belt, this stage was passed long ago and the prevalent model of evolution envisages the formation of folded mountain of the subduction margin. As the rising mountains pushed back the sea, the next zone to the west was similarly folded and uplifted. Thus, a progressive migration of orogeny, morphogeny and sedimentation were seen. The area is totally land now, remains dynamically active and the direction of evolution is the same.

The Cachar Valley

This valley lies between 24°10′ N and 25°10′ N latitude and 92°15′ E and 93°15 E longitude. The alluvial valley of this district is proximally formed by three kinds of hills which are as follows:

1. The precipitous southern face of E-W trending Barail ranges with an average relief of the order of 300 m or more to the North. The highest point is Kalapahar peak. The geological characteristics, including the dendritic and almost rectangular patterns being formed in the intensely-jointed sandstones strongly control the physiography and the drainage. The hills, said to had descended below zone of mainly flat-topped ridges. It, thus, formed a continuous bench dissected at the fringes from Kalain in the west to Kumbhirgram in the east. They too structurally consist of fens, terraces and very young lithified sediments, in which, the major components are sand, pebbles and boulders with some beds of clay.
2. A similar set of comparatively subdued ridges to the east marks the watershed with the Manipur river. The ridges tend to coalesce and divide along their length. The rivers like the Makri, the Irang and including the Barak itself, have considerable longitudinal courses before cutting across each other. The hills rise sharply till the crest of the watershed on the East of Irang River.
3. Further, the valley is limited by a group of low ridges to the south. They trend N-S and dispose like the outstretched fingers of a hand. It may be noted here that, the northern extremities of the hills of Mizoram are morphologically immature. Further, the presence of alluvial sediment in the intervening valleys, as well as, sharply entrenched channels at some places of certain tributaries of the Barak, suggest neotectonic uplift. Moreover, the Barak cuts the ridge at Badarpur; but, it merges again to the north to meet the northern hill range at Hilara. This septum, thus, separates the Cachar valley proper from the western plains of Karimganj and Sylhet (which is now in Bangladesh).

The next major ridge to the east is at Masimpur whose visible expression is limitedto the south bank. These two ridges seem to divide the Cachar valley into two sectors of distinct Geomorphic behaviour, *viz.*, the Silchar Block and the Salchapra Block.

4. The Silchar Block, lying between Lakhipur in the east and Masimpur in the west, varies in altitude from 18-20 m above MSL. The depositional valley of the Barak in this sector overlaps the erosional valley of the Sonai-Rukni rivers to the south near the village Palonghat. The alluvium, in this sector is usually 4-10 m thick and is typically underlain by laterised sandstone bedrock which is capped by iron hardpan.

Geomorphology of Cachar Valley

The River Barak comes down from the hills into the plains at village Lakhipur in Cachar District of Assam. Here, the river bed is *c* 15 m below the surrounding ground and the width of the channel is *c* 150 m. From Lakhipur, westwards to Nayagram, the course of Barak is nearly straight. The river begins to meander downstream of Nayagram, leaving cut-off palaeo-channel segments; and, thus, results in the formation of river-formed oxbow lakes/wetlands, locally called 'Anua', *e.g.*, Rupairbala Anua Beel on the south and Baskandi Anua Beel on the north. The present broad meandering loop was believed to have been formed before 1920 due to the shift of the course of River Barak from south to north and then to the present position accompanied by cutting-off of the two loops. Barak, then, flows straight south from village Govindapur No.2 to village Sonaimukh, where it meets the River Sonai coming down from Mizoram as River Tuirial. After this, Barak makes almost a closed loop and flows northwards to Kashipur. In this stretch, the Satkarakandi Anua Beel was said to have been cut-off before 1920. But, its counterpart, the Algapur Anua Beel, on the northern bank, was cut-off, perhaps, after 1930. From Kashipur to Masimpur, Barak flows west. But, evidences of northward shift are said to have been preserved in the channel scars and palaeo-loops near Khelma. However, there is said to be a geodynamic pressure on the river to migrate northwards as the southern hills evolved. Nevertheless, an equilibrium is said to be maintained due to the course coming-up against the low frontal spurs of the hills to the north. Significantly, the confluence of River Barak with River Sonai has shifted its location, during the period around 1986, (from its original confluence point around Sonaimukh) resulting in the formation of the latest 'Anua' (oxbow lake) called 'Dungripar Anua'.

Nevertheless, the floor of the valley in Silchar Block is made up of materials rich in clay which checks infiltration. It is, perhaps, the primary cause of waterlogging in the area. The sub-surface data available from boreholes, drilled by Geological Survey of India (GSI), Oil and Natural Gas Corporation (ONGC) and Central Ground Water Board (CGWB), revealed 3-5 m of clayey sediment at the top, underlain by laterised bedrock which is almost identical in composition. The zone of laterite is of comparable thickness (*c* 2-8 m); but, relatively fresh rocks are encountered below that. The rocks are, most commonly, shales and sandy shales of the Bokabil and Dupitila formations. The clayey nature of the strata could mean little unfavourable condition of exploitation of groundwater resources.

In contrast to Silchar Block, the Salchapra Block lies between the anticlinal Ridges of Masimpur on the east and Panchgram-Badarpur on the west. It also includes the Hailakandi Valley along the River Dhaleswari. The characteristics of the River Barak in Salchapra Block is markedly different from those in Silchar Block. It reveals both structural control and mobility.

In this stretch, in Salchapra Block, the River Barak generally flows westwards with much meandering. The nature and dimension of the meandering loops vary widely. There is a broadmeandering pattern, between the confluence of the River Jatinga with River Barak and River Katakhal with River Barak. The first loop has a length of *c* 4 km having a straight stretch of *c* 0.5 km. A comparison of different standard maps reveals no major changes in this segment of the river. To the contrary, appreciable changes are evident in the course of the river which lies between Salchapra and Panchgram. The course of Barak in this stretch is said to have been straightened between 1930–1972; there by, leaving the Salchapra and Katigora loops hanging. There had also been a down-cutting of 2-3 m along with the straightening process. Further, the influence of neo-tectonic activity on sub-surface structures is quite prominent, which could be due to the Great earthquake which had rocked Assam during 1950.

Incidentally, the northern half of Barak valley is a mosaic of 'Beels' representing the remnants of earlier shifts of the course of River Barak. Further, the constituent materials of the valley floor are rich in clay and there are identical problems of drainage congestion. The Tarapur-Ganirgram sector is said to be an example of such a problem.

The nature of the clay sequence and the occurrence of peaty layer indicates a phase in the history of Barak valley when a lake was formed and existed for sometime in the area. The nature of the floor makes this area prone to flooding; and, rainy season witnesses a vast sheet of water in this region.

Hailakandi Valley

This valley lies along the course of the River Dhaleswari, which is one of the major south bank Tributaries of the River Barak The River Tlawng descends from the Mizo hills and enters the plains of Barak valley as River Dhaleswari. After flowing for some distance in Barak valley, the River Dhaleswari was blocked by a blind dam near the village Krishnapur around Katlichhara; and, the main flow of the river was diverted through a channel (basin) called Katakhal which joins River Barak at the village called Katakhal near Salchapra. Incidentally, the Hailakandi Valley is bounded by linear ridges which tend to swerve towards the N-E at their extremities and the general ground level varies between 15–25 m above MSL.

The Katakhal basin is said to be an artificial re-suscitated old channel which now carries the main discharge of the River Dhaleswari while the original channel has now become moribund. The Katakhal flows along the eastern margin of the valley while the original river Dhaleswari flows along the western margin. However, from its take-off point, at village Ganjakhawri to village Muktachhara, the Katakhal flows along the older channel. As such, there are plenty of land marks of the shift of the River Dhaleswari. Nevertheless, the plains are said to be a composite floodplain of the rivers Barak and Dhaleswari around the village Algapur in the main valley. In fact, the shift of the Barak towards the north and that of Dhaleswari towards the west are, probably, related; and, both these appeared to be in response to the dynamics of deeper structures. Nevertheless, evidences of sinking of some portions are seen through the occurrence of upright tree trunks buried under 2-3 m of soil in the Chandipur TE situated towards the west of the so-called 'Jita-Nadi'

It is, therefore, quite likely that, such tilts could cause major changes in the river courses.

The 3rd principal segement in which the meridional ridges divide the Cachar valley, is the westernmost Badarpur Block situated downstream of Katigorah. In this region, Barak has a straight course confined between rocky banks at some places, *e.g.*, at Badarpurghat, where well-bedded Bokabil rocks are exposed. The depositional regime is restricted to a narrow strip on either side of the river. The thickness of the alluvial sediment here is *c* 1.5-2.0 m and the sand fraction is certainly more than that in the east. However, large proportions of clay are found here too. Further, the sediments are underlain by a zone of grey clay with some concentration of iron which, probably, represents the weathered bed rock.

Geo-environmental Considerations

A recapitulation of the inorganic evolution of the region (as a preamble of the study of geoenvironmental considerations), reveals that, the development of the trough in which the river Barak flows, represents both the beginning of the evolution of the Barak valley and the culmination of the evolution of the marine basin in which the sedimentary rocks of the Tertiary folded mountain belts were laid down. The geological processes continued with the modifications imposed by the results of each preceding event. The river itself appears to be antecedent to some of the features it cuts across. In fact, the total topographic pattern developed is in relation to the evolution of the youngest rocks and structures. It is believed that there was a period of undisturbed lacustrine condition before geological events which perhaps gave rise to high energy environment concomitant with rejuvenations and entrenchment, as revealed from the recent history of the Barak system recorded in the alluvium. However, during the immediate past historic period, the river had oscillated in some sectors as a probable response to neodynamic phenomena.

With regard to the problems of the present-day environment, the major threat of the Barak Valley is of 'floods'. The flood prone area is more or less restricted to the depositional valley regime. Some of the factors which contribute to the severity of the flood are as follows:

1. Occurrence of the ubiquitous clay layers under the valley floor

2. Seasonal rush of large volume of water from the catchment
3. Tendency of the river to change its course

It may be noted here that, spilling-over of the river bank or over-bank flow is a natural behaviour of a river, particularly, in its depositional stretches. In fact, it is in this pattern, how deposition takes place. Conversely, the erosional reaches of the river become waterlogged and flooded mainly due to factors like impervious clay and obstruction of drainage. In fact, problems are compounded by mainly erosional problems in the upper catchment, which results in both higher run-off and transport of heavy silt load.

A survey of the bank of the River Barak in Silchar Sub-division reveals the following:

In Lakhipur-Baskandi sector, the height of the river bank is 12-15 m followed by 10-12 m in Baskandi-Rongpur sector, culminating in 8-10 m in the Rongpur-Masimpur sector, which continues almost uniformly till Barak bifurcates into Surma and Kushiara around village Harinagar/Haritikar near Badarpur.

Notwithstanding the above, it is quite understandable that the sectors of Barak valley along the course of River Barak where the River Barak possesses a height ranging from 8-10 m, is severely flood-prone. Accordingly, the sectors beginning from around Rongpur to around Haritikar is highly flood-prone. A significant feature of the river in the Salchapra Block is that, the valley in this region is wide and floored with clay. The ground slopes away from the channel, especially to the north, where the terrain is a mosaic of cut-off loops. Height of bank does influence spilling-over. This area is a flood plain and the river has an inherent tendency to spread out.

In contrast to above, the flooding situation in Hailakandi valley is little different from that of Cachar valley. It is controlled by the variable discharge of the Rivers Dhaleswari.and Barak impinging on each other. It is further interfered by the congestion of the discharge due to conflicting geomorphological features of the two river systems.

Nevertheless, the area of flooding of the Badarpur Block is small being restricted to the depositional valley. However, here too, *e.g.*, in the upstream region of village Malua near village Haritikar, bank erosion is an appreciable phenomenon.

Measures of Flood Mitigation

Measures of amelioration of both incidence and damage caused due to floods could begin from the Upper part of the sub-basin and the hills around the valley. However, an appropriate Environmental Management Policy in this area would emphasise on two important aspects, *viz.*,

1. Maintenance of vegetal cover and an avoidance of construction and other interventions in the more fragile sectors. However, the very nature of the Cachar valley and the dynamics of its river systems dictate that only an adaptive strategy is likely to prove viable in this region. A major strategy of such an approach could be an emphasis on the control of settlement in the less flood prone and floodplain areas, *e.g.*, the Boalipar-Hailakandi stretch. Secondly, and perhaps, more important, would be to create flood cushions by spill basins (as impervious sub-surface condition may not facilitate aquifer cushions). The backswamps, including the remnants of the old left-out loops of the rivers, particularly of the Barak, serve now as the natural locations for spill basins. They could be further developed conjunctively for pisciculture and agriculture. The third strategy could involve easing of drainage congestion which could most effectively be achieved by developing better drainage, for which, the resuscitation of abandoned channels might provide a partial solution.

The foregoing analysis has tried to identify the problems and perspectives, in which, detailed and appropriate studies could be undertaken in the different sectors.

It could be concluded at this stage that, the dynamic nature of the river and the ecology of the valley are two very important factors which could be taken into consideration for any kind of developmental works in the region.

Watershed

The lowest area of the catchment where water accumulates to form a lake is called a 'lake basin'. Water may directly drain into the lake or pass through a system of streams and rivers and other depressions on its way to the lake. The spatial extent of the catchment, the amount of rain and the amount of rain falling on it, ultimately determines the volume of water entering the lake.

Watershed contributes to surface run-off to the water body. Watershed on a defined drainage channel includes all the land and water areas from which water drains out. It is marked by an elevated line which forms a division between two areas drained by two separate streams or bodies of water, *e.g.*, a small watershed of few hectares which drains into a small stream forming part of a larger watershed; which, in turn, forms a part of a still larger watershed, until the combined watershed becomes a major river basin draining millions of hectares of land. There could be sub-watersheds within a watershed, *e.g.*, a tribuatary into a lake/wetland having its own watershed, which is part of a large total drainage area to the lake/wetland. A lentic body reflects the size of the watershed, its topography, geology, land use, soil fertility, erodibility and vegetation.

Quality of water is degraded with increasing ratio of watershed area to lake area. In a large watershed, water from precipitation will interact and leach minerals from soil before being discharged into the lake/wetland. Lakes with small watersheds, maintained primarily by groundwater flow are known as 'seepage lakes'; while, lakes fed by inflowing streams or rivers are known as 'drainage lakes'.

Further, land use pattern has impact on the quality and quantity of water entering the lake/wetland. In the urban areas, the high proportion of impervious surfaces prevents the seepage of rainwater into the soil; and, there by, increases the rate of surface water flow to the lentic body. Moreover, high flushing rates from the urban areas could increase erosion of the watershed area providing sufficient force to mobilize and carry materials (minerals, etc.) along with the soil; thereby, affecting the water quality.

Soil types and their susceptibility to erosion is another important component of watershed. Further, the type and abundance of vegetation determines the extent of erosion. Areas with native undisturbed vegetation are less prone to erosion than areas with disturbed vegetation like agricultural lands.

Whether the watershed is a 'closed' (without any outlet) or 'open' (with outlet) system, is another aspect to be considered. In a closed system, all the water draining into a wetland/lake, stays within the lentic body.

In addition to the above, the water quality of the upstream region has an important role in the health of the lake ecosystem. Lakes situated in the upper end of the watershed may have different water quality than the downstream. Further, the downstream water quality may be affected, either positively of negatively, depending on the physical and biological characteristics of the upstream lakes.

Biogeography

The region has varied aspects of Biogeography.

The concept of South-East (SE) Asia as the 'Centre of origin' of the cyprinoids (Menon 1973, Banarescu and Nalbant 1982, Darlington 1957, Briggs 1979) has been revised; as there is, perhaps, no evidence to support it (Kottelat 1989). 'The area of greatest taxonomic diversity of a group of organisms is said to be the area of origin of that group of organisms.' This is, perhaps, the basis of the theory of centre of origin as proposed by Darlington (1957). Conversely, Matthew (1915) had postulated that, the most advanced species would be found at the centre of origin and the more conservative or primitive species farthest from it. However, all these hypotheses had been based on continental stability. Now, it is believed that, continents do drift. Kottelat (1989) had examined the freshwater fish composition of SE Asia and its environs. He had concluded that, there may not be a single centre of origin for the Ostariophysean fishes in the south, SE and East Asia. He had included the species of India, the Irrawaddy, and the Ceylon (now called Sri Lanka) as Indian fauna. However, SE Asian ichthyofauna included the ichthyofauna of Chao Phraya, the Mekong and the adjoining Islands. Interestingly, the Chinese fauna included the fishes of China and the Red River of the Tonkin Gulf.

MORPHOLOGY, MORPHOMETRY, HYDRODYNAMICS

Morphology

General Description

Limnological analyses of a lake or stream may sometimes require a detailed knowledge of morphology and morphometry, particularly, the depth and shoreline characteristics. These are sometimes critical to detailed analyses of physical, chemical and biological properties of freshwater. Accurate hydrographic maps of lakes and streams are rarely available with sufficient detail for a limnologist.

It is interesting to note here that, the line of demarcation between land and water is definite or indefinite depending upon a number of natural circumstances. Lake basins with steep incline of the bottom at the shore regions have margins which are less subject to changes than are those whose slope of bottom is so slight that even the smaller alterations in water level either expose or inundate peripheral areas to considerable extent. Hence, the margins may or may not be so definite, which, some of the maps may indicate. In lakes with bordering low-lying swamps, bogs or marshes, the margin may not only shift with changes of lake elevation, but it is also intangible at any particular time. Not infrequently, the lake basin is of such a form that one part of its margin is steep and definite while another part has extensive tracts virtually at water level. Amount of marginal slope, after all, determines the degree of definiteness of a lake margin.

General Description of Morphology of Lotic and Lentic Systems

(A) Lotic

The lentic and lotic systems are regarded as potential habitats for fishes. The fishes generally select the micro-habitats and not the overall habitat type. Most working lotic biologists use some more complex systems based on gradient, bottom type, flow rate, temperature and other water quality parameters, stream order, etc. Based mainly on origin and certain other parameters, lotic systems could be of the following types:

Headwater Streams (Rocky Creeks)

The headwater streams or the rocky creeks, range from tiny brooks which mark the recognizable beginnings of streams to rather sizeable, almost torrential creeks. They flow directly from the run-off and upland springs; and, the smaller brooks could be intermittent, *i.e.*, flowing only during the wetter parts of the year and drying completely in the late summer. Such streams usually have steep gradient in V-shaped valley and also nearly straight courses. Even the large rocky streams may have boulders and rocks or gravel bottoms with relentless swift currents and few pools where the water is little quiet. They usually have little visible vegetation.

Microhabitat

They do not have a diversity of micro-habitats. However, there are fast water flumes where the water is channelled by boulders and rock outcrops, interspersed with protected areas of quiet waters around large boulders. There may be falls with pools at the bottom and crevices between layers of rocks. Undercut banks and fallen trees and logs provide a different type of shelter. Sand bars may sometimes build up at the mouths of tributaries. Relatively few species live in the smaller headwater streams. Those streams which dry in the summer will not generally have any permanent resident; but, some fish may move-in when water is running. In segments, where water flows the year round, more species could be looked for in the still places where the rocks break the flow.

Headwater Streams (Marshy Creeks)

Marshy headwater streams drain low lying areas which are fed by springs. Some are slow, tortuously meandering brooks linked with alders. Others flow through grassy marshes. Sometimes, they are dammed by beavers. Much of the bottom is covered with flocculent organic debris from the vegetation in the stream and along the shore due to which the water is often stained brown. Moreover, submerged aquatic vegetation may form dense beds.

Microhabitats and Fishes

Micro-habitats, here, are somewhat more diverse than rocky creeks. The bottom may be clay, mud or organic debris. Much of the shoreline consists of peat or roots of swamp plants. Floods may wash-in a log here and there. Meandering streams often have undercut banks on the outsides of the curves and sandbars on the inner sides. Patches of vegetation provide shelter to the small fishes. Marshy creeks are somewhat more diverse than rocky creeks.

Further, marshy creeks have more micro-habitats and hence a bigger diversity of fauna. Their banks are often formed of the peaty remains of plants and by the roots of living plants. The streams may be shaded by grasses which hang-out over the water. As a result of this shading, along with their underground source, headwaters tend to be cool, rather stable environments. Such habitats are generally suitable for perches, killifishes, minnows, etc. Further, water of marshy creeks is characteristically stained little dark brown. As such, it could be difficult to see the fishes in the shadows as well as when they are under the roots of shore vegetation. In addition to the above, it may be noted here that, marshy creeks could also flow directly into swampy or marshy shorelines of lakes and ponds and their lower reaches could be the nesting ground of fishes, like perches, etc.which we could normally expect to find in the lentic systems.

Mid-reach Streams

These have alternating pools and riffles, the latter being a section of the stream where water current is deflected as it flows over cobbles and gravels; so that, the surface is permanently disturbed. They also vary in gradient and in bottom type. The alternating pools and riffles provide more micro-habitats where fishes can find their preferred conditions. There could be various kinds of shelters available. Aquatic macrophytes are abundant and diverse. Some parts of the streams could be shaded by overhanging vegetation.

Microhabitats and Fishes

Signs of erosion could.be seen which may lead to trees falling into the stream and deep pools tend to form around their roots. Streams which are subjected to flooding, some pools are nearly or completely cut off from the main stream. These pools come and go as the sand and gravel bars shift with heavy rain. Such semi-permanent pools are good places to see fish because they are free of currents. They are also often clear and the fish either are trapped here or can exit only through a narrow passage which they generally try to avoid unless they are seriously disturbed. Diversity in this stretch of the stream is more than that in the headwater stretch. Waterfalls are common in mid-reach streams. Even low falls and cascades over rocky outcrops and artificial cataracts over dams are fascinating. Falls and rapids are also suitable habitats for fish. The violently-stirred water rapidly comes into equilibrium with atmosphere; absorbing all the oxygen it can hold; and, at the same time, getting rid of undesirable gases, such as, carbon di-oxide and methane. Fishes which require high level of oxygen, tend to thrive in fast-flowing waters and downstream from waterfalls.

The fishes which live in streams having pool-riffle type of micro-habitats are very selective in their choice of dwellings. Some are definitely pool-dwellers while others are distinctly more at home in the riffles. While the pools could be the suitable habitats for cat fishes and other LGF (Large growing fishes), one could certainly look for the minnows and small perches in the riffles. The danid minnows could be found in the lower part of the riffles and in the nearby pools. Strangely enough, some of the loaches are riffle species in streams but they could also be found along the wind-swept shores of large lakes. In cooler streams, one could watch for trouts lurking under logs and undercut banks. Other species which live in both lakes/wetlands and streams usually prefer the areas where water movement is minimal.

Base-level Streams

These are usually slow-moving meandering streams with few riffles. They generally have sand bars on the insides of the curves and undercut banks on the outsides. They tend to be larger than the mid-reach streams and are subjected to periodic flooding.

Microhabitats and Fishes

Low gradient streams are often quite uniform for several kilometers. Only the depth, vegetation and bottom type may vary as the stream winds its way through meadows, farms, and occasional patches of forest. Sand bars and weed beds provide excellent fish habitat. In flat countries, these are subjected to periodic flooding. Therefore, a sandy bottom stream may have multiple channels, so that, it appears to consist of interwoven strands of water. Such streams are appropriately called 'braided streams'. Base-level streams usually have many logs and fallen trees which provide excellent shelter for fishes. Often one could find places where the banks are deeply undercut and the soil has eroded away from the tree roots.

Base-level streams, in general, are quite clear enough for fish watching although thick vegetation are generally seen. The pale spots at the bottom could be fish nests One could quietly make way into the weed beds to look for the minnows, killi fishes and the invertebrates.

Large Rivers

These are large navigable streams, many of which, might have been modified by flood control dams. Also, long stretches of their shorelines might have been modified by construction for protection against erosion. Pollution may be severe in some places. Many of these large rivers are inter-connected by canal systems. Further, there is perhaps no good definition of a river as opposed to a creek. It usually does not make much difference, from the standpoint of fish, how large the stream is, as long as, it has a variety of habitats and each habitat is extensive enough to be permanent through dry and wet seasons.

Microhabitats and Fishes

In general, the larger the river, more microhabitats are expected to be for fishes. In the open water, there may be sand bars, deltas around the mouths of tributaries and sharp dropoffs into deep water.

Large rivers are difficult places to watch for fishes. They are usually deep and turbid. Hence, it is hard to see except when they are in shallow water and very close to shore. However, some of the large rivers have substantial commercial fisheries of carps and cat fishes. Common fish groups in the large rivers include the Carps and minnows, silurids, bagrids, schilbeids, sissorids, etc.

Stream Survey

The streams (lotic system), like the lakes/wetlands (lentic system) may be mapped by various methods.

Deflection Angle Traverse (DAT)

In DAT, the direction of each successive axis line is established by measuring the angle which the last line makes with the forward prolongation of the preceding one. The same result is obtained, if, instead of a deflection angle, the angles which are formed by the two lines of course, without prolongation are used; but, deflection angles are mostly smaller in magnitude, easier to plot; and, on certain kinds of instruments, are easier to read.

In certain cases of mapping, the plain table, if available, can often be used to considerable advantage, particularly, if a telescope slidable with stadia crosswires is employed.

In addition to survey for mapping, the following data pertaining to habitat parameters of lotic systems could also be collected in the same exercise:

Selection of Habitat Parameters in the Study of Lotic System

1. Valley segment
2. Stream order
3. Total number of identified reaches
4. Reach type

5(a). Beginning co-ordinate
(b). End co-ordinate
6. Reach length
7. Micro-habitat type
Fall
Cascade
Riffle-pool
Run-sheet
8. Substrate type
Bedrock
Bouldeer
Cobble-gravel
Fines
9. Current flow/velocity
10. Bankful width
11. Bankful depth
12. Channel width
13. Channel depth
14. Wetted width
15. Wetted depth
16. Fish cover type
17. Total cover (per cent)
18. Riparian type
19. Riparian vegetation type
20. Signs of erosion

(B) Lentic

Water, together with its contained substances and energies, constitutes the immediate environment of aquatic organisms. Therefore, an analysis of the fundamental features of natural, uncontaminated waters is the necessary approach to an understanding of interrelationships, which can not be adequately determined until the things interrelated are themselves understood, atleast to some extent.

Study of interrelationships in a lentic system could begin with its morphology, which pertains to Length (L), Breadth (B) and Area (A) of the lentic system. They have bearing on wind action, waves, water movement and distributional pattern of depth of the water body. Morphology of a lake has relationship with the occurrence and growth of aquatic macrophytes and distribution and breeding ground of fishes.

Hydrographic Mapping

Accurate detailed hydrographic maps are basic requirements. Government and Non-governmental Organisations (NGO) have charted mainly big lakes, wetlands and rivers in such a way that, their maps could be tuned to limnological purposes. However, for the vast majority of inland waters, maps suitable for limnological works are still lacking.

Some of the morphological features of study with regard to various parameters are given below:

Geographical Co-ordinates

Latitude, Longitude and Bearings

A levelled surface, *e.g.*, the surface of a lake, is parallel everywhere to the mean surface of the earth and so it is essentially spherical. A level line is the arc of a circle whose radius varies according to the elevation above or below mean sea level. In the survey of small areas, level surfaces are assumed to be plains.

It may be noted here that, the equatorial plain bisects the earth normal to the polar axis. Further, the centre of the earth is the point of origin of the angular designations given to both latitude and longitude. Longitudes are described as west or east of any selected meridian which is called a primary meridian. Longitudes are true N-S lines. The directions of lines may be expressed as 'bearings' and azimuths. Bearings are measured, either clockwise or anti-clockwise, in quadrants and never exceed 90°. When the magnetic poles of the earth do not coincide with the geographic poles, the compass needle does not point to the true north; rather, it points along a line called a magnetic meridian. The angular deviation of the magnetic meridian from a true or geographic meridian is called a 'magnetic declination', which is east, if the compass needle points to the east of the true north; and, west, if the compass needle points to the west of the true north. The amount and direction of declination depends on both the geographic position of the observer and the year and day on which the observations are made.

Morphological Calculations

Some of the calculations related to the morphology of a lentic body is given below:

Maximum Length (L)

The maximum distance on the lake surface between any two points on the shore line; or, length of the line which connects two most remote extremities of the lake/wetland is called the maximum length. In fact, this length is potentially the maximum fetch or effective length for wind to act on the surface of the lake without land interruption. The form and position of this line must be such that it represents, as correctly as possible, the true open water length. This line may be straight, as in lakes of regular, ovoid form; but, often must be curved in form, as in instances of oxbow lakes or other lakes with irregular shape.

Measurement of maximum length of irregularly-shaped lentic bodies based upon a straight line connecting the most remote extremities and crossing portions of the mainland have little or no limnological value and be avoided. Some lakes are of such shape that, it is difficult to select one position for a maximum length measurement. In lakes of other shapes, *e.g.*, stellate form, there may not be a definite maximum length axis. In all such instances, the method and position of measurements be clearly specified in descriptions.

Maximum Width or Breadth (B)

The maximum distance between the shores perpendicular to the line of maximum length. The mean width (b) is equal to the surface area ($A_?$) divided by the maximum length (L).

$$\overline{b} = A_o/L$$

Maximum Effective Length (MEL)

It is the length of the straight line connecting most remote extremities of a lentic body along which wind or wave action occur without any land interruption.

MEL and Maximum Total Length (MTL) may be and often are identical. However, when islands or groups of islands are so located that they effectively interrupt the continuity of wave action and virtually divide the lentic body into two or more portions, MTL and MEL are different.

Maximum Effective Width (MEW)

It is the length of the straight line connecting the most remote transverse extremities of the lentic body along which wind and wave action occur without any land interruption.

Area (A)

The area of a lentic system is of vital importance in revealing the availability of water mass. Only in certain cases, it is feasible to determine the area of the water body by direct measurement in the field. Consequently, the methods for measuring area usually require a map of suitable scale and accuracy. Some of the common methods for calculating area of a lentic body are as follows:

1. By directly reading from the map using a mm graph paper.
2. By using a planimeter

In the first case, the map drawn on a tracing paper is placed over a mm graph paper and the number of squares (full or part) are counted along the shoreline boundary of the lentic body. Then, they are totalled and converted with the help of conversion table to give the area.

In the second case, a polar planimeter is an ingenious but relatively inexpensive instrument which permits an integration of the area of any plane surface regardless of its shape. A planimeter consists of two units, *viz.*, the pole arm unit and the trace arm-carriage unit. The tracer arm, containing the tracer point and the magnifying lens, connects the carriage unit. The latter consists of a dial, a vernier, a measuring wheel, a support wheel and a zero-setting wheel.

In practice, the operator traces the periphery of a plane surface with the tracer point in a clockwise direction. The planimeter units of the enclosed irregular area can then be read directly from the dial, measuring wheel and the vernier. Then, using the same pole and the tracer arm settings, as used for measurement of the unknown area, one can measure a region of known area. On a map, a square of side dimensions equal to 5 to 10 cm., made equivalent to some number of metres from the scale of the map, makes a convenient regular plane surface of known area. The area in question can then be determined by a direct proportional equation:

$$A/PU = A'/PU',$$

where,

A = Area of known plane surface

PU = Planimeter units of the known plane surface

A′ = Area of unknown plane surface

PU′ = Planimeter units of unknown plane surface

In using the planimeter, the carriage unit is gently placed, with the tracer arm, at the centre of the area to be measured. Place the pole weight or pin at some distance from the ball-socket joint of the carriage unit and connect both with the pole arm. The angle formed between the pole arm and the tracer arm be approx. 90 degrees. Never be allowed this to approach 180 degrees or a straight line or < 20 degrees; as the planimeter measures to the limit of its capacity at these angles.

Shore Line (SL)

The SL or length of shoreline (LSL) is the total distance of the shore of the lentic body on all sides.

The intersection of the land with permanent water is of nearly constant length in most natural lentic bodies. The LSL could, however, fluctuate widely in ephemeral lakes and in reservoirs and also in wetlands in response to variation in amount of precipitation and in discharge.

The LSL could be determined directly by measurement or from maps. Gadgets like chartometers, rotometers, etc., are inexpensive devices which measure the length of the lines by means of a tracing wheel whose revolutions are recorded directly on a graduated dial. The shoreline is traced from a starting point through the periphery of an area. This distance is then converted to real units by a simple proportional comparison to the distance measured on the scale of the map with some map measurer. Sometimes, the LSL could be determined directly from the field survey. Often, however, it is necessary to get this information from maps.

Direction of Major Axes

It is expressed in general compass direction, *e.g.*, NE-SW or NE-SSW.

Some of the pertinent informations related to lake morphology are briefly discussed below:

(a) Size of the Lake

Lakes differ in area from those barely larger than a pond to those of great size. For example, Lake Superior is the largest freshwater body on the globe which has an area >31,000 sq..miles. The Caspian Sea has an area of *c* 1,70,000 sq.miles. The Great Lakes of North America have a combined area of *c* 1,27,000 sq.miles. In India, the Lake Sone ('Sone Beel') is the biggest wetland in Assam and is one of the biggest in India too.

(b) Lake Elevation

Lentic bodies present a striking diversity even in elevation. While, as a whole, the elevation of surface is above sea level, exceptions are known, in which, the lake surface is far below it. The elevation of bottom of lake basin is, also, mostly, above sea level; although, exceptions do occur, in which, the lake surface is far below it. Lakes which conform to this rule, vary from those with slight altitude to high mountainous region, some of which may exceed an elevation of 3600 m.

The elevation of bottoms of lake basins is, also, generally, above sea level; although, exceptions do occur. Further, surface elevation of virtually all lakes are subject to variations from season to season and from year to year. These seasonal variations are often detectable by the increasing amount of exposed beach during the summer or during the dry season; although, care be taken not to mistake increasing extent of beach due to beach building. Moreover, unusual fall in water level sometimes expose shoal areas; thus, eliminating the principal rooted vegetation zones and temporarily robbing the lake of its most productive region.

In addition to the above, it may be noted here that, considerable change in surface level, annual or periodic, may have various other limnological effects, such as, increase or decrease of:

1. Outflow of water at outlets
2. Erosion and transportation of shore materials due to wave action
3. Formation of beach pools
4. Contact of water with earthy deposits on shores
5. Inflow of water from contributing streams and lakes
6. Available space for shallow water bottom organisms

Shorelines of Lentic (Lakes) and Lotic (Streams) Systems

A reasonably accurate hydrographic survey is essential for evaluation of the basic morphometric characteristics of lentic and sections of lotic water bodies. Such surveys could be performed with minimum of equipments; although the work could be facilitated and made more accurate when simple surveying equipments are available.

Often the general outline of the lentic or lotic system could be obtained from aerial photographs. The shoreline then is measured and formulated, section by section, until the entire area is circumscribed. It is sometimes easier to make shoreline surveys during the winter; because, during the rainy season, the margins may become swampy and not easily accessible.

Most shoreline surveys use the traverse method. In this method, a series of points are connected by straight lines of known lengths and angles from each other. Computerised surveying instruments are also available with automatic data collection capability. In other words, a series of connected straight lines whose lengths and angles have been determined by appropriate methods and instruments is known as a 'traverse'. Also, sophisticated methods of point location and survey are possible by the use of the Global Positioning System (GPS), in which, co-ordinates are determined from triangulation from satellites. The baseline data of the morphometric characteristics can, however, be obtained with a simple compass and measuring tape; but, at considerable sacrifice of accuracy.

Morphometry

General Description

Limnological analyses of a lentic or a lotic system very often require a detailed knowledge of morphometry; particularly, of the volume characteristics of the body of freshwater (FW). Depth analysis, including measurement of areas of sediments and of water strata at various depths, volumes of strata, and shoreline characteristics are often critical to detailed analyses of physical, chemical and biological characteristics of freshwaters. Further, for evaluation of different parameters, like signs of erosion, nutrient loading rates, chemical mass, heat content and thermal stability, biological productivity and effectiveness of growth, and for many other structural and functional components of the ecosystem, morphometric parameters are required. Management techniques, such as, loading capacity for effluents and the selective removal of undesirable components of the biota are also tremendously dependent on a detailed knowledge of the morphometry and water retention times in FW ecosystems. Further, it is a characteristic feature that, the morphometry of lakes and streams changes with time. Hence, even if bathymetric (depth contour) maps are available from governmental or other sources, their accuracy be checked carefully. It is, therefore, essential that, the rudiments of the construction of bathymetric map and the computation of morphometric parameters be understood before starting such works.

'Morphometry' is that branch of Limnology, which deals with the measurement of morpho- anatomical features of the basin of a lentic or a lotic system and its included watermass. It may be noted here that, certain fundamental conditions of biological productivity arise directly from the structural relations of inland waters which arise out of size and form interrelations, *e.g.*, other things being equal, greater the length of the shore line, greater is usually the productivity, which seems to be one of the important limnological relations concerning the form of the periphery of a lake/wetland. Thus, it becomes necessary for the limnologist to make various measurements of shoreline, area, depth, slope, volume, etc.; and, to establish from them certain ratios which serve as indices of differences among the various lentic systems.

Further, in order to make morphometric data valuable, measurements be based upon reliable hydrographic maps of appropriate size and construction. Other things being the same, the larger the map, the more accurate are the results. It may not be very useful to use the diminutive maps. Small maps, if otherwise suitable, be enlarged: (a) directly by some photographic process; (b) traced from a lantern slide projected on a screen; (c) by means of a pantograph; or, (d) be stored in a computer with softwares for doing such kinds of works. However, care be taken that, certain limitations inherent in small maps, are not accentuated in enlargements.

An account of the morphometric features of lentic and lotic systems are given below:

Morphometric Features of Lentic and Lotic Systems

(A) Lentic System

The depth contour map of a lake/wetland (particularly of a deep one) may be prepared through soundings.

Sounding Technique for Preparation of Depth Contour Map

The bathymetry of a lake/wetland basin is determined quite accurately from a continuous record of the basin contours with the help of an accurate sonar·device (also called depth sounder or fathometer). The depth sounder is moved across the lentic body at a constant speed along straight lines from known shoreline points. The continuous record shows small irregularities in bottom contours which easily could be missed by manual sounding methods. However, the disadvantages of depth sounders is that, most of them are inaccurate in water bodes which are < 2.0 m in depth.

Manual soundings of depth are made along transects between shoreline points with weighted calibrated lines; or, in shallow waters, with metered sounding poles. Where sediments are soft, the weights or ends of the poles could consist of a plate (30-40 cm in diameter) to prevent appreciable penetration into the sediment

The depth measurements along transects are then plotted precisely to scale within the map of shoreline. Points of equal depth at convenient intervals (*e.g.*, 1m, 2 m or 5 m) are interpolated between points of known depth in several directions. These points of equal depth are connected by contour lines (isopleths).

Important Terminologies in Lake/Wetland Morphometry

Maximum Depth (Z_m)

It is the greatest depth of the lentic body and is generally expressed in metres.

Mean Depth ($\overline{Z}$)

It is the volume of the lake divided by its surface area (at zero depth). In other words,

$$\overline{Z} = V/A_o$$

Relative Depth (Z_r)

It is the maximum depth as a percentage of mean diameter (Hutchinson, 1957). In other words,

$$Z_r = 50\,Z_{m.}\,\sqrt{\pi}/\sqrt{A_o}$$

Mean Depth-Maximum Depth Relation

It is a ratio of mean depth to maximum depth. It is expressed as a decimal value and it serves as an index figure which indicates, in general, the character of the approach of basin shape to conical forms.

Maximum Depth-surface Relation

It is the ratio of maximum depth to square root of the lake surface. Expressed as a decimal value, it is an indication of the relation of depth to horizontal extent.

Volume (V)

The volume of the lake/wetland basin is the sum of the values of the strata at successive depths from the surface to the point of maximum depth. The volume is approximated closely by plotting the areas of contours, as closely placed as possible, against depth; and, integrating the area of this curve by planimetry.

The volume of a lentic body can be determined by computing the volume of each horizontal stratum as limited by the several submerged contours on the hydrographic map and taking the sum of the volumes of all such strata. The volume can be estimated by summation of a series of truncated irregular cones of the strata. In other words, the volume could be calculated as follows:

$$V = h/3\,(A_1 + A_2 + \sqrt{A_1 A_2})$$

where,

h = Vertical depth (thickness of each stratum in metres)

A_1 = Area of the upper surface in m^2

A_2 = Area of the lower surface of the stratum in (m^2) whose volume is to be determined.

In short, for determining volume of the lentic body; at first, the surface area of the lentic body is determined; then, the area circumscribed by each submerged contour is determined. Then, computed the volume of the first (uppermost) horizontal stratum bounded by the surface and the plane circumscribed by first submerged contour using the above formula. Subsequently, volumes of succeeding strata are computed similarly. The sum of volumes of several strata is the total volume of the lake

It has been assumed commonly that, the increase in benthic area below an area of the lake surface resulting from the slope of the lake or stream bottom is negligible. This assumption could, however, be valid for very shallow lakes or shallow areas of a littoral zone where the slope is slight.

Volume Development (VD)

This index figure is used to express the form of the water mass or the form of the basin of the lentic body. This expression represents the ratio of the total volume of the lentic system to the volume of the cone whose area of base is equal to the surface area of the lentic body and whose height is equal to the maximum depth of the lentic system. VD may also be computed by the use of the simple formula:

$$VD = 3.(md)/mxd$$

where,

md = Mean depth of the lentic body

mxd = Maximum depth of the lentic body

When the value of VD approaches unity, the lentic system is close to the form of a cone whose height is the maximum depth of the lake/wetland; and, whose base is equal to the surface area of the lentic body.

Shore Development (SD) or Shoreline Development (D_L)

It is the ratio of the length of the shoreline (SL) to the length of the circumference of a circle whose area is equal to that of the lake. In other words,

$$D_L = SL/2\sqrt{\pi A_o}$$

where,

A = Is the area of the lentic body

As the shore line (SL) becomes more irregular, SD deviates more and more from the minimum SD value of one, *i.e.*, that of a circle. Many sub-circular and elliptical lakes have SD values of about two; while, lakes/wetlands of flooded river valleys have much large SD values. SD is of interest because it reflects the potential for development of littoral communities, which are usually of high biological productivity.

Volume Curves

These curves not only exhibit certain structural relations of the lake basin, but also make it possible to derive information from them by interpolation. A curve constructed by plotting depths along the ordinate and the superimposed volume above each submerged contour along the abscissa, is considered to be a convenient volume curve.

Depth-Area Curve (DAC) or Hypsographic Curve (HC)

Such a curve is a graphical representation of the relationship between surface area of a lake basin and its depth. This relationship may be expressed in absolute units of area (m^2, ha or km^2) which are at a given depth; or, in terms of per cent of lake area which is at a given depth. We could now plot the area of a given lake at each depth and likewise we could plot the area at each depth as a per cent of the surface area. In short, a DAC for sub-surface horizontal areas could be constructed by plotting depth along the vertical axis (ordinate) and the area along the horizontal axis (abscissa). Such a curve not only represents certain elements in the form of the lake basin but it also provides a means whereby areas at any depth level may be determined.

Depth-Volume Curve (DVC)

DVC is closely related to DAC and represents the relationship of lake volume to depth. The units may be expressed as volume units against depth or in per cent of lake volume above a specific depth.

Slope of Basin (SOB)

SOB of a lentic body is usually expressed in per cent although formulae are available for expressing the same result in degrees. In limnological works, there could be need for one or both of the two approaches to bottom slope measurement; *e.g.*, (a) determination of the slope between two submerged contours or isobaths, and (b) determination of the mean slope of an entire lentic body having many submerged contours. Nevertheless, slope between two adjacent contours may be calculated by the following formula:

$$\text{S or SOB} = (C_1 + C_2)/2 . 1/A$$

where,

S or SOB = Slope in per cent between the two contours

C_1 and C_2 = Lengths of the two contours involved

l = Vertical distance or interval between the two contours

A = Area of the bottom included between the two contours

Moreover, the mean slope of the entire lentic body could be calculated as follows:

$$S = [½(C_o + C_1 + C_2 + C_3 + \dots\dots\dots + C_{n-1} + ½ C)/n].D/A,$$

where,

S = Mean slope in per cent

$C_{o,}$ $C_{1,}$ C_2, C_3 are the lengths of the contours

N = Number of contours

D = Maximum depth of the lentic system

A = Surface area of the lentic body

Depth of Lentic Systems

Lentic bodies differ widely in depth. However, it is unlikely that even the deepest lakes would approach those of the oceans. Nevertheless, it is a striking fact that, a remarkable lake in southern Siberia, *viz.*, the Lake Baikal, has a maximum depth of *c* 1706 m. In North America, Lake Michigan is *c* 265 m; while,the Lake Ontario is *c* 224 m; and, so on. However, most lakes have a maximum depth of <30 m.

In contrast to the above, all wetlands, by definition, are 6 m or < 6 m in depth. Sone Beel, the biggest wetland in Assam, is 6 m in depth at the Full Storage Level (FSL). However, 'bottomless' lakes of local fame are said to be pure fictions.

Lake Bottoms

The term 'bottom' has been used here to include all parts of the bottom of a lake basin from the water's edge to the deepest region.

Lake Bottom Diversity

Tremendous amount of diversity prevails in the bottom of different kinds of lakes; as well as, in the individual lakes of the same type. All lakes of a particular region might have had the same origin; might have begun their history with essentially the same materials composing the basin; and, might exist under the same climatic conditions; yet, the bottom materials might be, and often are, markedly dissimilar in different lakes. It may be noted here that, every lake/wetland accumulates deposits on its bottom. The kind of bottom deposits and the rate of deposition are largely matters of local circumstances. Likewise, the distribution of materials, once they are in the lake, also depend upon local conditions. As such, two lakes of similar position, size, depth, age and original basin materials may now differ makedly with regard to their bottom deposits. Further, the nature of bottom deposits is a preeminently important feature in determining biological productivity. Different kinds of bottom deposits maintain different kinds of organisms which have some bearing on the biological productivity.

Sources of Bottom Materials

The principal sources of bottom materials are as follows:

1. The planktonic organisms which die and sink.
2. Plant and animal remains from the littoral zone
3. Wind-blown materials, both organic and inorganic
4. Silt, clay and other similar materials introduced by tributaries or washed-in directly by run-off water from adjacent slopes.

   Since these additions to the bottom are usually more extensive at certain times of the year, such as, spring floods and spring thaws, a definite stratification of the bottom may occur; so that, a transverse section of the deposits shows an alternation of the clay layer or other kinds of inwash with other layers of very different character representing an accumulation during other times of the year
5. Marl, which is largely calcium carbonate, produced by plants and animals or precipitated from the water by inorganic processes.
6. Remains of a blanket of floating algae.
7. Eroded materials from the shore zones

All these materials may contribute to the bottom of a lake. Further, newly deposited organic materials undergo decomposition; thus, finally, could change their character. Since, there is a continual addition of new bottom-making materials through the water, a transverse section of the bottom deposits shows a gradient varying from the topmost newly-deposited materials down to the deeper more chemically-stable substances. Those animals which normally live in bottom deposits, inhabit only in upper few inches of depth.

Classification of Bottom Deposits

A great diversity of bottom deposits occur in lakes. Different types of deposits are produced due to predominance of certain materials in the mixtures. However, we are yet to evolve a satisfactory classification of deposits. As such, there is much need for a critical study of bottom deposits.

Caspari (1910) classified bottom deposits in the Scottish lochs into three types, *viz.*,

1. Sand or grit
2. Clay
3. Crown mud

Further, three other types occurred sporadically and were listed as rarities, *viz.*,

1. Diatom ooze
2. Ochreous mud and
3. Calcareous deposits

Except in the following, the predominant component in each of these types is obvious from the name:

The brown mud in the Scottish lochs has an impalpable brown humus-like substance as its principal constituent which could result from the decay of vegetable matter. The ochreous muds could have a high content of limonitic iron which may give them a distinct red colour. The bottom deposits in Scottish Lochs may vary little from loch to loch. Hence, such a classification may not be considered to be broad enough to include all kinds of bottom. Extensive studies of the lake-bottom deposits of southern Sweden had been done by the Scandinavian limnologists, Lundqvist, *et al.*(1927). They had used a classification based on the relative proportion of clay, marl and fossil material in them.

Stratification of Bottom Deposits

Bottom deposits sometimes show a series of horizontal strata.

Alternation of conditions bringing about differences in deposition could result in such stratification. The clay layer, originating from the adjoining hills, is laid down during the spring, when inflowing streams carry their maximum load of eroded materials; while, the black ooze layer is deposited largely during the remainder of the year and almost wholly by processes going-on within the lake itself. Such a situation is said to be quite favourable for the study of the rate of deposition. Various circumstances may bring about some form of stratification; although, it could be such that, a lake may be so situated and so composed that detectable stratification is absent.

Historical Significance of Bottom Deposits

Bottom deposits constitute the historical record of a lentic body. The oldest deposits are at the bottom of the original basin and the most recent ones are at the top with intervening materials occurring in the order of their formation in the past. Study of the bottom materials revealed important information concerning ancient conditions within the lake itself and past conditions on the shore. Notably, deposits differ greatly in degree of disintegration. They could include even those which could be so completely reduced to yield little recognizable materials; or, could include partially preserved remains of plants and animals, which could provide much evidence of past events. For example, a stratum of sand in a bog lake deposit may possibly represent some period when the surrounding land area was denuded of its vegetation and wind-blown materials from the ancient glacial sand surroundings were, for a time, had greatly increased, until a new cover of vegetation had appeared.

False Bottoms

In northern bog lakes, a very striking phenomenon usually occurs, *viz.*, a false bottom formation. It is a mass of opaque, very finely-divided materials suspended in the water not far below the surface. When viewed from a boat, it may have the superficial appearance of being a definite, firm bottom; but, as a matter of fact, it is deceptive to the extreme.

Lake Elevations

Lakes present striking diversity even in elevation. As a concept, elevation refers to the surface of the lake above Mean Sea Level (MSL). However, exceptions are known in which the lake surface is far below the MSL. Lakes which conform to this general view, vary from those with small altitude to lakes situated in high mountains, some of which may exceed an elevation of 3600 m. Further, as a generalized rule, elevation of bottoms of lake basins is also above sea level, although exceptions do occur in this case also, *e.g.*., Lake Erie has a bottom which is above sea level; while, in Lake Superior (which is one of the Great Lakes), the bottom is *c* 122 m below the sea level.

Surface elevations of virtually all lakes are subject to variations from season to season and from year to year. Increasing amount of exposed beach during the summer or during the dry season could help in detecting these seasonal variations. However, care be taken not to mistake increasing extent of beach due to the beach building process or wave action on an exposed sandy shore for drop in level of water surface.

Lake Margins

The line of demarcation between land and water is considered to be definite or indefinite depending upon a number of natural circumstances. Lake basins with steep incline of bottom at the shore regions have margins which are less subject to changes than are those whose slope of shoal bottom is so slight that, even the small alterations in water level either expose or inundate peripheral areas of considerable extent. Hence, the margin may or may not be so definite as maps indicate. In lentic bodies with bordering, low-lying swamp, bog or marsh areas, the margin not only shifts with changes in lake elevation but it is also more or less intangible at any particular time. Not infrequently, a lake basin is of such a form that, one part of its margin is steep and definite while another part has extensive tracts virtually at water level. Amount of marginal slope may, however, determine the degree of definiteness of a lake margin.

High Water and Low Water Marks

High water marks (HWM) could easily be identified by ridges of debris; by certain kinds of bottom materials; by marks of erosion on headlands and elsewhere; and, also by others which are purely matters of local circumstances

Low water marks (LWM) are often less easily recognized but may sometimes be inferred from the positions of more permanent animal and plant zones of shallow waters. The nature of the margin has much to do with determining how clearly high and low water marks may be indicated.

Beach Zones

Exposed sandy beaches of big lentic bodies generally maintain three parallel environments, which have, to some extent, a certain degree of distinctness depending somewhat upon local conditions. These are:

The Inner Beach

That part, which extends from the water's edge, during periods of calm water, up the slope to the place where the surface of the sand ceases to be saturated with water and shows the first traces of drying, is called the inner beach. This zone is relatively narrow and exposed, in part, alleast, to the slight wash of the gentle waves of the calm weather.

The Middle Beach

That part which occupies space just beyond the inner beach and constitutes the region which is subject to the wash of the waves, only during the conditions of ordinary rough weather, is referred to as the middle beach. However, in calm weather, this part of the beach, while containing water at lower levels, shows a dry surface.

The Puter Beach

That part of the beach which extends from the middle beach to the outer limits of the beach proper, is called the outer beach. It is washed by waves only during the violent storms or during times of highest water levels. It is usually dry to some depth during the summer and is subject to certain amount of shifting due to wind action and could display a scanty encroachment of beach plants.

These zones have certain definite limnological relationships, one of which is the fact that, one or more of them may support definite biotic communities, *e.g.*, special assemblage of burrowing insects, etc.

Shore Cutting

On a shore, which faces the waves directly, *i.e.*, the crest of the on-coming waves, is almost parallel to the shore line. The final plunge of a wave lashes against the opposing land, thereby, loosening a certain amount of it. The amount of material loosened, depends entirely upon its nature and resistance. As water from a spent wave retreats down the slope of the beach, it carries and rolls a certain amount of loosened material into the lake, whence it may be transported to other locations by gravity, by undertows or by alongshore currents. Fine salt remaining in suspension for a considerable period of time, becomes scattered widely, ultimately, settling to the bottom. On high abrupt shores, which face the water, undercutting takes place, leading to the occasional fall of overhanging masses into the water.

Shore Building

It results from many processes, all of which produce additions to the original lake margins. A very common example is the familiar beach ridge which occurs along the outer margins of certain beaches as a result of wave action. Also, on exposed sandy beaches, a form of beach building is of common occurrence during the summer. In this process, waves coming on to the gently-sloping shoal, whose water depth is less than the depth of the wave influence, strike the inclined bottom; thereby, pushing and carrying ahead some of the sand. Each succeeding wave continues the pushing and carrying; ultimately, dropping sand at the water's edge. Under favourable conditions, the end result is the substantial increase in breadth of the beach.

Barrier Beach

Barrier beaches are ridges of sand formed in the face of the strongest prevailing wave action. In general, they are parallel to the shore and are located in shallow water some distance from the margin.

(B) Lotic System

Streams

Length (L)

The length of streams could be determined in two different ways:

By Direct Measurement in the Field

This may be available from the original field survey. In smaller streams, results of reasonable accuracy may be secured by rapid data recording along the bank or in the channel. Nevertheless, little approximate results may be obtained by walking and counting with the help of a pedometer.

Measurements from Maps of Suitable Character, Scale and Accuracy

This may be done satisfactorily by means of any one of the methods applicable for the measurement of lake/wetland shorelines.

Width (W)

Stream widths are usually measured in the field. The problem of measuring stream widths directly presents difficulties owing to the fact that great variations occur from changes of water level. Often the method and number of direct measurements must be matters of judgement and practicability depending on local circumstances and the needs of the work. For limnological purposes, some system of measurement is generally employed which could include all parts of the stream, from the narrowest to the widest. Such measurements could have dependable significance, if essential variations are depicted. In field works, stream width measurements may be based upon: (a) a selected water level, (b) general channel width, (c) high water level. In limnological works, the first-mentioned method is usually preferred. The common need is to determine a 'mean width' accurate within certain desirable limits.

However, measurement of stream width on ordinary maps of small scale is usually out of question. Dependable measurements can be made only on such maps which show particular sections of a stream on a very large scale and which have been drawn from careful and detailed field surveys.

Other Morphometric Considerations

Means of making other morphological measurements of streams are easily derivable from the methods applicable for the lentic systems. In order to have reliable data, parameters, like area, shorelines, volumes, longitudinal and transverse profiles, be based upon careful detailed surveys; and, the hydrogrphic maps be drawn on large scale. Shore development be computed on the same basis as applicable for lentic water bodies.

Hydrodynamics

The ecology of the surface run-off waters play an important role in the development of fisheries of streams because ecology determines the habitability and abundance of flora and fauna in its different sections. The main water precipitated on the earth flows into depressions and follows the course of least resistance seeking its own level. The local variations in run-off waters depend upon the following:

1. The nature of soil (porosity and solubility)
2. The slope gradient
3. Its vegetation
4. The climatic conditions, *e.g.*, temperature, wind, humidity, etc.
5. The volume and frequency of precipitation

Due to slope on land surface, the run-off canalizes itself quickly into small gullies, which, in turn, join to make larger channels and rivulets. The rivulets finally converge to form brooks, which eventually lead to larger streams.

A 'stream' may be defined as a mass of water with its load of suspended inorganic or organic materials flowing towards a lower level following the line of least resistance. The streams, depending upon their source, are either seasonal or perennial.

Large rivers are navigable streams. Most well-known rivers in the world, *e.g.*, the Mississippi-Missouri, Ohio, etc., have been modified by flood control dams with locks for commercial or recreational boat traffic; and, long stretches of their shorelines have been modified by construction of dams and embankments for flood mitigation. Many of the large rivers are interconnected by canal systems.

However, there is, perhaps, no complete definition of a river as opposed to a creek. However, from the standpoint of fish, it usually does not make much difference how large the stream is, as long as, it has a variety of habitats and each habitat is extensive enough to be permanent during both dry and wet seasons.

Notwithstanding the above, some of the pertinent aspects related to hydrodynamics of lotic system is given below:

Water Movements

Movements of water have numerous far-reaching limnological relations. Some amount of understanding is, therefore, necessary of the major physical features of water movement. The principal forms of movements of water are waves, currents and seiches.

Waves

These are mainly wind-produced. They occur on every body of water in forms and magnitudes depending upon various local conditions, *e.g.*, area of open water; direction, duration and velocity of winds; shape of shoreline; and, relative amounts of deep and shallow water. Waves could become formidable in size and action in the big lentic bodies; while, in the small ones, waves, although potentially present, are always minimal and not without their limnological significance. Incidentally, the greater the expanse of water over which the wind blows, the greater the potential wave height, wave length and wave velocity.

In open water, where a wave is free to move without modification, wave motion is essentially as follows:

Assuming that a wave is already formed and in action, each particle of water describes a vertical, circular path and theoretically returns to the position from which it had started. Therefore, the wave moves along, producing a vertical rise and fall at successive positions, but no horizontal movement of the water. Such a wave is sometimes called a 'wave of oscillation. Such a wave is modified by circumstances, though not infrequently, *e.g.*, the wind, when its velocity is such that it blows the crest of the wave over, causing it to fall forward; and, thus, produce a 'white cap', in which, actual horizontal motion of the water does occur. Similarly, as a wave from the open lake approaches the shallow water, certain changes occur, one of which is that, the wave height is increased and then the top pitches forward in a sort of forward rolling motion, forming 'surf'. This also results in horizontal motion or flow of the water. Surf is often explained as due to the drag of the bottom of the wave when it encounters the shoal, causing the top to overturn the bottom and thus fall over; although, such explanation is not always true.

The formula for computing the height of the waves, as quoted by Johnson (1919) from Stephenson, is as follows:

$$\text{Height (H)} = 1.5.\sqrt{\text{distance}}$$

In this formula, height was expressed in feet and distance in nautical miles. By distance, it is meant that, expanse of open water over which the wind blows. It is pointed out, however, that, for short distances, Stephenson's formula requires little modification.

Nevertheless, the formula for representing conditions in large reservoirs is as follows:

$$H = 0.025.\sqrt{f}$$

where,

H= Height of wave in feet.

F= 'Fetch' or clear distance (expressed in feet) over which the wind blows.

Waves, in which, there is a definite forward movement of water itself are known as 'waves of translation'. These may result from the effect of waves of oscillation coming onto a very gentle sloping shoal and forming breakers. They may also result from the sudden addition of considerable volume of water into the lake.

A natural stream usually has great transporting power in respect of material arising from disintegration.

It may be mentioned here that, water thrown-up on shore runs back down the slope; thereby, forming undertow. Often the downflowing water is met by the next expiring wave; so that, shifting shore materials are alternately swept or rolled back and forth.

Depth of wave action is a matter of considerable limnological importance; but, unfortunately, precise information as to the extent in different conditions is lacking. It is sometimes stated that, the maximum depth of wave action is equal to the wave length. According to other statements, the disturbance caused by waves extends to a distance below the surface equal only to the height of the wave.

In short, water movement could be summarized as follows:

Stream velocity is a function of the gradient of the bed. It varies from near static in pools in lower reaches of the rivers to higher rates of *c* 9 m or more per second. However, the maximum velocity of a flowing stream in its vertical section is generally found somewhere within the first third of the depth of the water from the surface. In shallow streams, the maximum velocity zone lies quite near the surface. The velocity is reduced at or near the surface because of surface tension and is diminished at the sites and the bottom of the stream bed due to frictional effects. The current velocity is not uniform in all parts of the transverse section of the stream. These parameters have a bearing on the rheotropic movement of planktonic organisms, young fish, etc.

The total volume of stream water passing at a point in a given period of time is called 'discharge'. The rate of discharge is proportional to the rate of flow and volume of water fed into a stream. The discharge of a stream is estimated by the formula

$$Q = W.\, D_m.\, V_m$$

where,

Q = Discharge in cubic feet/cubic metre per second

W = Channel width

D_m = Mean channel depth

V_m = Mean velocity in feet/metre per second.

A natural stream usually has great transporting power in respect of material arising from disintegration of sub-surface layers of hills and slopes of upper part of valleys. The movement of suspended matter is a function of competency, capacity and load. Competency relates to the maximum size of the sediments transported by a stream and is dependent upon its velocity and turbulence. The power of a current to transport varies as the square of the velocity. As such, rocks, boulders and large stones are physically rolled down by the torrents near the sources of the rivers. These large pieces of stones are gradually ground in their onward course into shingle, gravel, sand and silt which are transported by the main rivers towards the sea or partially strewn over the flat plains during the floods. The size of the materials, forming the bed of the river or what is carried by the stream, becomes smaller with the reduction of its current velocity.

Currents

Water current is as important as the waves in in-shore building activities. They are secondary to the surface currents which pass alongshore carrying and dropping materials. In the smaller inland lakes, these currents are largely determined by the prevailing winds. As these currents pass projecting headlands, they pick-up loose materials and transport them elsewhere. Since shorelines often present a series of sinuosities, along-shore currents tend to pass more directly from one head land to next, instead of following closely the concave shoreline. Having encountered a headland and picked-up a load of materials, the current loses some of its initial velocity when it reaches the other side; and, as a result, deposits along its path, some of the load of suspended material. Eventually, a sand split begins to grow out from some point on the more protected side of the headland; which, in turn, reaches the shore somewhere along the side of the bay, forming a bar which completely cuts off a small portion of the lake and establishes a sand-split beach pool. Incidentally, a step in the process of smoothing-out the concavity in the shoreline is completed. There is a general tendency of the along-shore current to follow more or less a straight course. However, the sand split does not ordinarily pass in a straight course from one headland to the next. Instead, it tends to approximate a sigmoid form. It, thus, cuts off only a narrow portion of the lake. Further, before the first sand split is completed, another may begin to form from the lakeward side of the first. However, it could repeat the history of the first in a general way somewhat further along. This process continues until the bay is completely obliterated and the shoreline extends more or less directly from the original headland to the other. Vegetation begins to invade the growing sand split. At first, certain sedges and small seedlings followed by shrub-like plants and finally forest trees. Beach-pools, often formed, serve as amazingly good habitat for aquatic biota. Such situations also provide splendid opportunities for studying ecological succession.

Currents in lakes are of three kinds, *viz.*, vertical, horizontal and returning (or undertow current).

True vertical currents seldom occur in inland lakes but may be present in large waters, *e.g.*, the Great Lakes. However, when present in inland lakes, they are said to be due to some unusual thermal, morphological and hydrostatic circumstances.

Horizontal currents are common in lakes. They are usually produced by wind and often modified by the shape of the shoreline and form of the basin. In large lakes, like the Great Lakes, the surface velocity could be *c* 5 per cent of that of the wind causing it; but, it is < 5 per cent in smaller lakes. In Lake Okoboji, Iowa, USA, a wind of 243.902 m per minute produced a current of 7.621 m per minute. Also, a thermal stratification could be established and only heaviest winds could disturb the water level below a depth a 20 m. Further, the ratio of wind velocity to water movement could diminish as the wind velocity increases and water velocity generally decreases with increase in depth. Lakes with large inflow or outflow or both could have areas of flow not due to wind action. This is particularly of those lakes which are expansions of a river system through which surface water drifts due to river current. Returning currents are formed when water is piled-up on an exposed shore as a result of an onshore wind. Such action raises the water level at that position. As a result, the excess water may return underneath along the bottom. The magnitude and duration of such currents depend upon the velocity and duration of the wind. Return currents may be set up by steady, vigorous on-shore winds which extend to the opposite side of the lake. During summer, a considerably deep lake is thermally stratified. However, the upper portion of the water (epilimnion) is approximately of the same temperature and hence of the same density. A strong steady wind of some hours' duration continually could drift the water upon the exposed shore. As such, the excess water may return down the slope of the basin to a considerable depth. There could be some exceptions under certain special circumstances. This downflowing returning current is diverted in a horizontal direction and flows towards the offshore after encountering the colder, denser water of the thermocline region. In such a circulation, the upper stratum of the epilimnion flows towards one shore; the lower stratum (the return current) flows towards the opposite shore and the stratum between the two (which is known as the 'shearing plane') is pratically without any motion. Based upon the use of floats adjusted to different levels, that such a return current may also occur in part beneath the thermocline (Whipple, 1927). If this is true, then, the hypolimnion (the region below the thermocline) is not so completely devoid of circulation, as is sometimes supposed. There are also evidences of sub-thermocline circulation, the nature of which is, as yet, little understood. During the Spring overturn and fall winters, the water is almost of uniform density from top to bottom. As such, return currents may extend even to the bottom of the relatively deep inland lakes. If wave fronts meet the shoreline obliquely, the water piled-up on the shore, may seek to re-establish the water level by flowing off laterally on the surface and along the shoreline. Various forms of these alongshore return currents, *e.g.*, lateral whirls or eddies, may result from the modifying circumstances of local shore and basin characteristics.

Tides

Tides are almost imperceptible in inland lakes and even in the Great Lakes. Lake Michigan is said to have a tide of *c* 5 cm. This virtually means that, tides in freshwaters are, so far as known, negligible phenomena in limnology.

Seiches

'Seiches' are oscillations which occur in water levels in lakes and along the sea coasts under certain circumstances. Forel (1895) made extensive study of these phenomena although these were known to some extent for atleast two centuries. A seiche consists of a local periodic rise and fall of the water level. It is an example of a standing wave, in which, the water particles do not travel in circular orbits; but, the advance and return of the particles are in the same path. Any influence, which produces a temporary, local depression or elevation of water level, may produce a seiche.

The following are the influences which most commonly produce seiches in lakes:

1. Winds, temporarily strong, which pile-up water on the exposed margin of the lake
2. Sudden changes in the barometric pressure over a portion of the lake area.
3. Earthquake
4. Landslides
5. Sudden very heavy rainfall at one end of a lake.

Of these, the first seems to be, by far, the most common cause, particularly, in the small inland lakes. A sudden vigorous summer storm, accompanied by high steady wind, which subsides as abruptly as it begins, piles-up water on the exposed shore; thus, making the water level high at that point and corresponding little low level on the opposite shore. When the blow is over, the lake surface begins to swing, with alternating rise and falls on each of the two opposite shores. The swing diminishes rapidly in amplitude and ultimately disappears completely.

Each lake has its own period of oscillation upon its essential dimensions. The amplitude, depending upon the dimensions of the lake and the intensity of the initial cause, may vary from a fraction of a centimetre in small lakes to one metre or more in large ones. In Lake Geneva in Switzerland, it is reported that the amplitude of a seiche may reach two metres. Seiches of considerable magnitude occur on the Great Lakes becoming most noticeable in the protected covers and bays. Seiches on Lake Erie sometimes expose, for a short time, the shallow water bottoms along the protected side of the peninsula, known as 'cedar point'.

In inland lakes, seiches are seldom seen because their amplitudes are very small. Seiches may cause a current to run-in and later run-out through narrow strips of narrowly contracted passages rarely where local circumstances provide the necessary settings.

The formula for computing the period of oscillation of a seiche in a lake whose basin has definite regularity of bottom is given below (Forel, 1895):

$$t = l/\sqrt{gh}$$

where,

t = Time of one-half oscillation in seconds

l = Length of axis of seiche in metres

g = Acceleration due to gravity (980 cm/sec^2}

h = Depth of water in metres.

For evaluating in the case of lakes having irregular basins, the formula is given below (Whipple, 1927):

$$t = 2l/3600.\sqrt{dg}$$

where,

t = Time of oscillation in hours

r = Length of lake (or length of axis of seiche in feet)

d = Mean depth in feet along the axis of the seiche

g = Acceleration due to gravity (32.16 ft/sec^2).

Forms of Seiche

The different forms of seiches are given below (Forel, 1895):

Longitudinal Seiche

Its axis corresponds with the direction of the long axis of the lake.

Transverse Seiche

Its axis lies in the direction of one of the shorter axes of the lake.

Both longitudinal and transverse seiches are further of four forms:

1. Uninodal: Having one node only
2. Binodal: Having two nodes
3. Plurinodal: Having several nodes
4. Dicrotic: Having two beats (shown as two peaks on a limnograph) due to interference of uninodal and bimodal seiches.

Sometimes, lesser known forms of water motion are considered as seiches, *e.g.*, the short period, back and forth flow of water through narrow channels in certain localities of very large lakes; and, the sub-surface seiches, a type which has been postulated as the cause of certain submerged currents in Lake Erie.

Sub-surface waves, sometimes produced in large bodies of water, occur where sub-surface water is denser than the overlying water. A strong localized wave starts as an impulse (wave) in the underlying layer of water which moves forward in the direction of the wind. As this wave moves along, the warmer lighter water just in front of it passes over the crest of the wave but in the opposite direction; thus, producing a surface current, opposite to the direction of the wind.

Sub-surface seiches usually arise from a temporary displacement of the thermocline due to the weight of the piled-up surface water on one side of the lake caused due to strong wind action. The superimposed layers of water are of such density that, the waters resist mixing. The underlying cold heavy water will be depressed under the region of temporarily accumulated surface waters and will rise to a corresponding degree under the opposite shore waters. When calm follows storm, the underlying heavy water begins to restore equilibrium by seeking its level again, swinging to the other side but going past its original level. Such seiches have been reported from many inland lakes.

HYDROGRAPHY, MORPHOLOGY, MORPHOMETRY AND HYDRODYNAMICS OF FRESHWATER BODIES IN BARAK DRAINAGE, MIZORAM AND TRIPURA

Information on the hydrobiological conditions of any water body is of prime necessity before endeavouring to utilize it as a productive fishery. It is an established fact that proper planning depends on the availability of reliable data. But, unfortunately, in the limnological and fisheries sector, there is an acute shortage of such data in the North-Eastern region of India.

In Barak Drainage

Contrary to such background, the naturally-formed lakes and rivers constitute great potential of fishery resource in India. Of the *c* 0.72×10^6 ha lentic coverage in India, the North-East region of the country, particularly Assam, portrays significant possibilities of such water bodies for productive fisheries. Besides lotic territories, the region is gifted with myriads of riverine and tectonic wetlands (locally called 'Beel',' Haor', 'Anua') which alone constitute *c* 81 per cent of the total lentic areas ($> 0.12 \times 10^6$ ha) in Assam.

In Assam, there are *c* 1392 number of wetlands having a total of *c* 22,896 number of fisheries of different categories. Out of these, the number of registered wetlands is only 394 (30.38 per cent) covering an area of *c* 70,000 ha; of which, *c* 19,000 ha is in good condition, *c* 15,000 ha is in semi-derelict condition and *c* 35,000 ha is in derelict condition (Govt. of Assam, 1980). The remaining wetlands are mostly unregistered and under the control of both government and private sectors. In this context, it is interesting to note that the district of Cachar includes the highest number of unregistered wetlands in Assam.

An account of some of the prominent lentic bodies in Barak drainage of Assam is given below:

Lentic Bodies

Some of the significant lentic bodies in this region are given below:

1. Beel (Perennial wetland): Sone Beel, Sat Beel, etc.
2. Haor (Seasonal floodplain wetland): Chatla Haor, Puneer Haor, etc.

3. Anua (River-formed oxbow wetland): Baskandi Anua, Satkarakandi Anua, etc).

An account of some of these is briefly given below:

(*a*) Sone Beel (Kar, 1990)

It is situated between 92 º 24′ 50″–92º 28′ 25 ″ E and 24º 36′ 40″–24º 44′ 30″ N within Karimganj district of Assam and falls in a valley geologically called syncline (Figure 4, Plate 1).

The physiography of the district consists of small hillocks intervened by wide low valleys. The hillocks have NE-SW and NE-SSW trend near the Barail range and N-S trend towards south away from the Barail range. Notably, Sone Beel, the biggest 'Beel' (wetland) in Assam is situated in between two hill ranges, *viz.*, the Badarpur-

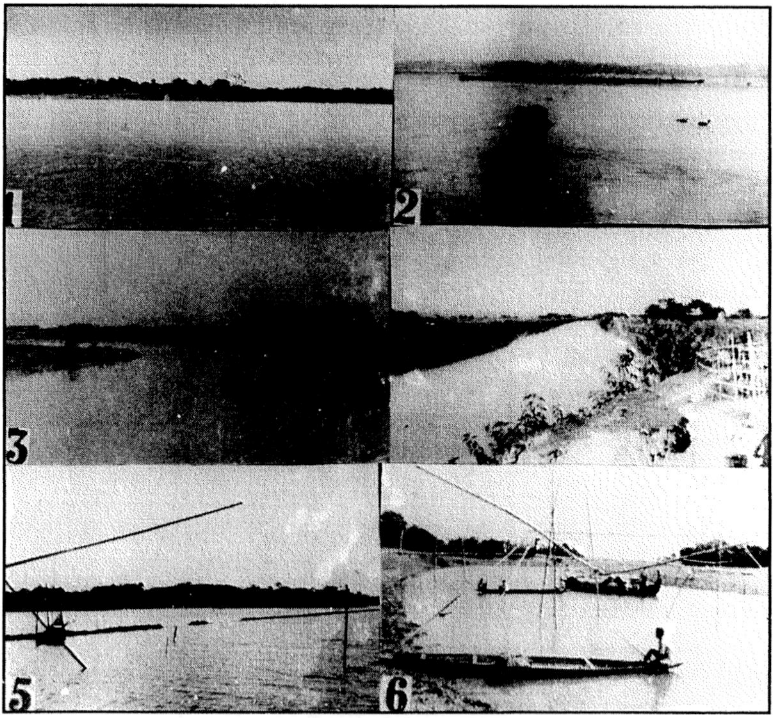

Plate 1: Sone Beel at FSL and DSL

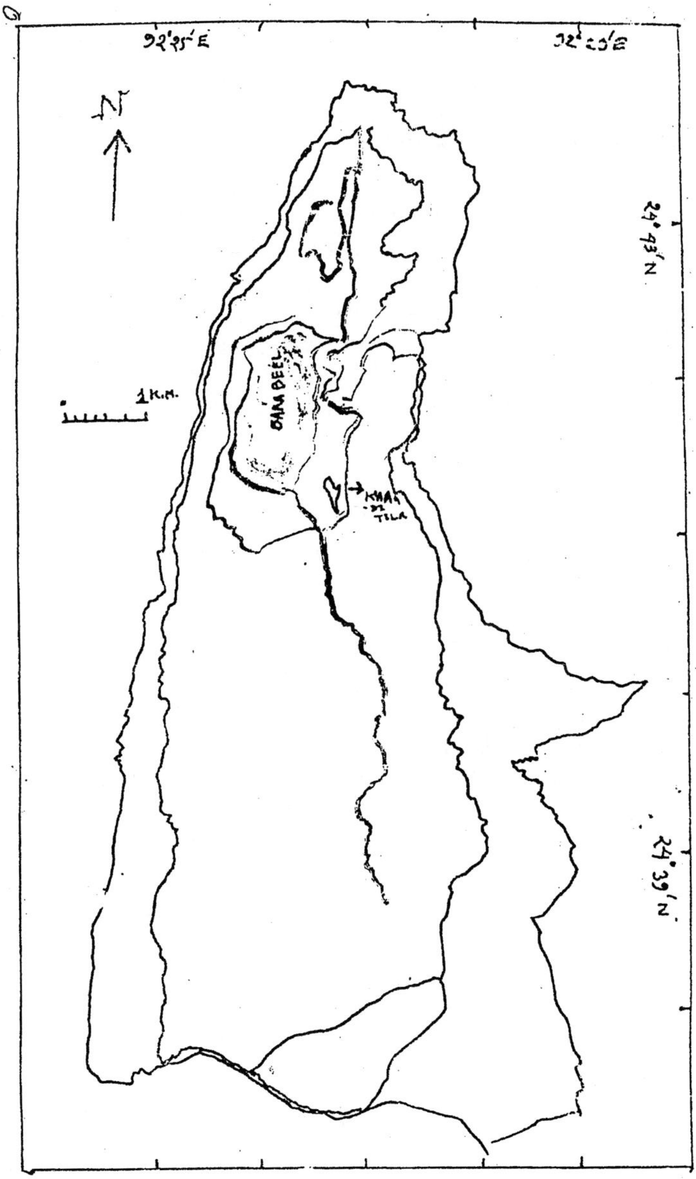

Figure 4: Boundary Contour Map of Sone Beel Showing FSL and DSL

Saraspur range and the Chowkirmukh-Dohalia range. In the East, the neighbouring structure is the Badarpur line of folding; while, there is the Chargola anticline towards the west. A typical geomorphological feature is the tightfoldedness of the anticlines represented by hillocks having very high dips of the sedimentary beds.

Information obtained from ONGC and GSI (Personal communication) reveal that, Cachar represents a type area of Surma sediments exhibiting only Tertiary deposits (70 million years old). Investigations into the rock samples of this wetland revealed that the hillocks around the weland, were, probably, formed after Tipam sedimentation. Most of the wetlands in this region, including the mighty Sone Beel, might have been originated after the Dupitila sedimentation during the Mio-Pliocene period.

Around Sone Beel, the soil in the catchment of the plains is generally loamy, but occasionally sandy or gravelly admixed with quartz. Conversely, hilly portion of the catchment consists generally of fine grain sand stones bearing many angiosperms and thus forming the evergreen forest.

The principal feeder of the wetland is the major inflow, the river Singla, which drains a total catchment area of *c* 46,105 ha. The wetland receives water from 12 minor inlets and many other canals flowing from both hills and plains, all of which together drain a total of *c* 18,941.9 ha of the catchment area of the wetland. Out of this, *c* 11,003.9 ha lies in the plains; while, *c* 7,938.0 ha falls in the hilly portion of the wetland. These, otherwise, form respectively 58.09 per cent and 41.91 per cent of the total catchment area of the wetland.

Further, the catchment of the wetland also includes the reserve forests (RF) of the State, notably, the Singla RF. During monsoon, the wetland receives some humic as well as inorganic and organic nutrients from the hillocks and cultivable areas, particularly around the swollen tail end of the wetland..

On the west, the wetland is connected to Karimganj town by a *c* 30.0 km railway line; whereas, Badarpur Junction is situated at a distance of *c* 24.5 km from the north of the wetland. The nearest town Hailakandi is situated at a distance of *c* 9.5 km to its east. The Block HQ Ramkrishna Nagar is situated at a distance of *c* 8.5 km from the south end of the wetland.

The maximum length (L) and breadth (B) of the wetland at Live Storage Level (LSL) were measured as 12.5 km and 3.9 km respectively. Interestingly, these values reduce to 4.07 km and 2.22 km respectively at its Dead Storage Level (DSL).

The area of Sone Beel at LSL was measured as 3458.12 ha; while, at DSL, the area diminished to only 409.37 ha. The length of the shoreline was measured as 35.4 km while the shore and volume developments were recorded as 1.69 and 0.15 respectively with mean depth of 0.29 m. The gross volume of the wetland was found to be 101.54×10^6 m^3.

Silt Islands (SI) were recorded on the northern and southern parts of the lake. Of these, the Gopikanagar SI (area 3.74 ha and 25 MSL) and Khagdi tila SI (area 3.31 ha and 21 MSL) were noteworthy. Interestingly, the wetland surface itself is situated 23 m above the sea level (MSL).

The wetland basin tended to become deeper from south to north. The contours in the west were found to be almost parallel and closer than their counterparts in the east.

Although the wetland did exhibit variable water level ranging from 0.07 to 5.69 m at FSL (Jun-Sep) and 0.02 to 2.08 m at DSL (Nov-Apr), the average depth of the wetland was found to vary from 0.18 m (Feb 1980) to 3.34 m (Sep 1979) in 1979-80 and 0.16 m (Jan 1981) to 3.38 m (Jun 1980) during 1980-81.

Although 12 minor inlets were found to exist in different parts of the wetland, the wetland is mainly fed with the major inlet, the river Singla. It originates as 'Thing Tlawng Lui' at a height of *c* 365.21 m MSL in Mizo Hills, from where, after traversing a meander course of *c* 62.75 km, it enters Sone Beel.

The major outflow (there being no minor outflow) of the wetland, the river Kachua originates from the northern-most end of the wetland. It drains out the wetland water into the mighty river Kushiara after covering a length of *c* 19.30 km. Although the river Kachua was blocked by a blind dam constructed by the Government of Assam in 1950-51, the dam was replaced by a lock gate in 1964 after experiencing navigational and fishery problems. The maximum outflow of the wetland was found to be 87.03 m^3 sec^{-1}. in Aug 1980; while, the maximum inflow was recorded to be 33.91 m^3 sec^{-1} in July

1980 during the entire period of investigations. Concomitantly, the minimum inflow and outflow of the wetland were recorded as 0.0027 $m^3 sec^{-1}$ and 0.087 $m^3 sec^{-1}$ (both in Feb 1980) during the period of study.

The wetland, as a norm, was found to exhibit a decreasing trend of its storage capacity from post-monsoon period culminating in winter to early summer whence it again gained its storage at the onset of monsoon.

(*b*) Sat Beel (Kar, 1998)

This wetland is situated in the village Rongpur of Silchar Sub-division in Cachar district of Assam. This Beel is an aggregation of seven wetland units closely applied to each other and they become a single sheet of water during monsoon.

The physiography of the locality consists mainly of plain land with a small forest along the right bank of Barak. The tree species are the 'simul' (cotton) and the 'jarul'. The village 'Rongpur' acquired its name because of the colourful red flowers bloomed in large number in the simul trees during its flowering season. Unfortunately, such simul trees had been largely deforested for settlement of villagers due to increase of population during the last 25 years. Concomitantly, the seven wetland units, of which Sat Beel is constituted, are, Gajaria, Chepta, Bardoloo, Kachudaram, Mokachakri, Koia and Kejua. The river Barak, which flows *c* 1.0 km away from the Beel, diverted its original course by *c* 2.0 km during the last 30 years due to erosion of soil in its right bank. Sat Beel is situated at *c* 25 m MSL. The geology is mostly Tertiary formation and is mostly fertile clayey loam. Dihing series beds are exposed near the river Madhura which flows also near the Sat Beel..

Sat Beel is situated *c* 2.0 km away from Silchar town. The VIP road from Silchar city to Silchar Airport passes by the side of Sat Beel. A railway station on the Silchar-Jiribam (Manipur) line is situated *c* 5.0 km away from the Beel.

(*c*) Chatla Haor (Kar and Barbhuiya, 2000 b)

It is situated between 93°15′ N to 24° 10′ E in the Cachar district of Assam. It was said to be a 'Beel' (perennial wetland) some decades ago having its waterspread area reaching Silchar town. Due to gradual siltation and eutrophication occurring naturally in the successional process, accelerated by human interference, today it has become a 'Haor' (seasonal wetland) and retains water for approximately six months in a year having practically no dead storage level. So, it is almost completely dry during the winter. Having a water spread area of *c* 1600 ha at the FSL, Chatla is considered as one of the biggest 'Haor' in Assam (Figure 5, Plate 2).

Around Chatla, the soil in the catchment is generally sandy-loam, but shore vegetation is thin. The Haor is drained by a number of small inlets (*viz.*, Jalengachhara, Baluchhara, Salganga) and an outlet (*viz.*, river Ghagra) which drains itself into the river Barak. The catchment of the Haor includes a small portion of the Innerline Reserve Forest. During monsoon, the Chatla, like other similar wetlands, receive some humic as well as inorganic and organic nutrients from the hillocks and surrounding cultivable lands.

The Silchar city is situated *c* 15 km away from the east of the wetland. Hailakandi town is situated *c* 35 km away from the western shore of the Haor. The Block HQ at Barajalenga is situated *c* 5 km away from the southern end of the wetland.

The maximum length (L), breadth(B), depth(D) and water spread area(A) of the wetland at FSL have been measured to be 10 km, 2.5 km, 5.5 m and 1600 ha respectively. Prominent Silt Islands (SI), *viz.*, Bairagitila and Harintila have been found to occur towards the eastern shore of the Haor. Other small SIs, *viz.*, Haltia, Diblia and Barshangan occur towards the SW side of the Haor.

Among the inlets, river Salgonga originates from the foot hills of the Mizo Hill range while the Jalengachhara and Baluchhara, which are mostly rheophilic in nature, flow down into the Haor from the Innerline RF. The only major outlet, river Ghagra, drains the water of the Haor directly into the river Barak traversing a tortuous course of *c* 14 km from the northern boundary of the Haor.

(*d*) Puneer Haor

It is situated *c* 38 km away from south of Silchar city near the village Dhalai along the Assam-Mizoram border. This Haor has a waterspread area of *c* 2.5 ha at FSL and *c* 1.3 ha at DSL. The maximum L, B and D of Puneer Haor at FSL have been found to be 1.5 km, 0.9 km and 2.5 m while its average depth was found to be 0.4 m. The Puneer Khal, flowing along the eastern shoreline of the Haor,

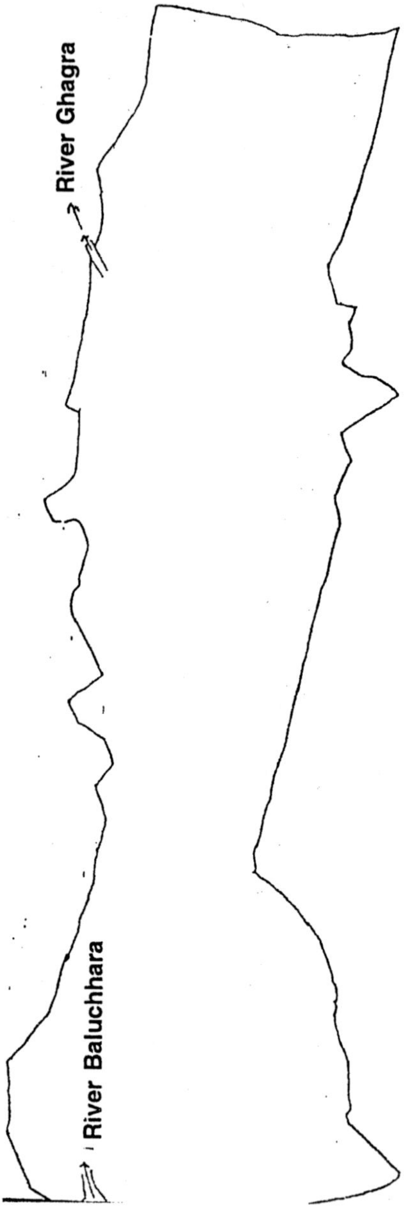

Figure 5: Boundary Contour of Chatla Haor at FSL

Plate 2: Chatla Haor at FSL

originates from Panchhara hill ranges and water from this khal spills over into the Haor at FSL. A drain from the adjoining Bhubandhar TE flows along the western shoreline of the Haor and water containing TE pollutants also spills over into the Haor at FSL.

The catchment of the Haor contains mainly trees and some amount of herbs and shrubs. The soil consists mainly of sand and gravel. A big portion of the catchment contains the tea plantation of the TE. There are also human habitations in the catchment. The catchment of the wetland includes a portion of the Dhalai RF. The Block HQ at Narsingpur is situated *c* 20 km away from the northern boundary of the wetland.

(*e*) Baskandi Anua

This oxbow wetland is situated between 24º 10′ N (latitude) and 93º 15′ E (longitude) in the Lakhipur Sub-division of Cachar district in Assam. It is said to have been formed due to change of course of the River Barak. Baskandi Anua borders villages Chandrapur Part II and Nijabaskandi in the north; villages Ratangram and Hazarigram in the south-west and village Chandrapur Part III in the south (Plate 3).

The Anua is situated near the Manipur range of hills. Towards the east, lies the Manipur valley while the Silchar sub-division is situated in the west. The North Cachar Hills is situated towards the north and the province of Mizoram lies towards the south. The catchment soil is found to be mainly sandy loam. Rain is the main

Plate 3: Baskandi Anua

source of water for the Anua. The wetland also receives water from the surrounding catchment having human habitation. The catchment vegetation includes herbs, shrubs and trees including a lot of bamboos.

Baskandi Anua is situated about 20 km away from Silchar city along the NH 53 which connects Silchar with Imphal. The Silchar-Jiribam railway line passes near the wetland. The Rural Development Block HQ is at Baskandi.

The L, B and A of Baskandi Anua have been found to be respectively 2.230 km, 205 m and 39.2 ha at FSL and 2.090 km, 190 m and 36.7 ha at DSL. The wetland basin tends to be deeper towards the southern side as compared to the northern. The Anua exhibits variable water level ranging from 0.25 m to 5.85 m at FSL (Jun-Sep) and from 0.14 m to 4.12 m at DSL (Nov-Apr).

(*f*) Satkarakandi Anua

This Anua lies between 92° 53′ E and 24° 1′ N in the Sonai Revenue Circle in the Cachar district of Assam. The wetland is situated at a distance of *c* 15.5 km away from Silchar city. River Barak flows towards the northern side of the wetland. The villages Satkarakandi and Dhanehary are situated on the eastern shore of the Anua. The villages Barahali and Satkarakandirpar are situated towards the south and west of the wetland. The NE and western shores of the Anua are protected by earthen embankments about 4 m in height as a preventive measure against flood (Plate 4).

The maximum L and B of the wetland were found to be *c* 1.75 km and 0.3 km respectively. The Anua was found to have an average depth of *c* 3.0 m at FSL. During monsoon, the Anua establishes connection with the river Barak through a channel on the northern side of the wetland which is, however, guarded by a sluice gate as a flood-mitigation measure.

(*g*) Sibnarayanpur Anua

This oxbow wetland is an abandoned segment of river Barak situated at a distance of *c* 32 km from Silchar city. It lies within the jurisdiction of Katigora Revenue Circle in Cachar district of Assam. It has a L, B and A of *c* 1.7 km, 0.7 km and 53 ha respectively. The catchment soil is mostly loamy. At present, it seems to have a connection with river Banaimulla (Plate 5).

Plate 4: Satkarakandi Anua

Plate 5: Sibnarayanpur Anua

Lotic Bodies

An account of some of the lotic bodies in Barak drainage is briefly given below:

Assam has a total river length of *c* 4500 km out of a total riverine length of *c* 1,12,650 km in India. The River Barak is the biggest river in South Assam constituting the Barak drainage. Barak is a part of the Ganga-Brahmaputra-Meghna Basin. The sub-basin Barak is surrounded by Barail, Patkai and Lushai Hills extending over an area of *c* 41,723 km^2 in the provinces of Assam, Manipur, Nagaland and Mizoram in India; and, *c* 751 km^2 in Myanmar upto Bangladesh border. Total catchment area of river Barak from its source upto the point of bifurcation is 25,866 km^2. Barak has a number of tributaries joining it in both its north and south banks. The principal north bank tributaries are the rivers Jiri, Chiri, Madhura and Jatinga; while, the south bank tributaries include the rivers Sonai, Rukni, Dhaleswari and Tuivai. Before entering Bangladesh, Barak bifurcates into two branches, *viz.*, the river Surma and the river Kushiara, which, after flowing some distance along the Indo-Bangladesh border, enters Bangladesh. The river Surma receives the river Baleswar while the river Kushiara receives the rivers Singla / Kachua and Longai coming down from Mizoram (Figure 2).

The geographical limits of Barak sub-basin lie between 22° 32′ N latitude and 94° 18′ E longitude. The gradient of River Barak from around Tipaimukh (at the border of Manipur and Mizoram) varies from 1 in 70 in the head reach. The Barak sub-basin, being situated at the route of SW monsoon, receives an annual rainfall of 2,198 mm.

The major types of soil in Barak sub-basin are alluvial (sometimes mixed with silt, sand and clay) in Cachar, Karimganj and Hailakandi and in the plains portions of Tripura province; while, it is laterite red or yellow in the hilly portions.

Based on our extensive field works, an account of River Barak and its principal tributaries in the Barak valley region of Assam are briefly described below:

River Barak

It is a 4th order river situated between the geographical ordinates N 24° 47′ 12.9″ and E 93° 2′ 9.1″. Its altitude was found to vary from 17-43 m MSL in our different study sites within the Barak valley region of Assam. After originating from Japvo peak in Nagaland, the first place it reaches the plains is the village Karong situated in Senapati district of Manipur at the Manipur-Nagaland border. It, then, flows through a tortuous course in Manipur, passes through places like Noney, Tipaimukh (at the junction of Manipur and Mizoram where the proposed Barak Dam is to be constructed), Jiribam, etc.; and, finally, enters the Cachar district of Assam through which it flows along a serpentine course. Ultimately it bifurcates into Surma and Kushiara at village Harinagar near village Haritikar at the Indo-Bangladesh border (Kar, 2003 a, 2005 a; Kar *et al.*, 2002 a) (Figure 2, Plate 6 and 7).

Observations in our study sites indicates the following:

The valley segment is mostly alluvial. The flow rate of the river ranged from 12-61 rev./10 sec; which, at the surface, revealed a range of 30-35 rev/10 sec. The Entrenchment Ratio (E/R) varied form 1.15-2.87. The reach type is mostly regime. The river bottom mainly consists of sand and silt as substratum which is also the case with river bank. However, bedrocks are found on the bank in some places. The bankful width and depth, channel width and depth and wetted width and depth are given below:

Parameter Type	*Bankful*	*Channel*	*Wetted*
Width (m)	430-490	340-410	150-270
Depth (m)	23-29	17-19	3-12

Depth, turbulence and some amount of overhanging vegetations serve as moderate cover and protection of the fishes in Barak. Depending on the contact of soil with water, Barak is of mesoriparian type having grass, shrubs and evergreen trees as riparian vegetation. Bamboos form one of the most dominant types of riparian vegetations.

The dominant texture of the riparian zone is mostly silt and fine grain sand stone with some amount of clay. The riparian land use pattern includes mainly human habitation and agriculture. Signs of erosion are common along the river bank.

Critically, our surveys revealed very turbid river water mainly because of heavy rainfall during monsoon. High turbidity indicates brining-in of eroded soil by the river. Inspite of its high depth, the

Plates 6 & 7: River Barak at Lakhipur and near Jirimukh in Cachar, Assam

river depicted the existence of backwater type pools having dense vegetation cover at some spots. Interestingly, bedrocks are found even at some spots in the mid-stream and downstream regions of the river. There is variation in water temperature and transparency with seasons. Turbid water is less productive although it gives cover to fish. Barak also has undercut bank.

River Sonai

It is a 4th order stream which originates from Mizo Hills as river Tuirial. The study sites lied between geographical limits of 24° 34′ 36.4″ N and 92° 53′ 47″ E at altitude ranging from 25-39 m MSL. The rate of flow of the river varied from 15-37 rev/10 sec. However, the surface water depicted maximum flow of 47 rev/10 sec by the flow meter. The valley segment is of alluvial type while the reach is of regime type. As such, the river is generally highly turbid during the rainy season and fishes are sometimes found gulping due to little asphyxiation. The bankful, channel and wetted breadth and depth are given below (Kar, 2003a, 2005a; Kar *et al.*, 2002 a), (Figure 2, Plate 8).

Parameter Type	*Bankful*	*Channel*	*Wetted*
Breadth (m)	70-110	50-85	27-50
Depth (m)	3.2-9.0	4.5-6.0	1.0-2.5

The fish cover is mostly turbulence and depth. Some amount of woody debris and overhanging vegetation also serve as fish cover. The river substratum is mostly fine sand and silt admixed with some amount of clay. Bedrocks are also found on the river bank at some spots. The river Sonai is of mesoriparian type with mostly bamboos, shrubs and medium size trees as riparian vegetation. Nevertheless, the dominant texture of the riparian zone is sand, and silt admixed with clay and fine grain sand stones. Signs of erosion are quite prominent in some places. The riparian land use pattern includes mainly human habitation and crop farming. A big portion of the riparian land is included in the RF.

The river Sonai is seasonally flooded due to intermittent rains. The river sometimes shows roots of plants along the bank. A peculiar feature portrays undercut banks on the outside of the curves and sand bars on the inner side. Patches of overhanging vegetation provides shelter for fishes. Reaches were selected both in the plainwater downstream region as well as in the rheophilic little upstream region.

More recent studies (Kar, 2005a) revealed small rivulets, *viz.*, Jorkhal and Nagakhal, joining the main river Sonai from the

Plate 8: River Sonai in Cachar, Assam

surrounding catchment. Gravels and cobbles formed the principal component of the substratum of these rivulets; which, along with higher transparency (as compared to the main river Sonai) and faster flow of water, probably, paved the way for some fishes to migrate locally from the main river to these rivulets.

The river Sonai joins the river Rukni (flowing down from Mizo Hills as River Serlui) around the village Palonghat and the combined flow, as river Sonai, joins the river Barak around the village Kaptanpur in Sonai Revenue Circle in Cachar.

River Jatinga

It is a 2nd order stream which originates from the North Cachar Hills in Assam and flows down the side of Jatinga village. It is one of the North bank tributaries of river Barak. The reach type is mainly pool-riffle to braided-type; while, the valley segment of the river is colluvial to alluvial type. Bedrocks, boulders, cobbles, gravels and fine sand form the main components of the river substratum. The study spots (selected reaches) of the river were situated within the geographical limits of 24° 57′ 49.4″ N and 92° 45′ 41.3″ E having an altitude of 19 m MSL on the average. The river is less entrenched (E/R: 3.06) possibly because the valley segment is V-shaped. Turbulence is the principal fish cover; while, small woody debris, overhanging vegetation and depth were also found to serve as fish cover. River Jatinga is of hydroriparian type having riparian vegetation consisting mainly of herbs, shrubs and trees. There is an abundant population of *Polygonum* spp., *Sesbania* spp., and *Bambusa* spp. Dominant land use pattern includes human habitation, some amount of agriculture (mostly 'Jhum'or shifting cultivation) and some tea gardens, many of which are abandoned today mainly due to transportation problems and insurgent activities. It is remarkable to note here that, not much soil erosion has been observed, which could be because of deforestation to a little lesser extent. However, recent earth-cuttings going-on in the gauge-conversion of the Indian Railways, from MG to BG is causing a lot of soil being deposited into the river bed sliding down from the surrounding hills leading to changes in the riverine habitat. Also, undercut banks are sometimes seen. Dominant riparian texture includes boulders and gravels. Flow rate of the river was found to be 56 rev/10 sec at the surface level and 23 rev/10 sec at a little deeper level; while, it was found to be 5 rev/10 sec at the deepest point where the machine could reach. The bankful, channel and wetted breadth and depth of the river Jatinga are given below (Kar, 2003a, 2005a; Kar *et al.*, 2002 a) (Figure 2, Plate 9).

Parameter Type	*Bankful*	*Channel*	*Wetted*
Breadth (m)	230	170	80
Depth (m)	12	8	4

Plate 9: River Jatinga in North Cachar, Assam

In river Jatinga, vegetation, depth and bottom types vary according to gradient. The river is braided-type at some spots mainly because of accumulation of sand bars and boulders by the water current. The river has many logs and fallen trees which provide excellent shelter for fishes. Also, there are spots where the banks are deeply undercut and soil has eroded from the bank.

Dhaleswari

This river is a 4th order river, which originates, as river Tlawng, from the Mizo Hills, around Lunglei. After flowing through a tortuous course through the Mizo Hills, this river enters the Barak valley region of Assam around a place called Ramnathpur (Gharmura), situated at a distance of *c* 30 km from Bhairabi in Mizoram. The river is called Dhaleswari from the point it enters Barak Valley region of Assam. After flowing for about 25 km in Barak valley of Assam, the river had been blocked by a blind dam and the main flow of the river had been diverted at a village called Karichhara (near Katlichhara) as river Katakhal which joins the River Barak, on its south bank, around Katakhal Railway bridge after passing through villages Matijuri, Mohanpur, etc. The original river Dhaleswari, as a dead river from village Karichhara, also joins the river Barak, on its south bank, around the place called Panchgram where the Cachar Paper Mill (of the Hindustan Paper Corporation) is situated. Incidentally, the Hailakandi town is situated on the bank of (now) dead river Dhaleswari. The study sites in these rivers were situated between the geographical limits of 24°57′ 48.9″ N and 92°45′ 41.1″ E having an altitude of 16 m MSL on the average. The reach type is regime while the valley segment type is alluvial. 'Run' and 'mid-channel type' pool were the dominant micro-habitat types. The E/R is generally 2.27. The river is of riparian type with riparian vegetation type varying from herbs, shrubs and trees including grasses and quite a dense population of bamboos. The riparian soil texture is largely silt and clay on both the banks while the riparian land use pattern is mainly human habitation, agriculture and tea gardens.The river substratum consists mostly of silt and fine sand with a large amount of clay. Signs of erosion are seen quite frequently. The water current was measured as 5.0 rev/10 sec on the average. There is not much natural fish cover and consisted mainly of turbulence, undercut banks and depth; but, not much overhanging vegetation. The total cover was estimated to be *c* 10 per cent (Kar, 2003a, 2005a; Kar *et al.*, 2002 a), (Figure 2, Plate 10).

A significant feature of river Dhaleswari in its upstream rheophilic stretch in Mizoram is the presence of micro-habitats with lot of bedrocks and boulders with fast water current. Such type of micro-habitats are considered as typical habitat for *Tor* species; and, this type of habitat are quite abundant in river Tlawng/Dhaleswari

Plate 10: River Dhaleswari in Cachar, Assam

in its stretch between Sairong and Bhairabi within Mizoram. This kind of sites are considered as 'ideal sites' for conservation of endangered *Tor* species (Mahseer fish) (Kar *et al.*, 2002b, 2002 c).

The bankful, channel and wetted breadth and depth of river Dhaleswari are as follows:

Parameter Type	Bankful	Channel	Wetted
Breadth (m)	350-390	280-330	140-190
Depth (m)	17-21	15-19	4-9

In Mizoram

The province of Mizoram (area 21,081 km^2) is a beautiful terrain (maximum height *c* 2,743.90 m MSL) situated in the southern part of North-East India bordering Bangladesh in the south-west and Myanmar in the east having the Tropic of Cancer passing through it. Inhabited by a population of 8,91,058 (2001 census), growth rate was found to be 21.18 per cent (1991-2001); while, density per sq.km.was found to be 42. *c* 85 per cent of the rural mass are engaged in various kinds of agricultural practices including fisheries. The province has *c* 24,000 ha of cultural fishery resources and *c* 6000 ha of riverine water spread area which is spread over *c* 1100 km^2 of riverine stretches. *Per capita* availability of fish per annum is only 3.01 kg. The major rivers of Mizoram are the Tuirial, Tlawng, Tuirini, Tuivai, Mat, Kolodyne (Chhimtuipui), Tuichong, Karnafuli and Serlui. Of these, the rivers Tuirial, Tlawng, Tuivai and Serlui join the Barak drainage at different points of their course. A small river, called, Tuirini, also joins the river Tuirial. The river Chhimtuipui originates in Myanmar, flows northward along the Indo-Myanmar border for some distance, then, takes turn towards south to flow back to Myanmar again. This river, further, as river Kaladan, is said to have joined the the river Karnafuli in Chittogong district of Bangladesh after flowing through Indo-Myanmar-Bangladesh border. The river Mat joins the river Chhimtuipui in Mizoram. The river Karnafuli, after originating from around Marpara region in the Mizoram-Tripura border, flows along the Mizoram-Bangladesh border; and, ultimately, joins the Bay of Bengal in Bangladesh (Figure 1, 2).

In addition to the lotic systems, there are few lentic bodies in Mizoram. The prominent among those are the following:

Lentic Bodies

1. Tum Dil Lake near Saitual
2. Rih Dil Lake near Champhai along Indo-Myanmar border
4. Palak Dil Lake near Tuipang along Indo-Myanmar border

4. Rengtlang Dil Lake near Parva along Indo-Bangladesh border.

An account of Tumdil Lake is briefly given below:

Tum Dil

It is one of the very few perennial wetlands in Mizoram situated at a distance of *c* 40 km away from Aizawl city. It lies between the geographical limits of 92-93° N and 22-24° E. It has a maximum length, breadth and depth of 0.73 km, 0.10 km and 6.5 m respectively. It has a waterspread area of *c* 12.0 ha at FSL. It receives water mainly during the monsoon months from the surrounding catchment. There is no major inlet or outlet of the lake except few drains. This lake is under the joint management of departments of Fisheries and Tourism of the Govt. of Mizoram. In Tum Dil, the land use pattern involves mainly pisciculture and horticulture. Among the practices in Animal Husbandry, it is mainly poultry farming with little amount of cattle farming. There is much amount of naturally-occurring farm forestry as well as much amount of support area of forests around these wetlands. However, there is not enough of grazing in view of not a big population of cattle being available. Concomitantly, the amount of pasture land is also limited. As in the case of other wetlands, not much effort has been seen in the reclamation of wastelands. The source of pollutants in these wetlands is mainly horticultural wastes and refuges as well as domestic sewage and effluents from the surrounding human habitations. The nature of pollutants include mainly the phosphate and nitrate compounds. However, in view of lesser number of human settlements, the amount of pollutants released is expected to be not much (Kar, 1998).

Lotic Systems

Among the many rivers, an account of river Tuirial is given below:

River Tuirial

It is a 3rd order stream which originates from around Serchip region of Mizo Hills and is formed by the union of two 2nd order streams, *viz.*, Chite and Zelphui. After following a tortuous course in Mizo Hills, it enters the Barak valley region of Assam as river Sonai and merges with another river called Rukni (river Serlui in Mizoram) around village Palonghat. The combined flow joins the river Barak in its south bank near the village Sonaimukh (Kar, 2003a, 2005a; Kar *et al.*, 2002 a) (Figure 2; Plate 11).

The study spots in this typical hill stream are situated in Mizoram between the geographical limits of 22º 23′ 45.5″ N and 92º

Plate 11: River Tuirial Around Village Tuirial (in Mizoram)

57′ 59.9″ E and at altitude ranging from 500 m- 920 m MSL. The reach type is cascade to pool-riffle type. The valley segment is mostly of colluvial type. Falls, cascades, riffles, eddy pools and mid-channel pools constitute the diversity of micro-habitat types in this river. The E/R of the river is generally 8.35. Tuirial is a mesoriparian type of river with riparian vegetation consisting of herbs, shrubs and trees. Herbs include different types of grasses and a rich population of bamboos; while, teak (*Tectona grandis*) forms a rich component of bank vegetation. There is also a very big abundance of *Saccharum* spp almost all along the river bank. The riparian soil texture is generally dominated by gravels and cobbles with some amount of boulders and few bedrocks. The riparian land use pattern consists mainly of human habitation, agriculture ('Jhum') and plantations. The water current is generally 28 rev/10 sec on the average. Fish covers include mostly turbulence and undercut bank; while, depth and vegetations also serve as fish covers in some spots. The total cover in some area could be approximately 30 per cent.The river substratum was found to consist mostly of gravels and cobbles and also some boulders and few bedrocks. Signs of soil erosion was seen to be quite significant in some portions with accumulation of heaps of silt and sand on the river bank. The bankful, channel and wetted breadth and depth are given below:

Parameter Type	*Bankful*	*Channel*	*Wetted*
Breadth (m)	92.0	74.0	38.0
Depth (m)	11.0	7.7	2.3

In Tripura

The province of Tripura is flanked by both hills and plains having four principal rivers flowing through its terrains, *viz.*,

1. River Monu
2. River Khowai
3. River Gomati
4. River Feni

In addition to the above, there are other little smaller rivers, *viz.*,

1. River Howrah

2. River Deo
3. River Muhuri

Of these, the rivers Monu and Gomati originate from the Longtarai Hills while the river Khowai originates from the Atharamura Hill ranges. The river Feni is said to also originate from the escarp of the Chittagong Hill tracts and flows down beside the village Shilachhari along the Indo-Bangladesh border. The river Howrah originates from the Baramura Hill ranges (Figure 2).

Notwithstanding the lotic systems mentioned above, there are a number of lentic bodies in Tripura. Some of these lentic bodies are called by the word 'Sagar', meaning 'tank', which are said to have been constructed by the Monarchs who had ruled Tripura (Tiperrah) once upon a time. Some of the lentic bodies are cited below:

Lentic Bodies

1. Rudra Sagar
2. Fatik Sagar
3. Amar Sagar
4. Kurti Beel

In addition to the above, there is a big reservoir, called 'Dombaru Reservoir', formed by damming the river Gomati at Mandirghat/ Tirthamukh region in South Tripura district. The expanse of this reservoir also extends to Gandachhara region of Dhalai District.

Although not many, there is a small oxbow wetland situated in Amarpur Sub-division of South Tripura district formed by slight change in course of river Gomati.

An account of some of the wetlands in Tripura is briefly given below:

Rudra Sagar

This is one of the prominent wetlands in Tripura situated within the geographical limits of 23° 28′–23° 33′ N and 91° 16′–91° 20′ E in the Melagarh Development Block of Sonamura Sub-division in West Tripura district, about 50 km away from capital city of Agartala. The wetland has a waterspread area of 788 ha at FSL which shrinks to 240 ha at DSL. However, it is said that, Rudra Sagar had a waterspread area of *c* 1000 ha during the late 1940s. The wetland has a total catchment area of *c* 13,336 ha which includes *c* 716a of small hillocks having human habitation, *c* 4942 ha of crop land, *c* 175 ha of marshy land, *c* 178 ha 'Jhum' land, *c* 466 ha of Govt. orchard land, *c* 215 ha of pasture land, *c* 1340 ha of Reserve Forest area and *c* 4534 ha of Scrub/bamboo/degraded forest. The wetland receives water through three major inlets covering a total length of *c* 41 km which includes Noachhara 28 km, Mayhranichhara 5 km, and Durlavnarayanchhara 8 km. The wetland has a depth of 7.5-8.0 m at FSL and 1.5-2.5 m at DSL while it was said to be 12-13 m at FSL and 5-6 m at DSL during the late 1950s. The air temperature ranges from 31-38°C during summer which could sometimes plunge down to 8°C during winter. The wetland soil is mostly sandy clay loam admixed with some amount of silt; thus, rendering it quite favourable for cultivation of paddy (Kar, 1999a).

Rudra Sagar has a rich diversity of biota consisting of aquatic macrophytes (AM), phytoplankton, zooplankton, macro-invertebrates, fishes, reptiles and birds. The AM includes, among others, *Trapa natans* L. var. *bispinosa, Hydrilla verticillata, Vallisneria spiralis, Nymphaea* sp., *Nymphoides* spp., *Nelumbo* sp., *Ipomoea* spp., *Wolfia* sp., *Azolla* sp., *Salvinia* spp., *Marsilea* sp., *Euryale ferox*, etc.The wetland has abundant population of *Eichhornia crassipes. Microcystis* sp., *Notoc* sp. and *Anabaena* sp., which serve as indication of eutrophication/pollution. However, in addition to the above, *Volvox* sp., *Chlorella* sp., *Euglena* sp., *Diatoma* sp., *Pinnularia* sp, *Navicula sp*, etc., are some of the other principal phytoplanktonic flora while the zooplanktonic fauna include forms like *Brachionus* spp., *Keratella* spp., *Daphnia* spp., *Ceriodaphnia* spp., *Moina* spp., *Cyclops* spp., *Diaptomus* spp., and so on. The macro-invertebrates include, among others, the annelids, arthropods and molluscs, *viz.*, earthworms, leeches, aquatic insects, like *Bellostoma, Ranatra,* aquatic beetles, aquatic spiders, etc.; and, molluscs like *Viviparus* spp., *Lamellidens* spp., *Pila* sp., etc.

The ichthyofauna of Rudra Sagar includes, among others, the following:

The Indian Major Carps (IMC) like *Labeo calbasu, L.gonius, L. rohita, Cirrhinus mrigala, Catla catla;* minor carps and minnows like *Amblypharyngodon mola, Salmostoma bacaila, Danio* sp., *Rasbora* sp. *Puntius* spp.; loaches like *Botia dario;* bagrids like *Mystus* spp., *Aorichthys* spp.; Silurids like *Wallago attu, Ompok* sp., schilbeids like

Eutropiichthys vacha; perches like *Anabas testudineus, Badis badis, Nandus nandus*; murrels like *Channa* spp; the eels like *Mastacembelus* sp., *Macrognathus* spp., etc. There are a good varieties of ornamental fishes like *Botia dario, Colisa* spp., etc. The wetland has introduced exotic species like *Cyprinus carpio, Ctenopharyngodon idella* and *Hypophthalmichthys molitrix*.

The amphibians and reptiles includes mainly the toads, frogs and water snakes.

Rudra Sagar serves as the habitat for feeding, breeding and nesting of resident aquatic birds like the little cormorant, Indian shag, darter, pond heron, little egret, large egret, adjutant stork, common teal, pintail, large blue kingfisher, common kingfisher, black-capped kingfisher, etc.; as well as, wintering ground for migratory aquatic birds like Indian white wagtail, grey wagtail, different types of swans, ducks, etc.

The Scheduled Caste (SC) community constitutes *c* 15.12 per cent of the total population of Tripura today. Demographic history of Rudra Sagar indicates that, about 1000 refugees, who had immigrated from erstwhile East Pakistan (now Bangladesh) were rehabilitated in this wetland during the early 1950s. They were allotted *c* 6.0 acres of land for cultivation in addition to land for homestead along with fishing rights in Rudra Sagar. They had formed a Fishermens' Co-operative Society called 'Rudra Sagar Udbastu Samavay Samiti' for their developmental activities. There are a number of villages along the periphery of this wetland under the jurisdiction of number of Gram Panchayats having a total population of *c* 35,233, of which, the population of SC community is 17,888. Today, as such, SC population comprises *c* 50.77 per cent of the total inhabitants around Rudra Sagar but most of them (78.75 per cent) lead a life below poverty line. Land use pattern includes Cultivation of 'Boro' paddy inside this wetland during winter, which could be one of the major causes of degradation of Rudra Sagar.

Rudra Sagar has a beautiful palace called 'Neer Mahal' situated somewhere at the centre of this wetland, which was constructed by the then Monarch of Tipperah, Maharaja Beer Bikram Manikya as a summer resort. This is still preserved as a Heritage site and tourist spot by the present Govt. of Tripura.

Kurti Beel

This wetland is situated within the geographic boundary of 92° 15′ E to 24° 45′ N in the Kadamtala Block of North Tripura near Churaibari on the Inter-state border of Assam and Tripura. It could be regarded as a classic example of a wetland which has reached almost the climax of siltation and eutrophication. It is situated *c* 30 km away from Dharmanagar town and *c* 110 km away from Silchar city along NH 44. The Beel has a maximum length, breadth and depth of 0.6 km, 0.3 km and 2.8 m respectively; and, a waterspread area of *c* 93.67 ha at FSL. Kurtichhara is the main inlet of the of the Beel while Kalanadi drains into the Thal river as the major outlet. The Beel also receives water from the surrounding catchment and domestic sewage through a number of small drains (Kar,1998).

Land use pattern involves mainly fishing, agriculture and horticulture. Practice of animal husbandry is not of much varied type except some amount of cattle farming and very little amount of poultry farming. There is neither much amount of farm forestry nor much support area for forests. However, enough of grazing and pasture lands are available particularly during dry season when the water level in the wetlands go down.

The source of pollutants in this wetland are agricultural wastes and domestic sewage and effluents from the surrounding human habitations. The nature of pollutants include mainly the phosphate and nitrate compounds.

Lotic Bodies

An account of river Gomati, as one of the principal rivers of Tripura, is given below:

River Gomati

It is a 3rd order river which originates from Tirthamukh-Mandirghat Hills of South Tripura district and is one of the biggest rivers in Tripura. After flowing a tortuous course through Jatanbari, Natun Bazar, Amarpur Udaipur and Melagarh, it enters Bangladesh. Our study sites in this river are located between the geographical limits of 23° 30′ 32″ N and 91° 28′ 38″ E at an altitude of 41 m MSL on the average. The reach type is regime while the valley segment is mostly alluvial. The micro-habitat types include riffles, runs, eddy pools and debris pools. Nevertheless, cascades are seen in a small stretch of the upstream portion. The bankful, channel and wetted breadth and depth are as follows:

Parameter Type	Bankful	Channel	Wetted
Breadth (m)	68.60	56.0	47.80
Depth (m)	5.78	3.71	0.71

The river has an average depth of 0.49 m during the dry season. The E/R is generally 1.43. Further, the river is of mesoriparian type with riparian vegetation consisting of mainly herbs and shrubs but not many tress.. A rich population *Enhydra fluctuans, Vetiveria zizanoides* and *Bambusa* spp. have been recorded. The river substratum generally consists of silt, sand and clay in the mid-stream and downstream regions while it is mostly boulders and cobbles with some amount of bedrocks in the small upstream portion. The riparian land use pattern mainly includes human habitation, fish farming and some amount of crop farming. The water current is generally 17.87 rev/10 sec on the average. Fish covers include turbulence, turbidity, large and small tress and tree-trunks, sometimes depth and some amount of overhanging vegetation. The total cover is generally 20 per cent of the area of the habitat. Signs of erosion is significant at some spots with accumulation of silt and sand (Kar, 2003a, 2005a; Kar *et al.*, 2002 a) (Figure 2, Plate 12).

Table 1 shows the Habitat Inventory of six major rivers of NE India flowing through three provinces, *viz.*, Barak valley region of Assam, the province of Mizoram and the province of Tripura. The Table revealed differences in valley segment among the different rivers. In the plain region of Barak valley and Tripura, the valley segment is alluvial; while, in the true hill stream stretches of Mizoram and in the upstream regions of the rivers in Tripura and Barak valley, the valley segment is colluvial. The latter makes the habitat colonisable for the rheophilic fishes, most of which have developed adhesive apparatus on the ventral surface of the body in order to withstand the fast torrent of the rivers. Further, colluvial valley segment provides a diversity of micro-habitats while the alluvial valley segment generally shows only regime type of reach. Moreover, the rivers are generally wider where there is alluvial valley segment. The fish covers in the studied rivers have been found to be mostly

Plate 12: River Gomati Around Village Tirthamukh/Mandirghat (Top) and at Amarpur (Bottom) (in Tripura)

turbulence and depth types, the former ranging from 2-15 per cent. The total cover among all these studied rivers have been found to range from trace (< 5 per cent) to moderate (5-20 per cent) during the surveyed period. In some of the studied rivers, *e.g.*, in river Tuirial,

Table 1: Habitat Characteristics of Some of the Rivers of N.E. India

Sl.No.	*Habitat Parameters*	*River Barak*	*River Jatinga*	*River Sonai*	*River Dhaleswari*	*River Tuirial*	*River Gomati*
1.	Stream order (Source of Map: Survey of India)	Fifth order	Third order	Fourth order	Third order	Third order	Third order
2.	Position	N 24° 46'-50" E 92° 1'-2"	N 24° 14'-57" E 92° 35'-45"	N 24° 34'-44" E 92° 53'-57"	N 24° 16'-57" N E 92° 31'-45"	23°34'-43" E 92° 40'-48"	N 23° 30'-32" E 91° 28'-38"
3.	Microhabitat type	Runs and backwater pools	Riffle, cascade, run, eddypool, mid-channelpool and backwater pool	Run, eddypool, trench pool, and backwater pool	Run, eddypool, backwaterpool and mid-channel-pool	Riffle, cascade, mid-channelpool, trench pool and backwater pool	Run, dammed pool, eddypool and debris
4.	Mean wetted width (m)	130	45	50	32	30	24.5
5.	Mean wetted depth (m)	9	1.6	1.4	2	1.5	1.71
6.	Cover type (%)	Overhanging vegetation cover (2-14). Depth cover (7-13).	Under cut bedrock (10), Overhanging vegetation (7), Depth (15) and small woody debris (2).	Overhanging vegetation (5-13), under cut bank (2-4), Depth cover (4-15) and small woody debris (1-2).	Depth cover (8), Overhanging vegetation (3) and undercut bank (2).	Overhanging vegetation (4-7), turbulence cover (7-25) and depth cover (3-5).	Floating vegetation (20), depth (1-25) and small woody debris (3).

Contd...

Table 1–Contd...

Sl.No.	*Habitat Parameters*	*River Barak*	*River Jatinga*	*River Sonai*	*River Dhaleswari*	*River Tuirial*	*River Gomati*
7.	Substrate type	Fine sand and boulder	Boulders, Cobbles, Gravels and Bedrocks	Fine sand, silt and bedrock	Fine sand	Gravels, Cobbles, Bedrock and fine sands	Fine sand
8.	Riparian land use	Human habitation and agriculture	Tea garden, human habitation and railway	Human habitation and agriculture	Human habitation	Human habitation, and Jhum cultivation	Human habitation, Protected forest, and agriculture
9.	Signs of erosion	Visible	Not much	Visible	Visible	Visible	Not much

the fish cover is dominated by turbulence and overhanging vegetation.

The substrate type in the hilly portion of the rivers is found to be dominated by gravels, cobbles, boulders and bedrocks while in the plain portion of the rivers, the substrate type mostly consists of fines. It is mesoriparian in all the studied rivers but the riparian vegetations vary from shrubs to trees in the different rivers. Presence of Tea Gardens (TG) along the course of some of the rivers, notably, the Barak and the Jatinga, is a matter of concern, in view of the fact that, TG washings possibly being discharged into the rivers. Signs of erosion have been predominantly visible in all the studied rivers.

Chapter 3

Physico-chemical Characteristics of Water

Introduction

Water and soil are the two basic subjects on which the dynamics of a water body and thereby the life of the biota in it depends. To a limnologist, fishery scientist and a fish farmer, study of water is a pre-requisite for the welfare of the fishes and to know their environment. Soil regulates various water quality parameters. Hence, study of soil is equally important. Further, agencies like Public Health Engineering Department (Water Works), Pollution Control Boards, Various Industrial Organisations, etc., require to analyse water/ wastewater samples. The methodology recommended by APHA (1995) is largely followed.

The purpose of physical and chemical examination of water is as follows:

1. To know the exact composition of the sample at the particular point of time of sample collection
2. To classify water with regard to the general level of gaseous and mineral constituents.
3. To ascertain the presence or absence of constitutents affecting various beneficial uses of water.
4. To determine the level of organic impurities

5. To determine the degree of clarity and to ascertain the nature of matter in suspension

After the analyses are over, the results may be interpreted to suit different purposes for different organizations (according to the aim of the work), *e.g.*, surveillance of water quality, effluent quality, etc.

Among the physico-chemical characteristics, the following parameters are generally required to be studied in a water body by a limnologist:

Physical

1. Temperature
2. Turbidity
3. Transparency
4. Colour
5. Suspended solids
6. Silt load

Chemical

1. pH
2. Dissolved Oxygen (DO)
3. Free Carbon di-oxide (FCO_2)
4. Total Alkalinity (TA)
5. Total Hardness (TH)
6. Specific Conductivity

In addition to the above, the following parameters are sometimes analysed:

1. Biochemical Oxygen Demand (BOD)
2. Chemical Oxygen Demand (COD)

Further, the radicals and metals are often analysed:

1. Phosphate (PO_4)
2. Nitrate (NO_3)
3. Metals like Iron, calcium, potassium, etc.

Estimation of some of the parameters, mentioned above, are given below:

Physical

(A) Temperature

In limnological studies, water temperature, as a function of depth, is often required. Water temperature readings are related to the study of pH, Dissolved oxygen, salinity, conductivity, etc. However, heated water discharges into the water body generally have detrimental effects.

Principle

Water temperature could be measured with the help of good quality mercury-in-glass precision thermometer (-5 to 50 × 1/10°C) having minimal thermal capacity to facilitate rapid equilibration. It be checked and calibrated (if required) against a precision thermometer certified by the National Bureau of Standards (ordinary thermometers may have errors upto 3°C).

Temperature of both air and surface water could be measured directly. However, bottom water temperature could be measured by collecting the water samples from various depths with the help of a 5 litre capacity Kemmerer sampler. Alternatively, water temperature at depths may be most conveniently and accurately measured with the help of a thermister or with the help of a less expensive reversing thermometer or thermophone.

(B) Turbidity

Suspension of particles in water interfering with passage of light is called 'turbidity'. The suspended particles may be clay, silt, finely divided organic and inorganic matter, plankton and other microscopic organisms. Turbid water should be unfit for drinking and is undesirable for food products, beverages, etc. Turbidity is said to be a measure of performance of water treatment plants.

The following methods are generally used for determining the turbidity of a water sample:

1. By Jackson's Candle Turbiditimeter (JCT)
2. By Electrical/Electronic turbiditimeters commonly called Nephelometers

The JCT is generally used for high turbid waters.

(*a*) Jackson's Candle Turbiditimeter

Principle

It is based on the transmittance of light from a flame of a 'standard candle' (made of bee's wax and spermaceti designed to burn @ 114-126 grains per hour) through the sample column of certain path length, such that, the flame becomes indistinguishable against background illumination. Turbidity is inversely proportional to the path length. In brief, it is based on comparison of intensity of light scattered by the sample and a standard reference under comparable conditions. Higher the intensity of scattered light, higher is the turbidity.

The lowest turbidity which could be measured with a JCT is 25 units (JTU). As such, indirect secondary methods are required for measuring turbidities in the range of 0-5 units. However, the results obtained with different types of secondary instruments do not match with one another because of fundamental differences in the optical systems; even though, the instruments are all pre-calibrated against JCT.

This method is not in much in use today mainly because of difficulties in getting the 'standard candle'.

(*b*) Nephelometric Method

Principle

Formazin polymer is the turbidity standard preference suspension for water. It is easy to prepare and is more reproducible in its light scattering properties. A given concentration of Formazin suspension having 40 NTU has approximately a turbidity of 40 JTU. Therefore, turbidity based on formazin will approximate units derived from JCT, but will not be identical.

Materials

Turbiditimeter: It consists of a light source for illumination of the sample and one or more photoelectric detectors with readout device to indicate the intensity of light scattered at 90° to incident light. The sensitivity of the instrument permits detection of turbidity difference of 0.02 NTU or less in water having turbidity <1.0 NTU. The instrument measures generally from 0-40 NTU. Various ranges are necessary to obtain both adequate coverage and sensitivity. However, meters having wider ranges are also available today.

Sample tubes: Clear colourless glass, scrupulously cleaned both inside and outside without scratch. The tubes be sufficiently long, so that, they are not required to be touched where light strikes.

Reagents

(*a*) Turbidity-free water: It is obtained by passing distilled water (DW) through membrane filter having pore size <100 mm. If filtration does not reduce turbidity, DW itself may be used.

(*b*) Stock Turbidity Suspension:

(*i*) Solution I: It is prepared by dissolving 1.0 g of Hydrazine Sulphate in DW and diluted to 100 ml in a volumetric flask

(*ii*) Solution II: It is prepared by dissolving 10.0 g of Hexamethylene Tetramine in DW and diluted to 100 ml in volumetric flask.

(*iii*) In a 100 ml volumetric flask, 5.0 ml of Solution I is mixed with 5.0 ml of Solution II. Allowed to stand for 24 hours at 25 ± 3°C. This be diluted to 100 ml mark of the flask and shaked properly. Turbidity of this suspension is 400 NTU.

NB: Stock solutions be prepared monthly.

(c) **Standard Turbidity Suspension**

10 ml of Stock Turbidity Suspension (as prepared above) be diluted to 100 ml with turbidity-free water. The turbidity of this suspension is defined as 40 NTU. Such suspensions be prepared weekly.

Procedure

1. Turbiditimeter calibration: Generally, these meters are kept calibrated by the manufacturer before being sold out. Further, manufacturer's operation manual could be useful in calibration, if required.
2. Measurement of turbidities of < 40 NTU: The sample is thoroughly shaked. The air bubbles are allowed to escape. The sample is poured into the turbiditimeter tube and the

turbidity is read directly from the scale or from the calibration curve.

3. Measurement of turbidity of > 40 NTU:The sample is diluted with turbidity-free water until its turbidity falls between 30.0 and 40.0 NTU. Now, the turbidity of the original sample is computed from the turbidity of the diluted sample and dilution factor. The Stock Turbidity Suspension of 400 NTU be used for continuous monitoring. High turbidities determined by direct measurement are likely to differ appreciably from those determined by dilution technique.

Calculation

$$NTU = \frac{A \times (B + C)}{C}$$

where,

A = NTU found in the diluted sample

B = Volume of the dilution water

C = Sample volume taken for dilution.

Interpretation of Results

The turbidity readings be reported as follows:

Turbidity Range (NTU)	*Record to the Nearest NTU*
0-1	0.05
1-10	0.10
10-40	1.0
40-100	5.0
100-400	10.0
400-1000	50.0
>1000.0	100.0

(C) Transparency

It is a measure of the depth to which one may see into the water. Obviously, this is variable with the day condition and the eyesight of the observer.

The Secchi Disc is a simple device used to estimate this depth; and, this parameter could also be called 'Secchi Disc visibility'. The secchi disc is a simple device used to estimate this depth. It consists of a weighted circular plate, 20 cm in diameter, with the surface painted with opposing black and white quarters. It is attached to a calibrated line by a ring at the centre, so that, when held by the line, it hangs horizontally. The procedure for determining the secchi disc visibility is as follows:

Procedure

1. The secchi disc is slowly lowered into the water body until it disappears. This depth be noted.
2. The disc be lowered few more cm and then slowly be raised again until it re-appears. This second reading of depth be also noted.
3. The average of these two readings is taken as the final Secchi disc visibility depth or, in short, transparency.

Chemical

(A) Hydrogen-ion-concentration (pH)

pH is the negative logarithm of hydrogen-ion-activity. pH of most natural waters falls in the range of 4.0–9.0. However, majority of the water are slightly basic because of the presence of carbonates and bicarbonates. A departure from the normal pH could be due to influx of acidic or alkaline wastes, mainly from the industries. pH adjustments of effluents as well as water treatment plants is a common practice. Where as acidity and alkalinity are measures of total resistance to pH change or buffering capacity of a sample, pH represents the free hydrogen-ion-activity not bound by carbonate or other bases.

Hydrogen-ion-concentration, expressed as pH, may be determined by the following methods:

1. Colorimetric method
2. Electronic method

(*a*) Colorimetric Method

Colorimetrically, pH is determined on the spot with the help of a Hellige Comparator using BDH (Qualigens) indicator solution and colour discs of the ranges from 2.0–10.5.

However, the colorimetric method is said to suffer from interferences due to colour, turbidity, salinity, colloidal matter, oxidants and reductants. Indicators themselves may alter pH of the poorly buffered liquids unless pre-adjusted to nearly the same pH as the sample. Hence, colorimetric method is suitable only for rough estimation.

(*b*) Electronic Method

This method involves operation of an electronic pH meter using a standard 'glass electrode' along with a 'reference calomel electrode' which produces a change of 59.1 milli volt (mv) per pH unit at 25°C. Glass is an ion-exchange material with preference for H^+ ions which are adsorbed on one surface and their positive charge is transmitted through the glass by Na^+ ion displacement. The H^+ ions are desorbed on the other side. Therefore, glass is a membrane specific for H^+ ions. Glass electrode consists of a glass bulb with dilute HCl in it and is connected to the internal circuit through AgCl.

Materials Required

1. Electronic pH meter with temperature compensation
2. Glass electrode
3. Reference electrode
4. Magnetic stirrer

Standard Solutions (Reagents)

1. Standard buffer prepared by dissolving required amount of substances in DW having conductivity <2.0 micro-siemens at 25°C.

   NB: The buffer solutions be preserved in polythene or pyrex bottles and the solutions be replaced every month.
2. Saturated Potassium-Hydrogen-tartrate solution to be prepared by dissolving finely crystalline 5-10 g of Potassium-Hydrogen-tartrate in 100-300 ml of DW at 25°C in a glass stoppered bottle. The clear solution be decanted and preserved by adding a thymol crystal
3. Saturated KCl solution to be prepared by dissolving AR grade KCl in DW.

Procedure

1. The pH meter is connected to 230 V AC having good earthing and switched on; and, kept as such for few minutes to warm up.
2. Before use, both the glass and reference electrodes be kept dipped in DW for atleast 24 hours

   NB: The tips of the electrodes be kept immersed in water when not in use.

At the time of using the electrodes, they are rinsed with DW and wiped with tissue paper and never with hand.

pH Measurement

1. The electrodes are connected to the pH meter.
2. Buffer solutions of pH 7.0 and 4.0 or 9.2 are prepared by grinding and dissolving standard buffer tablets (of respective pH) in DW.
3. The electrodes are washed with DW and dried with tissue paper.
4. The temperature of the buffer solution of pH 7.0 is noted.
5. The temperature control switch of the pH meter is adjusted according to the temperature of the buffer.
6. The electrodes are dipped into buffer solution of pH 7.0.
7. The meter will display some reading which may be nearly 7.0.
8. The 'Standardise Control' of the meter is operated to make the meter reading exactly 7.0.
9. The buffer solution of pH 7.0 is removed from the table and the buffer solution of pH 4.0 is brought in.
10. The electrodes are washed and rinsed as before.
11. Temperatue of the buffer solution of pH 4.0 is noted.
12. The temperature contol switch of the pH meter is adjusted according to the temperature of the buffer.
13. The electrodes are dipped into buffer solution of pH 4.0.
14. The pH meter could display exactly 4.0 or could display some reading which is nearly 4.0.

15. If it is nearly 4.0, the standardize control of the meter is operated to make the meter reading exactly 4.0.
16. The buffer solution of pH 4.0 is also removed from the table.
17. The sample water of unknown pH is brought in
18. The electrodes are washed and rinsed as before
19. The electrodes are dipped into the water sample.
20. The pH of the water sample is recorded directly from the pH meter.
21. The procedure is repeated for other samples.

(c) Dissolved Oxygen (DO)

It refers to the oxygen which is dissolved in water. It plays a key role in the dynamics of the water body particularly in the respiration of the aquatic biota. DO content in water results from:

1. Photosynthetic activity of the green plants.
2. Diffusion gradient at the air-water interface
3. Wind-driven mixing

There are broadly two methods for the estimation of DO in water:

1. Titrimetric (idiometric)
2. Electrometric (Electronic)

(a) Titrimetric Method: Alsterburg Azide Modification of Winkler's Method

Principle

In this method, the amount of DO is computed in mg/lit by the quantity of liberated iodine from a mixture of manganous sulphate and alkaline iodide sodium azide neutralized by a known volume of 0.025 N sodium thiosulphate solution. In other words, there is formation of brown hydrated oxide of manganese (IV) by oxidation of manganese (II) hydroxide by the DO when a solution of potassium iodide –sodium azide is added to the sample

$$MnSO_4 + 2\,KOH \rightarrow Mn(OH)_2 \downarrow + K_2SO_4$$

$$2Mn(OH)_2 + O_2 \rightarrow 2\,MnO\,(OH)_2 \downarrow$$

(Brown ppt)

The floc of Manganese (II) hydroxide is said to act as a gathering agent for the DO and settles down. On subsequent acidification by concentrated sulphuric acid, the hydrated oxide of manganese (IV) reverts to the original state of manganese (II) while oxidizing the iodide quantitatively into iodine.

$$2\,MnO\,(OH)_2 + 2\,KI + H_2O \rightarrow Mn(OH)_2 + 2\,I_2 + 2\,KOH$$

Manganese
Oxyhydroxide

The liberated iodine is said to be equivalent to the original DO content of the sample; The amount of liberated iodine is estimated by titration with standard solution of sodium thiosulphate using starch as indicator:

$$I_2 + 2\,S_2\,O_3 \rightarrow S_4O_6 + 2\,I$$

Interference

The Alsterburg azide modification of Winkler method, in which, Sodium azide is used, is applicable to most natural waters. Interference by higher concentrations of Fe (III) is reduced by this Alsterburg modification with azide playing the pivotal role.

Sampling Procedure

Collection of water sample be done very carefully, so that, the sample does not mix with air nor get agitated. Samples from any depth in streams, lakes, wetlands or reservoirs need special precautions to eliminate changes in pressure and temperature.

Surface water samples be collected in narrow mouth glass stoppered BOD bottles of 300 ml capacity with tapered ground glass pointed stoppers and flared mouths. The bottle be overflown 2 or 3 times its volume and the stopper replaced without air bubble being allowed to enter. Samples from depths of > 2.0 m are collected by Kemmerer type of sampler. The bottle is filled by overflowing for nearly 10 seconds but without turbulence or formation of bubbles.

Sample Preservation

DO in water be determined immediately after collection of sample. However, samples supposed to have low DO in them, may be stored by adding manganous sulphate-alkaline iodide sodium azide and sulphuric acid, shaked properly and be kept away from strong sunlight.

Procedure for DO Determination

Reagents Required

Manganous Sulphate

It is prepared by dissolving 364 g of $MnSO_4.H_2O$ in DW. Then it is filtered and diluted to 1 litre. This solution is not to produce colour with starch when added to acidified solution of KI. Alternatively, 18.2 g of $MnSO_4.H_2O$ is dissolved in DW, filtered and made upto 50 ml.

Alkaline Iodide Sodium Azide Solution

It is prepared by dissolving 500 g of NaOH or 700 g of KOH and 150 g of KI or 135 g of NaI in DW. This be diluted to 1 lit with DW. Next, 10 g of NaN_3 (Sodium azide) is dissolved in 40 ml of DW; and, this be added to 950 ml of the solution prepared above. Alternatively, if less quantity of the reagent is required, it is prepared by dissolving 25g of NaOH and 7.5g of KI in DW. This be diluted to 50 ml with DW. Next, 1 g of NaN_3 is dissolved in 4 ml of DW; and, this be added to 45 ml of the solution prepared above.

Starch Indicator

To 0.25g of starch powder, a few ml of cold water is added and is rubbed to a paste. This is added, with constant stirring, to 1 litre of boiling water, so slowly that, boiling never ceases. Boiling be continued for an additional minute or two and then allowed to cool. It is preserved in a glass bottle. A fresh solution be prepared as soon as mould appears; or, when a strong colour is not given by very dilute iodine solution.

Concentrated Sulphuric Acid

To be of AR grade.

Stock Sodium Thiosulphate (Hypo) Solution (0.1 N)

It is prepared by dissolving 24.82 g of $Na_2S_2O_3.\ 5\ H_2O$ in boiled and cooled DW. It is then diluted to 1 litre.

Standard Sodium Thiosulphate Titrant (0.025 N)

It is prepared by diluting 250 ml of stock solution to one litre boiled and cooled DW. Alternatively, 1 lit of 0.025 N hypo solution is prepared by dissolving 6.2047 g (approx. 6.3 g) of Sodium thiosulphate (AR grade) in 1 lit of DW. This standard Sodium thiosulphate solution (which is a secondary standard) is to be standardized with a primary standard (say, Potassium di-chromate solution) before every set of titration for DO estimation, for which, any standard book of chemistry dealing with volumetric analysis could be consulted.

Procedure for Estimation of DO

1. To the 300 ml/250 ml of sample collected in 300 ml/250 ml BOD bottle, is added 2 ml of $MnSO_4$ solution followed by 2 ml of alkaline iodide–sodium azide reagent.
2. Stoppered carefully and immediately to exclude air bubbles.
3. Mixed the contents by inverting the bottle atleast 15 times. When the brown ppt. settles leaving a clear supernate above the manganese hydroxide floc, shaked again and allowed the floc to settle for the second time.
4. After atleast two minutes, settling is supposed to have produced minimum 100 ml of clear supernate. Now, the stopper is removed carefully; and, immediately 2ml of concentrated sulphuric acid is added in such a way that it flows down the neck of the bottle. Re-stoppered and inverted gently, until dissolution of the above floc is complete. A brown colour solution is formed in the bottle.
5. This brown coloured solution is titrated against standard 0.025 N sodium thiosulphate solution (popularly called 'hypo') taken in a 50 ml burrette. The brown colour goes on fading until a straw coloured solution is obtained near the end-point.
6. Now, 2-3 drops of starch indicator solution is added. An intense blue coloured solution is obtained.
7. The titration is continued and hypo solution is added drop by drop from the burrette. The blue colour goes on fading, until the solution becomes completely colourless at the end-point.
8. The burrette reading is noted.
9. The titre value is the amount of DO present in mg/lit in the sample water.

The data may be tabulated as follows:

Obs. No.	IR of Burrette	FR of Burrette	Difference	DO in mg/lit.
1				
2				
3				

IR = Initial Reading

FR = Final Reading.

(b) Electrometric (Electronic) Method of DO Estimation by Dissolved Oxygen Meter

The Dissolved Oxygen (DO) meter (manufactured my M/S Systronics, India) is generally a battery operated portable instrument for estimation of DO and determination of temperature of the sample water at the same time. DO is measured in ppm (mg/lit) over a range of 0-20 ppm with 0.1 ppm resolution and temperature is indicated in degrees Celsius from–5 to + 45 degrees Celsius. The DO measurement is automatically temperature compensated for solubility of oxygen in water and permeability of the probe membrane. Salinity compensation is manual. Measurements are displayed by a 3 digit LCD.

Principle

The probe supplied with the unit (instrument) is a special Clark-type membrane covered polarographic sensor. A thermistor probe is also supplied for the automatic temperature compensation and measurement. A thin, permeable membrane stretched over the sensor isolates the sensor elements from the environment; but, allows oxygen to enter. When a polarizing voltage is supplied across the sensor, oxygen, which has passed through the membrane reacts at the cathode, causing a current to flow.

Panel Features in Brief

1. The following principal switches: Power, Salinity, Zero, Set zero, Mode, DO meas/Cal, Temperature
2. LCD display.

Power Supply

This unit is battery-operated. Hence, it does not require any mains outlet. It uses one 9 volt small battery, *e.g.*, Everyday 216, Hi-Watt 6F22, etc. Correct polarities be ensured while inserting fresh batteries.

Operating Steps

Probe Preparation

The DO probe consists of five main parts, *viz.*, main stem, electrode cover, sleeve, membrane, and bakelite nut

The main stem is made of Perspex with gold electrode at lower end and silver ferrule fully surrounding the gold. The upper end carries a connecting cable and a connector. The main stem has threads; so that, electrode cover along with membrane assembly could be fitted to it by means of bakelite nut.

The electrode cover is a thin cylinder which fits into lower half of main stem. A Teflon, oxygen permeable membrane is held on to electrode cover by means of a rubber sleeve.

Electrolytic Solution

7.5 per cent solution of KCl is prepared by dissolving 7.5 g of AR grade KCl in 100 ml of DW. AgCl crystals are slowly added until until AgCl stops getting dissolved.

Assembling of the Probe

At first, the bakelite nut is slided over the probe cover. Next, a piece of the membrane (1″ × 1″) is kept over the tip of the probe. Then, the sleeve is pushed fitted over the electrode cover in such a way that the membrane is flat and tight. There be not any wrinkles on the membrane. The membrane be removed and re-fitted if any wrinkles are detected. A few drops of electrolyte solution is poured and checked for any leakage through the membrane or near the sleeve. Now, the electrode cover (probe cover) be filled completely with the solution. It is kept vertical and the main stem of the probe is slowly inserted into the probe cover which is filled with the solution. The excess solution be let out. Next, the bakelite nut over the main stem is tightened. It is to be ensured that there is no air bubble. In the event of any bubble being detected, the probe is opened and re-filled, till no bubbles are visible.

The gold cathode should always be bright and untarnished. It be wiped with a clean lint-free cloth or hard paper. Any form of abrasive or chemical be never used.

Preparing the Instrument

1. The 'Mode' switch be set at 'Set Zero'
2. The meter be switched 'on'.
3. 'Set 20' control be adjusted until the meter displays 20.0.
4. The Mode switch be switched to 'zero' and the 'zero control' be adjusted to read 0.0. (This is electrical zero setting of the meter).
5. The prepared DO probe be attached to the probe connector.
6. 15 minutes be allowed for the probe to be polarized.

Air Calibration

Experience has shown that, air calibration method is quite reliable; yet, far simpler than the other methods. The air calibration method is given below:

1. The probe be placed in moist air. For this, the probe could be wrapped loosely in a damp cloth taking care that the cloth does not touch the membrane. About 10 minutes be waited for temperature stabilization.
2. The mode switch be changed to 'Temp'and the temperature be read and recorded.

   Table (1st Table) showing solubility of oxygen in freshwater against temperature be consulted, from the Manual supplied with the instrument, in order to determine the calibration value. Also, the altitude and the atmospheric pressure correction factor be determined from a different Table (2nd Table) in the same Manual. Calibration value from 1st Table be multiplied with correction factor from the 2nd Table. An example is given below:

   Temperature be assumed as: 25°C

   Altitude be assumed as: 1000 ft

   From Table 1, the calibration value for 25°C is, say, 8.4 ppm

   From Table 2, the correction factor for 1000 ft is about 0.96

   Therefore, the correct calibration value is: 8.4 ppm × 0.96 = 8.1 ppm

3. The 'salinity' knob be set to 'zero'
4. The mode switch be brought to 'cal'.
5. The 'CAL control' be adjusted until the meter reads the correct calibration value as indicated in step No. 2 above. Two minutes be waited to verify the calibration stability. Re-adjustment be done, if necessary.

The DO meter is now calibrated. Calibration be checked periodically.

DO Measurement

1. With the instrument calibrated, the probe is dipped in the water sample (whose DO is to be measured) and the sample water be constantly stirred. Stirring be continued with a stirrer during the process of measurement of DO in order to prevent depletion of DO near the membrane of the probe.
2. The 'salinity' control switch be set to the salinity value of the sample under test. Sufficient time be allowed for the probe to stabilize.
3. The DO value be read and recorded.

(D) Free Carbon di-oxide (FCO_2)

< 10 mg/lit of FCO_2 is said to be, generally, present in surface waters. However, ground waters may contain greater amount of FCO_2.

FCO_2 may be estimated by:

1. Nomographic Method
2. Titrimetric method

The nomographic method gives a closer estimation of FCO_2 when the pH and alkalinity are determined immediately after sampling. The pH be determined with an electrometric pH meter calibrated with standard buffer solution in the pH range of 7.0-8.0. The total alkalinity be determined titrimetrically.

The titrimetric method of estimation of FCO_2 be done using phenolphthalein as an indicator.

The titrimetric method is discussed below in detail:

Titrimetric Method of Determination of FCO_2

Principle

FCO_2 reacts with Na_2CO_3 or NaOH to form $NaHCO_3$. Completion of the reaction is indicated by development of pink colour at the end point (pH 8.3). In short, FCO_2 is estimated by neutralizing the dissolved CO_2 with 0.022 N aqueous sodium hydroxide solution using phenolphthalein as an indicator.

Sampling and Storage

Surface water sample may be collected by dipping a collecting bottle with double lid just below the water surface; and, after the bottle is filled-in, its inner lid be closed while the mouth of the bottle is still under water. Deep water sample may be collected with the help of a water sampler bottle. More precisely, sample water be collected by rubber tubing discharging at the bottom of a 100 ml graduated cylinder or Nelson's tube. The sample water be allowed to overflow several times. The tube be withdrawn while the sample water is flowing. The measuring cylinder be flicked (jerked) to throw off the excess sample water above 100 ml mark. Some loss of FCO_2 may be expected in storage and on transit of the sample, even with a careful collection technique. Hence, field determination immediately after sampling is desirable. Alternatively, the sampling bottle be completely filled with the sample water and be preserved at a temperature lower than the room temperature and FCO_2 be estimated as soon as possible.

Reagents

1. Preparation of 0.1 N stock NaOH solution: It is prepared by dissolving about 4 g of AR grade NaOH pellets in 1 lit of DW. It be standardized against 0.1 N H_2SO_4 using phenolphthalein as an indicator. 100 ml of this 0.1N solution diluted to 440 ml with DW gives N/44(0.022N)NaOH.

Alternatively, 0.4735 g (accurately) of NaOH (96 per cent pure) dissolved in 500 ml of DW gives 0.022N NaOH.

Procedure

1. Sample water is collected following the above procedure and 100 ml be taken in a conical flask.

2. 10 drops of phenolphthalein indicator solution be added.
3. If the sample turns pink, it indicates that FCO_2 is absent.
4. If the sample remains colourless, it be titrated rapidly with 0.022 N NaOH solution (taken in a 50 ml burrette), stirring gently with a glass rod until a definite pink colour is produced which persists for 30 seconds when viewed through the depth of the sample.

   NB: For checking loss of FCO_2 during titration, a second sample be taken. After adding 10 drops of phenolphthalein indicator, the full amount of the titrant used in the first titration be rapidly poured-in. If the sample turns pink, no FCO_2 is said to be lost. If it remains colourless, additional amount of the titrant be added until pink colour appears. The result of the second titration be considered.
5. The burrette reading be recorded.

The data be recorded in the following Table:

Obs. No.	Initial Burrette Reading (ml)	Final Burrette Reading (ml)	Difference (ml)
1			
2			
3			

Calculation

mg/lit $FCO_2 = (A \times N \times 44{,}000)$ / ml of sample

where,

A = ml of NaOH titrant

N = Normality of the NaOH used above

(E) Alkalinity

Alkalinity of water is its capacity to neutralize a strong acid to a designated pH; or, stated in another way, it is the quantity and kind of compounds present, which collectively shift the pH to the alkaline side of neutrality. Although the alkalinity of natural waters is generally due to bicarbonates, it is usually expressed in terms of $CaCO_3$. The three kinds of alkalinity are indicated as OH^- (Hydroxide) normal CO_3^{--} (Carbonate) and HCO_3^- (Bicarbonate). The three are summed-up as Total Alkalinity (TA). Carbonate and Bicarbonate are common to most waters because carbonate minerals are abundant in nature. The presence of OH^- can usually be attributed to water treatment or to contamination. The expected TA in nature usually ranges from 45-200 mg/lit of $CaCO_3$.

Alkalinity is significant in interpretation and control of water and wastewater treatment processes. For industrial wastes, measurement of alkalinity could indicate change in quality, if the source of the sample is known to have generally stable levels of alkalinity.

Principle

Hydroxyl ions present in a sample as a result of dissociation or hydrolysis of solutes are neutralized by titration with standard acids. Titration to pH 8.3, *i.e.*, decolourisation of phenolphthalein indicator, show complete neutralization of OH^- and half of CO_3^{--} while, titration to pH 4.4, *i.e.*, sharp change from yellow to pink of methyl orange indicator indicates alkalinity due to OH^-, CO_3^{--} and HCO_3^-.

$OH^- + H^+ \rightarrow H_2O$

Titration to pH 8.3

$CO_3^{--} + H^+ \rightarrow HCO_3^-$

$HCO_3^- + H^+ \rightarrow H_2O + CO_2$ Titration to pH 4.4
(from CO_3 or
original HCO_3)

The end-point may be determined empirically by titration and is that pH at which the derivative of ΔpH/Δ ml titrant is the greatest.

Materials Required

50 ml burette, 250 ml Erlenmeyer flask, ordinary filter paper (for watching and comparing the colour during titration), 250 ml reagent bottle (for storing 0.02 N H_2SO_4), 60 ml dropping reagent bottles: (2 Nos for storing phenolphthalein and methyl orange indicator solutions), 0.02N H_2SO_4 solution, phenolphthalein indicator solution, methyl orange indicator solution.

Preparation of 0.02 N H_2SO_4: 0.28 ml of AR grade conc. H_2SO_4 added to 500 ml of DW gives 500 ml of 0.02 N H_2SO_4. This being a secondary standard, it is to be standardized with 0.02 N Na_2CO_3 using phenolphthalein as an indicator.

Sampling and Storage

Sample water be collected in polythene or borosilicate glass bottles and be stored at a lower temperature. The bottles be filled completely with sample water. Samples be analysed without delay preferably within a few hours. Samples be kept sealed until ready for analysis because loss of CO_2 results in conversion of HCO_3^- to CO_3

Procedure

1. 100 ml of the sample water (collected following the standard procedure) is taken in a 250 ml Erlenmeyer flask
2. 4 drops of phenolphthalein indicator solution be added to it.
3. If the solution remains clear, '0' ppm of TA be recorded
4. If the solution becomes pink, it be titrated with 0.02 N H_2SO_4 (taken in a 50 ml burette), until a clear solution is obtained.
5. Volume (ml) of 0.02N H_2SO_4 required in phenolphthalein titration be recorded.
6. Next, 2-4 drops of methyl orange indicator solution is added to the same solution in the Erlenmeyer flask
7. Titrated again with the same 0.02N H_2SO_4 until the solution turns pink orange.
8. Volume (ml) of 0.02 N H_2SO_4 required in the methyl orange titration is also recorded.

The data be recorded in the following table:

Phenolphthalein titration

Obs. No.	*Initial Burrette Reading (ml)*	*Final Burrette Reading (ml)*	*Difference (ml)*
1			
2			
3			

Methyl orange titration

Obs. No.	Initial Burrette Reading (ml)	Final Burrette Reading (ml)	Difference (ml)
1			
2			
3			

Result

1. 10 times the volume (ml) of 0.02N H_2SO_4 required in phenolphthalein titration = phenolphthalein alkalinity (PA) in ppm (mg/lit)
2. 10 times the volume (ml) of 0.02 N H_2SO_4 required in methyl orange titration = methyl orange alkalinity (MOA) in ppm.
3. Total Alkalinity (TA) is the sum of PA and MOA when both have values to contribute to the same.

(F) Conductivity

Conductivity of a water sample is its ability of carrying electric current which depends on total concentration of ionized substances, mobility of each of such ions, their valences, etc. On the other hand, specific conductivity means conductivity of a segment of a particular solution having a length of 1 cm and a cross sectional area of 1 cm^2. Organic compounds contribute very little to conductivity because they are mostly non-ionic. Freshly prepared DW has conductivity of 0.5–2.0 micromhos/cm which may increase to about 4.0 micromhos/cm on storage due to absorption of CO_2 from air. Conductivity reflects the characteristics of the water supplied.

Conductivity of a flowing stream could be recorded continuously *in situ* along with other parameters, *e.g.*, DO, pH, and temperature. However, the monitoring equipments are required to be frequently checked to avoid electrode fowling.

There are different kinds of conductivity bridges manufactured by various companies. The 'Systronics Digital Direct Reading Conductivity meter Model No.304' is designed for measuring the specific conductivity of a solution by using a conductivity cell. It enables to measure the conductivity, without manual balancing and the specific conductivity is read directly on a digital panel.

Principle

A conductance cell and a Wheatstone bridge is used for measuring the conductivity. The conductance of a solution increases with temperature at a rate of approx. 2 per cent per degree Celsius. KCl has a lower temperature co-efficient of conductivity and hence is used in the cell. For precise works, conductivity be determined at 25 C

The main feature of Systronics Digital Conductivity meter 304 is that, it avoids the conventional method of manual balancing by employing a self-balancing ratio-transformer bridge technique. A highly stable oscillator powers the bridge. The current through the conductivity cell passes through one winding of the radio transformer. The other winding output is fed to a selective high gain amplifier, the output of which drives a current in the third winding, in such a way, that, the flux in this winding opposes the flux in the first winding. This goes on until both the fluxes cancel each other and consequently the detector output will be almost zero. The output of the amplifier, now, is a measure of the conductivity of the solution being measured.

Materials Required

1. Conductivity meter with conductivity cell
2. Beaker
3. Stirrer
4. KCl

Operating Procedure

(a) Calibration of the Instrument

All the controls are marked on the instrument.

The instrument be switched-on. The indicator lamp glows.

The CAL/READ switch be kept in CAL position and the CAL control be adjusted to full scale reading (100) on the meter.

After thoroughly cleaning, the 'cell' (conductivity cell) be immersed in the standard KCl solution, say, 0.1 N KCl which has a specific conductivity of the order of 0.01412 millimhos (14.12 micromhos)/cm at 30°C.

The CAL/READ control be put in the READ position. The meter needle should deflect.

The range control be turned in the counter-clockwise direction starting from 300 micromhos range to get a range where the needle will show maximum deflection. The maximum deflection of the needle could be obtained in the range of 0-30 micromhos.

The standard Table supplied with the instrument be referred to in order to determine the conductivity of the solution at 30°C.

Then, the cell constant control be adjusted, so that, the meter reading corresponds with the Table value. In other words, the meter reading be adjusted to read 14.12 micromhos.

Now, the instrument is calibrated for measuring the conductivity of the unknown sample.

The standard 0.1 N KCl solution is prepared by dissolving 7.459 g of dried AR grade KCl in 1 lit of DW.

(b) Determination of Conductivity (with the help of Systronics Conductivity meter Model 304)

1. The instrument be switched-on.
2. The 'scale selector' knob be brought to 200 millimhos position (Maximum).
3. The 'Read/Cal' knob be brought to 'Cal'
4. The 'Cal' be operated to display 100 in the display panel.
5. The cell be dipped in N/10 KCl (specific conductivity of N/10 KCl at 30 C is 14.12 micromhos/cm).
6. The 'scale selector' knob be brought to 20 millimhos position.
7. The 'cell constant' knob be brought to 0.95 position.
8. The 'Read-Cal' knob be brought to 'Read' position. The meter is supposed to display a value which is little less than 14.12.
9. The 'cell constant' knob be adjusted, so that, the meter reads 14.12.
10. The 'Read-Cal' knob be brought to 'Cal' position.
11. The cell be rinsed with DW

12. Next, the cell be dipped in sample water.
13. The 'Read-Cal' knob be brought to 'Read' position.
14. The 'scale selector' knob be rotated anticlockwise from 200 millimhos position (highest value) gradually stage by stage to the 20 micromhos mark (lowest value), till the meter displays 1.0.
15. Now, the 'scale selector' knob be rotated clockwise stage by stage and to be stopped at that point where we get the highest value.
16. This value be recorded as the conductivity of the sample water.

(G) Hardness

'Hardness' is defined as the total concentration of Calcium (Ca^{++}) and Magnesium (Mg^{++}) ions expressed as Calcium carbonate ($CaCO_3$). However, if other polyvalent metals, *e.g.*, Fe, Al,Mn, Sr, Zn are present in significant amount, they could be included in reporting hardness.

When hardness is numerically greater than the sum of carbonate (CO_3^{--}) and bicarbonate (HCO_3^-) alkalinity, the amount of hardness equivalent to Total Alkalinity (TA) is called 'carbonate hardness' and the excess is known as 'non-carbonate hardness'. When hardness is equal to or less than total of bicarbonate and carbonate alkalinity, all the hardness is carbonate hardness and there is no non-carbonate hardness.

Methods of Determination of Hardness

There are two methods of determination of hardness, *viz.*,

1. Hardness by calculation
2. Hardness by EDTA titration

(a) Hardness by Calculation

The accurate method of determination of hardness is by computation from results of Ca^{++}, Mg^{++} and other hardness producing cations. Hardness, in this case, be reported as 'hardness (Ca, Mg)' or 'hardness (Ca, Mg, Sr, Fe, Al)'.

(b) *Hardness by EDTA titration*

It is rather a rapid method of analysis. It measures calcium, magnesium; but, may be applied with appropriate modifications to any kind of water.

Principle

Ethylene-di-amine-tetra acetic acid (EDTA) and its di-sodium salts form a soluble chelated complex, when added to a solution of certain metal ions. If a small amount of Eriochrome Black T dye is added to an aqueous solution containing Ca^{++} and Mg^{++} ions at a pH of 10.0 ± 0.1, the solution will become wine red in colour. If EDTA is added as a titrant, Ca^{++} and Mg^{++} will be complexed. When all such Ca^{++} and Mg^{++} are complexed, the solution will turn blue from wine red. This is the end-point. A small amount of Mg-EDTA is added to sharpen the end-point; but, at the same time, introduces a sufficient amount of Mg to obviate 'blank correction'. A time limit of five minutes is set as the duration of titration to minimize any tendency towards $CaCO_3$ precipitation.

Reagents

Buffer solution: 16.9 g of NH_4Cl be dissolved in 143 ml of Conc. NH_3. 1.25 g of Mg-EDTA be added and diluted to 250 ml with DW. Alternatively, 1.79 g of di-sodium EDTA dehydrate (AR grade) and 780 mg of $MgSO_4.7H_2O$ be added in 50 ml of DW. The solution be stored in plastic or resistant glass container to prevent loss of NH_3 or pick-up CO_2. They be not stored for more than a month and be discarded, if 1 or 2 ml fails to produce pH of 10.0 ± 0.1.

Odourless and stable buffer be prepared by mixing 55 ml of conc. HCl with 400 ml of DW and then adding 310 ml of 2-amino-ethanol with stirring. 5 g of Mg-EDTA be added and diluted to 1 lit with DW.

Indicator: Solutions of indicator, called Eriochrome Black T (earlier called Solochrome Black T) are unstable and give poor end point. Dry mixture of the dye and NaCl are said to be satisfactory. Alternatively, the said dye be prepared by dissolving 0.5 g of the dye with 4.5 g of hydroxylamine hydrochloride in 100 ml of 95 per cent Ethyl or Isopropyl alcohol.

Standard EDTA titrant (0.01 M): 3.723 g of AR grade di-sodium EDTA dehydrate be dissolved in DW, diluted to 1 lit. This be checked by standard calcium solution. This be stored in polythene or pyrex glass bottles because EDTA extracts cations from soft glass periodically. This be standardized by standard Ca^{++} solution.

Standard Calcium solution: 1 g of anhydrous $CaCO_3$ (special reagent grade, low in heavy metals, alkali and Mg) be dissolved in 1: 1 HCl in a 500 ml Erlenmeyer flask. 200 ml of DW be added, boiled for few minutes, cooled and a drop of methyl red indicator be added. The intermediate orange colour be adjusted by adding 3 N NH_4OH or 1:1 HCl, as required. It be transferred quantitatively to 1 lit capacity volumetric flask and made upto the mark.

1 ml of this standard calcium solution = 1 mg of $CaCO_3$.

Titration Procedure

The sample volume be such that, it requires less than 15 ml of EDTA (Titration time limit is 5 minutes).

25 ml of sample water be diluted to 50 ml with DW.

1-2 ml of buffer be added to adjust pH to 10.0 to 10.1

1-2 drops of indicator be added.

The titrant be added slowly with constant stirring until the last reddish tinge disappears. The last few drops be added at 3-5 seconds interval.

The end-point, which is blue in colour, be compared against day light or fluorescent lamp.

A blank be run containing identical amount of buffer and indicator, *i.e.*, equivalent amount of DDW be taken and be added to identical amount of buffer and indicator; and, be titrated as done before.

Calculation

Hardness (EDTA) as mg/lit $CaCO_3$ = $(A \times B \times 1000)$/ml of sample

where,

A = ml titre (EDTA) used

B = mg $CaCO_3$ equivalent to 1 ml of titrant.
(That is, 1 ml of 0.01 M pure EDTA = 1 mg of $CaCO_3$)

Physico-chemical Characteristics of Water of Sone Beel in Assam (Kar, 1990)

In Sone Beel wetland, the water was found to exhibit interesting trend in its physico-chemical features. Turbidity, silt load, temperature, hydrogen-ion-concentration, dissolved oxygen (DO), free carbon di-oxide (FCO_2), total alkalinity (TA) and specific conductivity were the parameters investigated which are briefly discussed below:

Water Colour and Turbidity

The turbid water of the Beel was observed to possess the colour of tea and milk mixture caused by the ingress of rain water during monsoon. During the post-monsoon period, the water gradually started becoming clearer and pale green; but in some places, it still continued to remain muddy. During the following winter period, the water is generally pale green while during summer, it is dark green indicating positive changes in biological activity.

The turbidity of the Beel water generally fluctuates from 24.21 to 185.54 TU at the surface and 20.56 to 183.36 TU at the bottom level.

Silt Load

The silt load of the inlet water, measured at the Beel mouth was found to vary from 27.0 to 350.0 mg/lit. Concomitantly, the values for the outlet water was found to fluctuate usually from 9.0 to 88.1 mg/lit. The overall result indicated that more amount of silt is retained and deposited in the Beel and lesser amount is expelled through the outlet thereby resulting in overall siltation of the Beel.

Temperature

The surface water temperature generally lies between 18.7 to 32.3°C while the bottom water temperature generally range from 19.2 to 32.1°C. The fluctuation trend is usually almost identical with high values during the warm season followed by low values during the cold months.

pH

The pH values, when measured with electronic pH meter at the laboratory after collecting samples from the field, generally vary from 6.67 to 7.90 at the surface level and 6.61 to 7.79 at the bottom level. Low pH values are usually recorded during the monsoon period and higher values during the post-monsoon days.

Dissolved Oxygen (DO)

DO values generally fluctuate from 2.61 to 5.96 mg/lit in the surface water and 2.43 to 5.84 mg/lit in the bottom water. Low values are usually observed during the rainy days while comparatively higher trend is found in the dry days exhibiting more or less an identical trend at varying depth levels. DO saturation was found to vary from 35.42 to 70.40 per cent at the surface and from 32.34 to 70.31 per cent at the bottom levels.

Free Carbon di-oxide (FCO_2)

FCO_2 was found to vary from 1.75 to 11.8 mg/lit at the surface water level while it fluctuated from 0.90 to 14.55 mg/lit at the bottom water level. Higher values are generally found during the rains but lower values during the post-monsoon days.

Total Alkalinity (TA)

TA generally lies between 28.5 to 74.31 mg/lit at the surface water 25.0 to 76.01 mg/lit at the bottom water. Higher values are usually recorded during the dry season while lower values during the rainy season.

Specific Conductivity (SC)

It usually ranges from 51.98×10^{-6} to 89.09×10^{-6} mhos/cm at the surface water level and 52.5×10^{-6} to 156.66×10^{-6} mhos/cm at the bottom water level. Nevertheless, higher values to the tune of 393.0×10^{-6} mhos/cm could sometimes be recorded at certain spots around fish landing centres.

Physico-chemical Characteristics of Water of Chatla Haor in Assam (Kar and Barbhuiya, 2000 b)

The following are the results of analysis of the physico-chemical characteristics of water of Chatla Haor:

1. Temperature °C: 33
2. Turbidity (NTU): 83.27
3. pH: 6.09
4. FCO_2: 7.59 mg/lit
5. TA: 83.39 mg/lit
6. Conductivity: 142.91 micromhos/cm

Chapter 4

Physico-chemical Characteristics of Soil

Soil is the natural medium for growth of biota in both land and water; and, is a heterogeneous mixture. The physical parameters of soil in any area determine the different types of biota it can sustain. The methodology recommended by Jackson (1973) are largely followed.

Collection of Soil Sample

The importance of having a representative sample can well be appreciated. As such, the collected sample be a representative of the area of collection. An area could be treated as a single sampling unit, only if it is appreciably uniform. Variations in slope, colour, texture, growth of biota and management be taken into account while collecting soil sample; and, separate sets of composite samples need to be collected from each area. The procedure for collection of soil is briefly described below:

1. The study spot be selected and demarcated into a plot of 50×50 m^2 area.
2. The plot be divided into a convenient number of quarters or sampling areas of almost equal dimensions. For simplicity, the plot be demarcated, physically or mentally, into 10 equal quarters, each of length (L) 25 m and breadth

(B) 10 m (area: 25 m × 10 m = 250 m^2). Each quarter be numbered serially from 1 to 10.

3. Depending on the nature of the work, two/five quarters are selected randomly through lottery (which be *c* 15 per cent of the total area of all the quarters) and soil samples are collected only from these selected two/five quarters. Before collecting soil sample, the surface litter, pebbles, etc., are removed from the collection site.
4. From each quarter, one soil sample is collected from two desired depths in the same spot, *viz.*, (a) one sample from a depth of 0-15 cm and (b) another sample from a depth of 15-30 cm.

   Alternatively, depending on the nature of the work, the soil samples could be collected from three different depths, as given below, one sample from each depth:

   (a) 0-6", (b) 6-15", (c) 15-24"
5. In each sampling spot, a 'V'-cut be given in the soil using a sampling tool (locally called 'Kodal'). After reaching the desired depth, the soil is collected, mixed thoroughly and debris cleared. *c* 1.0 kg of soil sample of each depth is stored in a polythene bag and labelled.
6. If the soil samples of each depth are found to be almost homogeneous, then, all the soil samples of respective depth of all the quarters be mixed thoroughly. For example, say, there are five selected quarters. In each quarter, there are two samples from two depths. So, in total, there will be 10 samples, 5 of respective depth. Then, each of the 5 samples of respective depths are mixed. As such, at the end, there are only two thoroughly mixed samples of respective depths (*viz.*, one sample from 0-15 cm depth and one sample from 15-30 cm depth), which are kept in two separate polythene bags and brought to the laboratory. Each bag contains about 2 kg of freshly collected soil.
7. From each of these 2 kg of freshly collected stock soil, about 1 kg of soil is finally retained by quartering technique in the following manner:

   (a) The stock soil (*c* 2 kg) is spread on a piece of paper.

   (b) The debris are removed and the soil aggregates and lumps are softly broken. No grinding is done.

(c) With a spade, the entire stock soil (c 2.0 kg) is partitioned into 4 quarters. Two opposite quarters are retained and the other two quarters are discarded. The soil samples in the retained two quarters are again mixed thoroughly, debris are cleared; and, again, with the help of the same spade, the soill mass is again partitioned into 4 quarters; the two opposite quarters be retained and the other two be discarded. In this way, repeated quartering is done with each of the two stock soil samples till c 1.0 kg of each of the stock soil sample of respective depth (0-15 cm and 15-30 cm) are retained. In some laboratories, in the quartering process, 3 quarters are retained and 1 quarter is discarded.

8. The soil sample, so collected, is dried at room temperature or in shade.
9. About 500 g of dried soil sample is stored in a polythene bag. Each bag is labelled and brought to the laboratory for analysis.

The bottom soil in a water body is generally influenced by the water column above it and could also be *vice versa*. In a water body, particularly a lentic one (*e.g.*, a Beel), random sampling of soil is done by operating an Δ 232.25 cm^2 Ekman dredge at randomly-selected spots covering variable depths. These samples are dried in shade, powdered, labelled and brought to laboratory for chemical analysis.

The following physico-chemical parameters of soil are generally analysed:

(A) Physical

1. Temperature
2. Texture
3. Colour
4. Moisture content
5. Bulk density
6. Infiltration rate.

(B) Chemical

1. pH
2. Conductivity
3. Organic carbon
4. Available Nitrogen
5. Available Phosphorus.
6. Available Potassium.

The analysis of the above-mentioned parameters are briefly described below:

(A) Physical

Soil Temperature

Main Materials Required

Soil thermometer, note book.

Procedure

1. The soil thermometer be inserted into the soil upto a certain depth (*c* 3 ") at the study point.
2. It is required to wait for *c* 1 minute till the alcohol/mercury within the thermometer becomes steady.
3. The soil temperature be recorded from the dial of the thermometer.
4. The temperature be recorded at different study points.
   NB: The thermometer be not pressed too hard into the soil.
5. In the case of bottom soil in a water body, the thermometer is inserted into the collected soil immediately after collection and the temperature is recorded.

The data be recorded in the following Table:

Locality:	Date:	Time:
Site No./Sample No		Temperature (°C)

Soil Texture

Main Materials Required

Seives of different mesh size (mm); 0.02, 0.2, 2.0, 5.0, 7.0; soil sample, graph paper.

Procedure

1. The soil sample be sieved through a series of different meshes, as indicated above, beginning 7.0 mm (stone); then, proceeding gradually to 5.0 mm (gravel); and, 2.0 mm (coarse sand); 0.2 mm (fine sand); and, ending with 0.02 mm.
2. This exercise would separate the particles of different sizes.
3. With the help of a balance, the quantity of soil of each particle size be weighed.
4. The data be put in a textural triangle.

The data be recorded in the following Table:

Locality: Date: Time:

Site No.	Weight of the Different-sized Particles in the Soil Sample (g)				
	7.0 mm	5.0 mm	2.0 mm	0.2 mm	0.02 mm

Soil Colour

Soil exhibits a variety of colours. Soil colour may be inherited from the parental material (*i.e.*, lithocentric) or sometimes it may be due to soil-forming processes (acquired or genetic colour). Colour of soil renders valuable key for the identification of the soil type.

Main Materials Required

1. Soil auger (for collection of soil sample).
2. Munsell's soil colour chart.
3. Bucket for keeping the collected soil: 1 × 10 lit capacity.
4. White foolscap paper: 1 sheet.
5. Cardboard sheet: 50 cm × 50 cm: 1 No.

Munsell's Soil Colour Chart

It is a booklet formed out of Complete Edition of Munsell's Book of Colour (*vide* US Department of Agriculture Hand Book No. 18: Soil Survey Manual) and consists of seven soil colour charts. It is popularly called 'Soil Collection Display 199'. Each soil colour chart (in the form of loose leaflet) consists of different standard colour chips, systematically arranged according to Munsell's Notations. The arrangement is by three simple variables which combine to describe all colours and are known in the Munsell's System as 'Hue, Value and Chroma'. The 'Hue'notation of a colour indicates its relation to Red, Yellow, Green, Blue and Purple. The 'value' notation indicates its lightness. The 'Chroma' notation indicates its strength (intensity). The colours displayed on the individual soil colour chart are of constant Hue designated by a symbol in the upper right hand corner of the card. Vertically, the colours become successively lighter from the bottom of the card to the top by visually equal steps; their value increases from bottom upwards. Horizontally, they increase in chroma to the right and becomes greyer to the left. The value notation of each chip is indicated by the horizontal scale across the bottom of the chart.

Procedure

1. The soil sample be collected following the standard procedure.
2. The dried soil sample be spread uniformly over a card board sheet and the same be held directly behind the apertures separating the closest matching colour chips.
3. The particles of the soil be matched with the colour chips of the Munsell's Soil Colour Chart.
4. After matching the sample soil colour with the colour chip(s) of the Munsell's Chart, the colour name and the Munsell's Notation be ascertained from the chart.
5. The field observations be recorded in the following Table.

Locality:	Date:	Time:			
Sample No.	Hue	Value	Chroma	Colour	Expression
Example:					
1.	5 YR	5	6	Yellowish Red	5 YR 5/6

Soil Moisture (mainly for land soil)

Soil moisture is the water present in the space between the soil particles. Soil moisture influences the physical, chemical and biological characteristics of the soil. Soil moisture is the per cent of moisture (on weight basis) in a soil sample at any given time. It shows general moisture content of the soil. Soil moisture helps to understand the water movement and aeration of the soil.

Main Materials Required

Soil digging tools, polythene bags, beakers, hot plate, weighing balance.

Procedure

1. Soil samples be collected in polythene bags following the standard procedure.
2. A 100 ml beaker be weighed empty (W_1).
3. From the stock soil sample, *c* 100 g of fresh soil be poured out into the pre-weighed beaker.
4. The beaker with the soil be now weighed (W_2).
5. The weight of the soil be determined by subtracting $W_2 - W_1$.
6. The beaker with the soil sample be now kept on a sand bath and then be placed on a hotplate maintained at 105°C (approx.).
7. The soil sample will be seen to be getting dreid gradually.
8. After the soil sample is dried, the beaker with the soil be removed from the hotplate and cooled.
9. The beaker with the soil sample be now weighed again.
10. After this, the soil sample in the beaker is heated, cooled, and weighed repeatedly till constant weight is obtained (W_3).

Calculation

Per cent moisture content be calculated as follows:

Weight of the empty beaker = say Y g

Weight of the empty beaker + fresh soil sample = say X g

Fresh weight of the soil sample = X – Y = say A g

Final weight of the beaker + dry soil (after getting constant weight) = say Z g

Dry weight of soil sample = Z–Y = say B g

Therefore, weight of the moisture in the soil = A–B = say M

Moisture % = {[Weight of moisture (M)]/[Dry weight of soil sample (B)]} × 100

Bulk Density (BD) (Mainly for land soil)

The bulk density or volumetric mass of the soil is the weight of the oven dry soil per unit volume. It includes the weight of the solid soil particles and the pore space in the soil. The dry mass, conversely, is the mass of the dry soil which has no pore space. Bulk density is expressed as grams per cubic centimeter. It is said to increase with the depth of the soil due to the compaction occurring from the pressure on the top layer and also due to the low organic matter content. Bulk density of soil is influenced by density of the mineral particles, quantity of organic matter, soil compaction and burrowing activity of the soil animals. In other words, BD is said to be the specific gravity of the soil with its pores. BD helps in determining the degree of compaction of the given soil. BD also indicates the soil aeration status.

Main Materials Required

25 ml volumetric flask, Oven, weighing balance, soil sample.

Procedure

It varies according to the nature of the site, disturbed or undisturbed. Procedure for both the kinds of sites are described below:

(a) *For Disturbed Site*

Portion of the earth which is regularly ploughed is called 'disturbed site'.

1. The soil sample be oven-dried.
2. A 100 ml volumetric flask be weighed (W_1).
3. This pre-weighed volumetric flask be filled with soil sample collected by standard procedure.
4. The flask be now tapped and tapping is continued till the level of soil poured-in reaches the 100 ml mark of the volumetric flask.

Note: It is assumed that, by packing the flask with soil followed by tapping the flask, compactness of soil in the nature is represented.

5. Weight of the volumetric flask with the soil be noted (W_2).
6. Weight of the soil be determined by subtracting W_2–W_1.
7. The flask be now emptied and water be poured-in upto the 100 ml mark of the volumetric flask.
8. The volume of the water (say × ml) be noted by pouring it out into a measuring cylinder.

Calculation (with a model example)

Weight of the volumetric flask (W_1) = say 71.0 g

Weight of the volumetric flask + soil (W_2) = say 203.5 g

Volume of water in the volumetric flask (X) = 100 ml

Weight of the soil (W_2–W_1 = Y) = say 132.5 g

Now, we know that 1 ml = 1 cc

Result

Bulk Density (BD) in g/cubic centimeter =

(Weight of the soil in the volumetric flask)/(Volume of the water in the volumetric flask)

= Y/X = 132.5/100 = 1.325 g/cc

(b) For Undisturbed Site

Portion of the earth which has not been ploughed for some years is referred to as undisturbed site.

1. Soil sample from an undisturbed plot of land be collected by following the standard procedure.
2. External surface of the soil clod be coated with wax, *i.e.*, the soil clod be immersed in liquefied wax.
3. The volume of the clod be determined by finding out the volume of water displaced by it. The method is given below:
   (*a*) A known volume of water be taken in a beaker (V_1).
   (*b*) The wax-coated clod be put in the water in the beaker.
   (*c*) The increase in water level be marked and noted.

(*d*) The clod be removed and water be added upto the marked level.

(*e*) The volume of water upto the increased and marked level be noted now (V_2).

(*f*) The difference between V_2 and V_1 (V_2–V_1) gives the volume of the clod.

4. The external wax coating from the clod be carefully removed, the clod be broken and the soil be dried in a pre-weighed empty beaker (W_1).
5. The beaker with the soil be weighed (W_2).

Calculation and Result (with a model example)

Volume of water taken (V_1) = 100 ml

Volume of water after inserting the clod into it (V_2) = 127 ml

Weight of the empty beaker (W_1) = 46.7 g

Weight of the beaker + soil (W_2) = 102.0 g

Volume of the soil clod (V_2–V_1= say X ml) = 27 ml

Weight of the soil (W_2–W_1 = say Y g) = 55.3 g

Bulk Density = Weight of the soil/Volume of the soil = Y/X = 55.3/27= 2.05 g/cc.

Water Infiltration Rate (mainly for land soil)

Main Materials Required

1. Two iron rings: height 25 cm, thickness 2 mm diameter: 1st ring 30 cm, 2nd ring 60 cm
2. Scale
3. Soil sample.

Procedure

1. The two rings be inserted (hammered-in) about 10 cm inside the soil
2. Both the rings be filled with water upto the brim.
3. The system be left as such for *c* 10 minutes.
4. The height of water within the inner ring be measured after *c* 10 minutes.

5. The system be left as such again for *c* 10 minutes.
6. The height of water within the inner ring be measured again after *c* 10 minutes.
7. In this way, the height of the water column within the inner ring be measured repeatedly at an interval of *c* 10 minutes till constant height (depth) of the water column inside the inner ring is obtained.
8. A graph be plotted with depth (height) of the water column in the Y-axis and time interval in the X-axis.

The data be recorded in the following Table:

No. of Obsvns.	Height (cm) of the Water Column Inside the Inner Ring After Each 10 Minutes				
At 10 min interval	1st 10 mins	2nd 10 mins	3rd 10 mins	4th 10 mins	5th 10 mins

(B) Chemical

The air dry (in shade) soil is normally passed through 2 mm sieve before chemical analysis. Before sieving, the soil clods be lightly crushed in wooden mortar and pestle. Plant residues, gravel and other foreign matters retained on the sieve be discarded.

Soil pH

Main Materials Required

pH meter, soil sample, Buffer tablets, beaker, stirrer, DW.

Procedure

1. 20 g of powdered soil sample be taken in a 100 ml beaker.
2. 40 ml of DW be added to it.
3. The suspension be stirred at regular intervals for 30 minutes.
4. The electrodes of the electronic digital pH meter be dipped into the supernatant (suspension) just above the soil in the beaker; and, care be taken, so that, the electrode does not touch the soil below the supernatant.
5. The pH value be recorded.

6. The suspension be stirred well every time just before the electrodes are immersed.

The data be recorded in the following Table:

Locality:	Date:	Time:
Sample No.		Soil pH

Soil Conductivity

Main Materials Required

Conductivity meter, beaker, KCl, DW, soil sample.

Procedure

1. 20 g of powdered soil sample be taken in a 100 ml beaker.
2. 40 ml of DW be added to it.
3. The suspension be stirred at regular intervals of one hour. Alternatively, the soil suspension in the beaker be kept on an electrical shaker for one hour.
4. The conductivity cell of the electronic digital conductivity meter be dipped into the supernatant (suspension) just above the soil in the beaker; and, care be taken, so that, the electrode does not touch the soil below the supernatant.
5. The conductivity value be recorded.
6. The suspension be stirred well everytime just before the electrodes are immersed.

The data be recorded in the following Table:

Locality:	Date:	Time:
Sample No	Conductivity (micromhos/cm)	

Organic Carbon (OC)

Method of Walkley and Black (1934).

Principle

The OC is estimated by titration, in which, the organic carbonaceous matters are oxidized by a known volume of a standard dichromate solution and then titrating back the volume of dichromate left unused with a standard solution of ferrous (ammonium) sulphate using diphenylamine as indicator.

Main Materials Required

1(N) Potassium di-chromate (AR) solution: 49.4 g of AR grade $K_2Cr_2O_7$ be dissolved in 1 lit of DW.

0.5 (N) (approx.) Ferrous Ammonium Sulphate: 196 g of the hydrated salt be dissolved in 1 lit of DW containing 20 ml of H_2SO_4.

Diphenylamine indicator: 0.5 g of diphenylamine indicator be dissolved in 20 ml of DW and 100 ml of Conc. Sulphuric acid.

Concentrated Sulphuric acid (sp.gr. 1.84) containing 1.25 per cent Silver Sulphate.

Orthophosphoric acid (85 per cent).

Required glasswares like beakers, conical flasks, volumetric flasks, measuring cylinders, burettes, pipettes, test tubes.

Procedure

1. 1.0 g of the soil (ground and sieved through 0.2 mm sieve) be taken in a dry 500 ml conical flask.
2. 10 ml of 1 (N) $K_2Cr_2O_7$ be pipetted-into the conical flask and be swirled a little.
3. The conical flask be kept on an asbestos sheet.
4. Then, 20 ml of H_2SO_4 (containing 1.25 per cent of Silver sulphate) be added into the conical flask and the flask be swirled again 2 or 3 times.
5. The conical flask be now allowed to stand for 30 minutes.
6. Then, 200 ml of DW be added into it.
7. After adding 10 ml of Phosphoric acid and 1 ml of diphenylamine indicator, the contents in the conical flask be titrated with Ferrous ammonium sulphate till the colour changes from blue violet to green.
8. Similarly, a blank be run without soil.

NB: If more than 7 ml of Ferrous ammonium suphate solution is consumed, the whole procedure of determination be repeated with a smaller quantity of soil.

Calculation and Result

Organic Carbon (per cent) = [10(B-T)] × 0.003 × 100/ (B–weight of soil)

where,

B = Volume of Ferrous ammonium sulphate solution required for blank (B)

T = Volume of Ferrous ammonium sulphate solution required for Test soil sample(T)

Available Phosphorus (AP)

It indicates the amount of phosphorus available in the soil. AP could be estimated by two methods, *viz.*, (a) By Olsen's method (Olsen *et al.*, 1954); and (b) by Dickman and Bray's method (Dickman and Bray, 1940).

The Dickman and Bray's method is briefly described below:

Principle

The available phosphorus content in the soil is evaluated by rapidly extracting it with suitable reagents. In this method, the orthophosphate and molybdate ions are said to condense in acidic solution to give molybdo-phosphoric acid (phospho-molybdic acid), which, upon selective reduction (say, with Stannous chloride), produce a blue colour due to molybdenum blue of uncertain composition.

Main Materials Required

1. Klett-Summerson (USA) Photo-electric colorimeter (say model No. 800-3)
2. Dickman and Bray's reagent: 15 g of ammonium molybdate be dissolved in 300 ml of DW and be warmed to about 60C; be cooled and filtered, if necessary. To this, 350 ml of 10 (N) HCl be added followed by an excess 50 ml of 10 (N) HCl and made to 1 litre. The normality of the HCl used be adjusted correctly by titration.

3. Stannous chloride solution: 10 g of Stannous Chloride (crystalline) be dissolved in 26 ml of conc. HCl by warming and stored in an amber-coloured bottle. This is the 40 per cent $SnCl_2$ stock solution. Just before use, 0.5 ml be diluted to 66 ml with DW.
4. Glasswares like beakers, conical flasks, volumetric flasks, measuring cylinders, burettes, pipettes, test tubes

Procedure

1. 1.5 ml of soil extract be pipetted into a 25 ml volumetric flask
2. 5 ml of Dickman and Bray's reagent be added to it.
3. The neck of the flask be washed down and the contents in the flask be diluted to *c* 22 ml.
4. To this, 1 ml of diluted Stannous chloride solution is added, shaken and made upto the 25 ml mark.
5. The intensity of the blue colour be measured just after 10 minutes. This is important, as the colour starts to fade away after the stipulated time.
6. Preparation of Standard curve for Phosphorus: 0.439 g of AR grade Potassium di-hydrogen Orthophosphate (KH_2PO_4) be weighed accurately (after keeping it in hot air oven at 40°C for 1 hour) and be dissolved in about 500 ml of DW. To this, 25 ml of 7 (N) H_2SO_4 be added and made upto 1 litre with DW. This gives a 100 ppm stock solution of Phosphorus (100 microgram of P per ml). From this, a 2 ppm solution is made. For preparation of standard curve, different concentrations of P (1,2,3,4,5 and 10 ml of 2 ppm solution) are taken 25 ml volumetric flasks. The colour is developed by following Dickman and Bray's method and the reading taken with the help of a Klett-Summerson Photoelectric colorimeter against 660 nm. The standard curve is plotted by considering the colorimeter reading in the Y-axis and the amount of P(in ppm) in the X-axis.
7. The concentration of Phosphorus in the soil water is extrapolated from the standard curve.

Available Potassium (AK)

The readily exchangeable and water soluble Potassium (K) determined in neutral ammonium acetate extract represent the available potassium in the soil. The estimation is done with the help of a flame photometer (Jackson, 1973).

Principle

The amount of potash is estimated with the help of a Flame Photometer by comparison with a standard curve of known Potassium oxide concentration determined by the same photometer.

Main Materials Tequired

1. Flame photometer
2. Digital pH meter
3. Neutral normal ammonium acetate: Equal volumes of 2(N) acetic acid and 2 (N) Ammonium hydroxide be mixed and the pH adjusted to 7.0 with acid or ammonia
4. KCl (AR grade).

Procedure

1. 5 g of soil be shaken with 25 ml of neutral normal Ammonium acetate (pH 7.0) for 5 minutes and the contents be filtered immediately through a dry filter paper.
2. First few ml of the filtrate be rejected.
3. K is estimated in the extract with the help of flame photometer after necessary calibration of the instrument.

Preparation of Standard Potassium Curve

1. A stock solution of 1000 ppm K be prepared by dissolving 1.9084 g of AR grade KCl (dried at 60 C for one hour) in DW and the volume be made upto 1 lit.
2. Required portions be diluted with the Ammonium acetate solution to give 10-40 ppm of K.
3. After attaching the appropriate filter and adjusting the gas and air pressure, the flame photometer reading be set at zero for the blank (Ammonium acetate) and 100 for 40 ppm K.

4. The curve be obtained by plotting the readings of the flame photometer against the different concentrations (10, 15, 20, 25, 30, 35 ppm) of K.
5. K content in the sample be estimated through extrapolation.

Physico-chemical Characteristics of Soil of Sone Beel in Assam (Kar, 1990)

The values of various physico-chemical characteristics of soil of Sone Beel in Assam are given below:

1. Temperature (C): 19.9–32.3
2. pH: 5.09–5.99
3. Conductivity (µmhos/cm at 25°C): 47.42–322.08
4. Organic Carbon (per cent): 0.25–1.74
5. Available Phosphorus (mg/100 g): 0.15–1.93
6. Available Potassium (mg/100 g): 1.6–24.8.

Chapter 5

Plankton Communities

'Plankton, the living fraction of material, found in water and move passively by wind or current, is composed of bewildering varieties of microscopic organisms. The 'phytoplankton', which are predominantly autotrophic, are the primary producers of organic matters in the aquatic habitats. The nutritionally dependent animal component constitutes the 'zooplankton'.

Phytoplankton

The phytoplankton, though stands on the baseline of many food webs, in the aquatic environment, is, in turn, dependent on the activities of other microbial organisms, notably bacteria, which convert organic material into inorganic nutrients required by the plants.

The phytoplankton, found in this region, could, broadly, be classified into the following groups:

1. Cyanophyta
2. Euglenophyta
3. Chlorophyta
4. Chrysophyta (which includes the Diatomes)
5. Pyrrophyta

Zooplankton

The zooplankton, which are the primary consumers in the ecosystem, include many varieties. While some spend only a part of their lives drifting at the surface, others are found to extend to the deeper waters than the phytoplankton do. Many of them make daily sojourns between layers.

The zooplankton, which occur in this region, could broadly be classified into the following groups:

1. Protozoa
2. Rotifera
3. Cladocera
4. Copepoda
5. Ostracoda

Methodology for Plankton Study

The analysis of plankton is done after collecting plankton samples generally in 50 ml polythene tubes from both surface and bottom waters of the water body. Samples from the surface water could be collected by filtering 100 litres of the water from the waterbody covering the study area using 1 litre capacity tumbler and operating from a canoe. Bottom water plankton samples be collected with the help of a 5 lit capacity Kemmerer water sampler. A plankton collecting net, made of nylobolt bolting silk with mesh size of various diameters (according to the aim of the group to be studied, but generally of 55 μm), is used to filter the water sample for collecting plankton. At the time of collection, extreme care be taken to disturb the water as less as possible. The collected plankton samples be preserved in 2-5 per cent (v/v) aqueous formalin solution on the spot and transferred to the laboratory where the samples are to be centrifuged and the plankton volume be recorded in ml. Each plankton sample be then diluted to 100 ml with DW and mixed thoroughly. A sub-sample of 1 ml be then drawn from it with the help of a wide-mouth pipette and poured into a Sedgewick-Rafter Plankton Counting Cell. While pouring, due care be taken to ensure equal distribution of planktonic organisms in the entire counting cell. Quantitative estimation of the plankton (unit/litre) be done with the help of a stereoscopic microscope (Kar, 1990). Limnoplankton are identified after various authoritative works, notably, Ward and Whipple (1918), Smith (1950), Pennak (1953), Needham and Needham (1962), Prescott (1969), Michael and Sharma (1988), Battish (1992), Fritsch (1965) and so on.

An Account of Limnoplankton of Sone Beel (Kar, 1990)

47 different forms of phytoplankton belonging to five groups, as indicated above, have been recorded, till date, in Sone Beel. Of these, the Chrysophyta included the maximum number and Pyrrophyta, the least. The phytoplankton density in the Beel varied from 48-5308 (average 1027) units/lit. The minimum population is generally recorded during the rainy season and maximum during the dry season. The cyanophytes and the euglenophytes generally exhibit prolific growth during the Spring and Summer with occasional abundance during monsoon. The Chlorophytes usually show higher occurrence intensity during Summer and early monsoon with infrequent richness during Winter and Spring. The Chrysophytes are generally recorded throughout the year. The Pyrrophytes were found to register their presence mainly during the Autumn.

19 different forms of zooplankton, belonging to five groups, as indicated above, have been recorded, till date, in Sone Beel. The zooplankton density varied from 6-380 (average 49) units/lit. Low density is generally recorded during February-March and high density during November-December. In Sone Beel, a considerable portion of zooplankton population is considered to succumb to the day-in and day-out fishing operations; thus, leading to their poor population in the Beel. Nevertheless, the Copepods have been found to occur throughout the year. On the other hand, the rotifers were quite abundant mainly during early monsoon and autumn. However, the Cladocerans generally exhibit abundance during Autumn and Winter.

An Account of Zooplankton of Chatla Haor Wetland (Kar and Barbhuiya, 2004)

Studies conducted in *c* 1600 ha Chatla Haor in Cachar district revealed the occurrence of 18 species of zooplankton consisting of 2 species each of Protozoa and Copepoda, 6 species of Rotifera and 8 species of Cladocera. Occurrence of *Arcella* sp among the protozoans and *Brachionus calyciflorus* among the rotifers, indicate eutrophy of the wetland. Two protozoans, *viz.*, *Arcella* sp and *Paramoecium* sp, represented *c* 11.11 per cent of the total zooplanktonic taxa in the wetland. Verma and Dalela (1975) had reported *Arcella* sp in eutrophic waters. Six rotifers constituting 33.33 per cent of the total zooplanktonic taxa were also identified; of which, *Filinia* sp and *Lecane* sp were found to be abundant. The identified Copepods included the *Cyclops* sp and the *Diaptomus* sp and they represented 11.11 per cent of the total zooplanktonic taxa. The total zooplanktonic count was found to be 68± 45 units/litre (Kar and Barbhuiya, 2004).

Chapter 6

Aquatic Macrophytes

The aquatic macrophytes (AM) are interesting forms of plant life. As compared to the terrestrial plants, the aquatic plants remain comparatively less affected by the climatic and seasonal changes. Consequently, it is more feasible to study the plants for the greater part of the year. Further, they also exhibit striking adaptations to fit in the aquatic regime.

Methodology

A Δ 1 m^2 wooden quadrat is generally used to determine the macrophytic density. The submerged and emergent macrophytes are collected with the help of a barbed iron hawser. The floating ones are usually grabbed with hand. After collection, the AM are segregated, the total number and size range are noted species-wise and the wet biomass are determined on the spot by weighing them together wet with a double pan balance. The sampling operations are carried out in both littoral and limnetic zones of the Beel at different randomly-selected spots. The data are analysed to evaluate the seasonal succession and phyto-sociology of the AM communities. Herbaria for different plants are prepared and transferred to the laboratory for further works (Kar, 1990).

Hooker (1872), Bentham and Hooker (1862-1883), Prescott (1969), Biswas and Calder (1984) and other standard literature could be consulted for identification. It is essential to consult the Botanical

Survey of India to confirm the identifications. Shannon-Weaver (1949) function be used to study the equitability and diversity status of the AM and chi-square test (in a computer programme) be used to test their association with one another.

An Account of Aquatic Macrophytes of Sone Beel (depicting species composition, phyto-sociology, standing crop and seasonal succession) (Dey and Kar, 1989a)

AM was found to exhibit a heterogeneous assemblage of 23 species in Sone Beel. The list of important AM species is given below:

1. *Alternanthera sessilis*
2. *Azolla pinnata*
3. *Cynodon dactylon*
4. *Echinochloa stagnina*
5. *Eichhornia crassipes*
6. *Eleocharis acutangula*
7. *Euryale ferox*
8. *Hydrilla verticillata*
9. *Hygrorhiza aristata*
10. *Ipomoea aquatica*
11. *Jussiaea repens*
12. *Nechamandra alternifolia*
13. *Nymphaea nouchali*
14. *Nymphoides cristatum*
15. *Nymphoides indicum*
16. *Oryza sativa*
17. *Polygonum flaccidum*
18. *Sagittaria trifolia*
19. *Salvinia cucullata*
20. *Scirpus eriophorum*
21. *Trapa bispinosa*
22. *Vallisneria spiralis*
23. *Vetiveria zizanoides*.

Association of different species of AM, forming phyto-social units, are generally encountered in the wetlands. In Sone Bee!, among the floating varieties, *S.cucullata* and *A.pinnata;* and, *N.cristatum* and *N.indicum* were found to form distinct phyto-social units. Among the emergent varieties, *E.acutangula* and *S.trifolia;* and, *S.eriophorum* and *E.stagnina* were found to be associated together. Among the submerged types, *H.verticillata, V.spiralis* and *N. alternifolia* were found to make an association. Interestingly, *T. bispinosa,* the floating form, also, exhibited association with this group. Further, *E.crassipes* and *E.ferox* were found to remain solitary and thus formed mono-specific unit.

AM biomass was found to vary from 0.58–21.90 kg/m² (average 2.48 ± 0.82) having the maximum in December and the minimum in May. In addition to water level (WL) and amplitude of flooding (Welcomme, 1979), the growth and distribution of the AM was found to be influenced by water quality of the wetland.

E.crassipes was the sole perennial species in the wetland, followed by *H.verticillata* and *T.bispinosa,* occurring during most of the months of the year. During dry season, the emergent varieties (like *E.stagnina, E.acutangula, S.eriophorum, O.sativa, S.trifolia* and *P.flaccidum*) and the submerged types (like *H.verticillata, T.bispinosa, V.spiralis*) generally succeed, flourish and show abundance at less WL when the wetland exhibits a decreasing trend in its depth. It indicates an indirect relation of AM biomass with WL (r =–0.130 ± 0.442. P>0.05) during this period. Higher ranges of conductivity at this time was found to sustain (Pearshall, 1938) a rich biomass of *T.bispinosa.*

The wetland water, during dry season, in general, portray high DO produced by photosynthesis at rich insolation, in which, FCO_2 is consumed and shows a fall. Concomitantly, a direct relationship of AM biomass with DO ($r = 0.500 \pm 0.340, P < 0.05$) and an inverse with FCO_2 ($r = -0.780 \pm 0.178, P < 0.05$) was recorded in the study. The pH and TA, which were found to depict rise, portrayed their direct relationship ($r = 0.850 \pm 0.126, P < 0.05; r = 0.022 \pm 0.454, P < 0.05$) with AM biomass. *H. verticillata* and *V. spiralis,* as indicated earlier, were found to be closely (P < 0.05) associated during this period (X^2, 34.67). Also, a close (P < 0.05) association between floating *T.bispinosa* and submerged *V. spiralis* (X^2, 24.15) and between submerged *H.verticillata* and emergent *S. eriophorum* (X^2, 31.89) was discernible during the period.

With the onset of monsoon, the floating varieties, *viz.*, *N.nouchali*, *N.cristatum* and *N.indicum* generally occur in their flowering stage. Most of the emergent varieties encountered during winter, spring and summer, get submerged and decayed during monsoon. However, the littoral and sub-littoral zones of the Beel, during monsoon, are generally found to be moderately infested with emergent varieties like *C. dactylon* and *V.zizanoides*. *H.verticillata*, among the submerged varieties, are, sometimes, found to occur. Onset of monsoon, as stated earlier, cause decay of AM, thus, rendering poor AM biomass with corresponding low DO, pH and TA but high FCO_2 during the periods. Concomitantly, a direct relation of AM with DO ($r = 0.940 \pm 0.053$, $P < 0.05$), pH ($r = 0.160 \pm 0.443$, $P > 0.05$) and TA ($r = 0.530 \pm 0.320$, $P < 0.05$) were recorded. Significant ($P < 0.05$) phyto-social association between *C.dactylon* and *E.crassipes* (X^2, 12.44), *C.dactylon* and *N. indicum* (X^2, 10.21), *H.aristata* and *E.crassipes* (X^2, 13.87), *H.aristata* and *N.cristatum* (X^2, 18.96) were recorded during this season. Howe+ver, none of the AM species formed significant phyto-social relation with *E.ferox*, possibly, due to its thorny body; and, the latter, thus, formed a monospecific unit.

Almost all the monsoon varieties, as stated above, also occurred during the autumn; but, with profuse flowering restricted mainly to *N. cristatum* and *N. indicum*. With the re-appearance of dry winter, most of the emergent varieties gradually start re-appearing with the decrease in water level of the wetland. *Euryale ferox* was seen to occur only for a limited tenure during the period from Spring-beginning to mid-Summer. Since the plant is generally considered by the local fishermen as an impediment to fish movement and hindrance in fishing, they are usually removed from the Beel at the end of Summer.

High species diversity among the AM species were evident in this wetland. And, the level was found to be high biomass estimate of $H' = 2.015$; expected value, $E(H') = 2.014$; variance of $H' = 1.431$ E-3) in Sone Beel.

An Account of the AM of Chatla Haor Seasonal Floodplain Wetland (Kar and Barbhuiya, 2001)

23 species of AM could be recorded, till date, in the 1600 ha (at FSL) Chatla Haor situated in the Cachar district of Assam. These could be classified as follows:

5 free floating, 4 rooted floating, 2 submerged and 12 emergent. Of these, 6 AM species could be found throughout the year. These are: *Azolla pinnata, Eichhornia crassipes, Salvinia cucullata, Trapa bispinosa, Jussiaea repens* and *Cynodon dactylon.*

J. repens showed flowering during March-May, while *Nymphaea nouchali* bloomed during June-August followed by *Nymphoides cristatum* and *N. indicum* which exhibited profuse flowering during September-October. *Ipomoea aquatica* depicted significant growth of population during July-August. *Hydrilla verticillata, Vallisneria spiralis* among the submerged varieties and *Alternanthera sessilis, Cyperus platystylis, Echinochloa stagnina, Eleocharis acutangula, Enhydra fluctuans, Scirpus eriophorum, Sagittaria trifolia* among the emergent varieties succeeded at a lesser water level during the dry season. The floating varieties, *viz., A. pinnata, E. crassipes, S. cucullata* were recorded throughout the year and found to be associated with each other. Further, among the submerged varieties, *H. verticillata* and *V. spiralis* were found to be associated.

Significant phyto-social association were also observed among *Nymphaea* sp and *Nymphoides* sp; and, among *Eleocharis acutangula, Scirpus eriophorum* and *Echinochloa stagnina.*

An Account of the AM of Baskandi Anua River-formed Wetland (Dhar *et al.,* 2004)

16 species of AM have been recorded in the 39.2 ha (at FSL) Baskandi Anua which belong to 6 free-floating (*Azolla pinnata, Echhornia crassipes, Salvinia cucullata, Lemna pausicostata, Pistia stratiotes, Wolfia* sp); 2 rooted submerged (*Hydrilla verticillata, Vallisneria spiralis*); 6 rooted with floating leaves (*Nymphaea nouchali, Nymphoides indicum, N.cristatum, Trapa bispinosa, Euryale ferox, Nelumbo nucifera*); and 2 rooted emergent (*Jussiaea repens, Muradania nudiflora*). Of these, 6 AM species were found to occur throughout they ear. These are *Azolla pinnata, Eichhornia crassipes, Salvinia cucullata, Trapa bispinosa* and *Jussiaea repens.*

It has been observed that the AM start their seasonal cycle from the germination period in winter while emerging to a vegetative stage in summer and then to monsoon. This was followed by the flowering and fruiting during late monsoon and post-monsoon. Studies conducted during the period 1998-2000 revealed wet AM biomass to be ranging from 4.4 to 11.4 kg/m^2. Wet biomass was observed to be higher during monsoon and post-monsoon reaching trough value during winter with a concomitant increasing trend during summer. Significant positive correlation of AM biomass with water temperature (r= 0.1820) has been recorded during the study period.

Chapter 7

Ichthyodiversity and Ichthyogeography with Special Reference to Barak Drainage (Assam), Mizoram and Tripura

Fish constitutes almost half of the total number of vertebrates in the world. They live in almost all conceivable aquatic habitats. *c* 21,723 living species of fish have been recorded out of 39,900 species of vertebrates (Jayaram, 1999). Of these, 8411 are freshwater species and 11,650 are marine. India is one of the Megabiodiversity countries in the World and occupies 9th position in terms of freshwater Megabiodiversity (Mittermeier and Mittermeier, 1997). In India, there are *c* 2500 species of fishes; of which, *c* 930 live in freshwater (FW) and *c* 1570 are marine (Kar, 2003a). This bewildering ichthyodiversity of this region has been attracting many ichthyologists both from India and abroad. Concomitantly, North-Eastern region of India has been identified as a 'Hotspot' of Biodiversity by the World Conservation Monitoring Centre (WCMC, 1998). This rich diversity of this region could be assigned to certain reasons, notably, the geomorphology and the tectonics of this zone. The hills and the undulating valleys of this area gives rise to large number of torrential hill streams, which lead to big rivers; and, finally, become part of the Ganga-Brahmaputra-Barak-Chindwin-Kolodyne-Gomati-Meghna system (Kar, 2000 a).

Based on IUCN categories, the CAMP Workshop (CAMP, 1998) for FW fishes has identified certain fish species which has attained threatened/endangered status. At the same time, there has been little study with regard to details of endemism and species richness in NE India. As such, a detailed study related to Germplasm inventory, evaluation and gene banking of freshwater fishes would not only help us to land at a concrete decision on the above-mentioned aspects; but, would also contribute to fulfilling India's obligation to CBD with special emphasis on Articles 6 and 8 (UNEP, 1992). Further, development of database on the biological parameters is a pre-requisite for preparation of detailed Fish Inventory. Genetic characterization and Gene Banking is a step forward towards further confirmation of the species at the molecular level.

River is the basic storehouse of water to meet our country's demand for water for various uses. Indian continent is singularly blessed with unparalleled riverine resources, harbouring one of the richest fish wealth in the world. Out of *c* 2500 species of fishes occurring in India, *c* 450 species of fishes occur in the Indo-Gangetic plains (Kar *et al.*, 2002 c). Menon (1974) had listed 207 species of fishes from the Gangetic plains which belong to 29 families and 82 genera. According to another estimate, the Gangetic system alone harbours not less than 265 species of fishes (Yadava and Ravish Chandra, 1994),

Structural characteristics of the lotic environment are closely associated with the occurrence of fish species. Importance of habitats and the relationship between fish and habitat are of major concern to fishery biologists. A common use of fish habitats indicates the physical and chemical characteristics of the environment excluding biological attributes. Habitat features have been identified as major determinants in distribution and abundance of fishes from earlier times (Shelford, 1911). Fish species diversity is associated with habitat complexity with regard to depth, flow and substrate types. Recently, fish assemblages have been used as indicators of environmental degradation (Scott and Hall, 1997), ecosystem health in streams (Karr, 1991) and environmental stress (Cairns *et al.*, 1993).

There have been extensive studies on the Freshwater (FW) fishes of India, notably, Hamilton (1822), Day (1878, 1889), Shaw and Shebbeare (1937), Hora (1920), Misra (1959), Menon (1974, 1999), Dey (1973), Jayaram (1981, 1999), Sen (1982), Sen (1985), Talwar and

Jhingran (1991), Datta Munshi and Srivastava (1998), Nath and Dey (2000), Dey and Kar ((1989 b, 1990), Kar and Dey (1992, 2000, 2002) Kar and Barbhuiya (2000 a, b; 2004), Kar *et al.* (2002 a, b, c, 2003 a, 2004), Kar (1984, 1990, 2003 a, b; 2004, 2005 a, b, c). But most of them are concerned with taxonomy (Biodiversity), biology and aquaculture.

The ichthyofauna of North-East region of India has elements of the Indo-Gangetic region; and, to a some extent, of the Myanmarese and South-Chinese regions (Yadava and Chandra, 1994). Ghosh and Lipton (1982) had reported 172 species of fishes with reference to their economic importance while Sen (1985) reported 187 species of fishes from Assam and its environs. Sinha (1994) compiled a list of 230 species of fishes from NE India. Nevertheless, Nath and Dey (1997) recorded 131 species of fishes from the drainages in Arunachal Pradesh alone. Sen (2000) comprehensively compiled a list of 267 species of fishes from NE India. Further, according to Sen (2000), of the *c* 806 species of fishes inhabiting the freshwaters of India (Talwar and Jhingran, 1991), the NE region of India is represented by 267 species belonging to 114 genera under 38 families and 10 orders. This is *c* 33.13 per cent of the total Indian FW fishes. Of the 267 species, Cypriniformes dominates with *c* 145 species followed by Siluriformes (72), Perciformes (31), Clupeiformes (7), Anguilliformes (3), Cyprinodontiformes (3), Osteoglossiformes (2), Synbranchiformes (2), Syngnathiformes (1) and Tetraodontiformes (1). Kar (2003 a) reported the occurrence of 133 species of fishes through a pilot survey conducted in 19 rivers spread in Barak drainage (Assam), Mizoram and Tripura. Kar (2005 a), further, reported the occurrence of 103 species of fishes through an extensive survey conducted in six principal rivers in Barak valley (Assam), Mizoram and Tripura.

Notwithstanding the above, studies on the fish assemblage structure and their habitat requirements in the streams of North East India are scanty. Few initiatives have started in South India (Arunachalam *et al.*, 1988). A novel approach related to the study of Fish assemblage structure and species inventory and habitat requirement of individual ichthyospecies in the assemblage in the four principal rivers in Barak valley region of Assam, and in one river each in the provinces of Mizoram and Tripura have been undertaken and results reported by Kar (2005 a, c) as indicated above.

Methodology

General survey of the Fish Biodiversity was done using standard procedures (Armontrout, 1990). Also, NBFGR Manual (2000) was consulted for studying the habitat parameters Headwater to downstream studies were based on River Continuum Concept (Vannote, *et al.,* 1980). Spatial heterogeneity of river channel across small to large spatial scales (Forman and Godran, 1980), longitudinal (upstream vs downstream) and lateral (stream margin/ mid-channel) dimensions were studied.

Fish samples were collected through experimental fishing using caste nets (dia.3.7 m and 1.0 m), gill nets (vertical height 1.0 m- 1.5 m; length 100 m -150 m), drag nets (vertical height 2.0 m), triangular scoop nets (vertical height 1.0 m) and a variety of traps. Camouflaging technique was also used to catch the fishes. Fishes have been preserved at first in concentrated formaldehyde in the field itself and then in 40 per cent formalin. Fishes have been identified after standard literature (Day, 1878, 1889; Shaw and Shebbeare, 1937; Misra, 1959; Menon, 1974, 1999; Talwar and Jhingran, 1991; Jayaram (1981, 1999). Yield statistics were extrapolated (Dey and Kar, 1990; Kar, 1990) from daily catch statistics recorded at the landing stations (FAO, 1974) while the trend and cyclic variations were constructed by applying 12 months moving average method (Coxton and Cowden, 1950; Kar and Dey, 2000).

Some Concepts Related to Fish Community Ecology

The River Continuum Concept

It was originally proposed for aquatic invertebrates. It predicts that, there is an upstream-downstream gradient of changing physical conditions and associated biotic changes (Vannote *et al.,* 1980). It has been postulated that, species assort themselves along environmental gradients, like pH, temperature, etc. In general, it has been found that, species diversity increases from upstream to downstream regions in streams. According to Lowe McConnell (1975), this is due to the fact that, habitat diversity also increases along the stream gradient in the same fashion. However, this happens in such a way that, most of the upstream habitat types also occur in the downstream sections. In addition, detritus and plankton richness also seems to increase downstream which may lead to greater diversity in the higher trophic levels. Certain studies on

African stream fish communities (Toham and Teugels, 1997) revealed the occurrence of distinct pattern in the longitudinal zonation of species with a sequential shift in species compositions and also an increase in species diversity along the stream gradient. Further, spatial segregation of fishes is said to be influenced by environmental and physical characteristics of the stream like width, depth, current velocity and substrate type (Toham and Tuegels, 1998); as well as, certain biotic characteristics like canopy cover and the occurrence of the species instream (dead wood, leaves and aquatic plants). These are known to influence habitat features like depth and stream current (Angermeier and Karr, 1983) and also said to enhance invertebrate production (Benke *et al.*, 1984) which could serve as fish food.

The Habitat Guild Concept

There is diversity of habitats in stream ecosystems. Each of them harbours diverse and distinct types fish communities. The Habitat Guild Concept attempts to explain this observation. It implies that, species-habitat associations are strong and are forged by adaptations of the fish communities to their surrounding habitat; as well as, interactions between each other (Taylor, 1996). Such a view has been tested and found to be true in comparative studies of fish communities in riffle and pool habitats in some of the rivers of Oklahoma and Arkansas. It was found that, these communities differ in a number of attributes, like species richness, composition, abundance patterns, and responses to changes in environmental gradients (Taylor, 2000). However, riffle and pool communities are linked by a continuous gradient; at a very local scale; and, species may use both these habitats at different parts of their life cycles as well as at different seasons (Schlosser, 1982). Nevertheless, some of the juvenile fish may prefer riffles to pool habitats as they provide the required kind of food and flow patterns necessary at a particular stage of their growth; whereas, the same species may prefer pool habitats at later parts of their life cycles. Thus, generalized concepts, like the Habitat-Guild concept may not hold true at all levels of observation and would depend on the relative abundance of pool and riffle habitats within a river stretch. There have been some observations on size-related difference in habitat use by many species where smaller individuals of some species occur in faster-flowing waters than their larger conspecifics (Toham and Tuegels, 1997).

Tropical Fish Communities

The tropical Asian ichthyofauna constitutes a substantial part of the total lotic fish community. The Indo-chinese peninsula supports 930 species of native FW fishes which belong to 87 families (Kottelat, 1989). A number of tropical Asian FW fish are shared with African riverine ecosystems both at the family as well as at the generic level. Cyprinids, certain Siluriform catfishes, Channids, Mastacembelids, and Notopterids are shared between the two regions. At the generic level, *Anabas, Clarias, Garra, Labeo,* and *Mastacembelus* occur in both African and Asian rivers. Certain fishes, like Tilapia, which have been introduced into India from Africa, have become widespread all over the southern Asia. They have also replaced the native population in some places due to their dominance. There is large-scale abundance of Cyprinids and Balitorids in Asia in contrast to predominance of Characids and Cichlids in Africa. Incidentally, research on the ecology of the tropical fish communities is limited (Dudgeon, 1995). Further, there have been studies on fish diets and resource partitioning in certain Sri Lankan Hill streams (Wikramanayake and Moyle, 1989). Niche' segregation is dependent on seasonality, diet and habitat utilization as was revealed from their studies. Also, there are morphological segregation and specialization in these fish communities.

Classification of the Freshwater Fishes of the Indian Region (with particular emphasis on Nort-East India)

Classification of the Freshwater (FW) Fishes of the Indian Region after Jayaram (1999), with particular reference to Barak drainage, Mizoram and Tripura, is given below:

Super class: Pisces

Class: Osteichthyes

Sub-class: Actinopterygii

Super-order: Telesotei

Order: Osteoglossiformes

Family: Notopteridae

Genus: *Notopterus* Lacepede

Genus: *Chitala* Fowler

Order: Anguilliformes

Family: Anguillidae

Genus: *Anguilla* Schrank

Family: Ophichthidae

Genus: *Pisodonophis* Kaup

Order: Clupeiformes

Family: Clupeidae

Genus: *Hilsa* Regan

Genus: *Gudusia* Fowler

Order: Cypriniformes

Family: Cyprinidae

Genus: *Hypothalmichthys* Bleeker

Genus: *Securicula* Gunther

Genus: *Salmostoma* Swainson

Genus: *Aspidoparia* Heckel

Genus: *Barilius* Hamilton-Buchanan

Genus: *Chela* Hamilton-Buchanan

Genus: *Esomus* Swainson

Genus: *Danio* Hamilton-Buchanan

Genus: *Rasbora* Bleeker

Genus: *Bengana* Gray

Genus: *Amblypharyngodon* Bleeker

Genus: *Ctenopharyngodon* Steindachner

Genus: *Cyprinus* Linnaeus

Genus: *Tor* Gray

Genus: *Neolissochilus* Rainboth

Genus: *Osteobrama* Heckel

Genus: *Chagunius* H.M. Smith

Genus: *Puntius* Hamilton-Buchanan

Genus: *Semiplotus* Bleeker

Genus: *Cirrhinus* Oken

Genus: *Catla* Valenciennes

Genus: *Labeo* Cuvier

Genus: *Schizothorax* Heckel

Genus: *Schizopygopsis* Steindachner

Genus: *Crossocheilus* Kuhl and van Hasselt

Genus: *Garra* Hamilton-Buchanan

Family: Psilorhynchidae

Genus: *Psilorhynchus* McClelland

Family: Balitoridae

Genus: *Homaloptera* van Hasselt *emend* van der Hoeven

Genus: *Balitora* Gray

Genus: *Acanthocobitis* Bleeker

Genus: *Nemacheilus* Bleeker

Genus: *Schistura* McClelland

Family: Cobitidae

Genus: *Botia* Gray

Genus: *Acantopthalmus* van Hasselt

Genus: *Somileptus* Swainson

Genus: *Lepidocephalus* Bleeker

Order: Siluriformes

Family: Bagridae

Genus: *Rita* Bleeker

Genus: *Batasio* Blyth

Genus: *Chandramara* Jayaram

Genus: *Mystus* Scopoli

Genus: *Aorichthys* Wu

Family: Siluridae

Genus: *Ompok* Lacepede

Genus: *Wallago* Bleeker

Genus: *Silurus* Linnaeus

Family: Schilbeidae

Genus: *Ailia* Gray

Genus: *Pseudeutropius* Bleeker

Genus:*Clupisoma* Swainson

Genus: *Eutropiichthys* Bleeker

Genus: *Silonia* Swainson

Family: Pangasiidae

Genus: *Pangasius* Valenciennes

Family: Amblycipitidae

Genus: *Amblyceps* Blyth

Family: Sisoridae

Genus: *Bagarius* Bleeker

Genus: *Gagata* Bleeker

Genus: *Nangra* Day

Genus: *Erethistes*

Genus: *Hara* Blyth

Genus: *Conta* Hora

Genus: *Laguvia* Hora

Genus: *Glyptothorax* Blyth

Genus: *Exostoma* Blyth

Genus: *Pseudecheneis* Blyth

Genus: *Sisor* Hamilton-Buchanan

Family: Clariidae

Genus: *Clarias* Scopoli

Family: Heteropneustidae

Genus: *Heteropneustes* Muller

Family: Chacidae

Genus: *Chaca* Gray

Family: Olyridae

Genus: *Olyra* McClelland

Order: Mugiliformes

Family: Mugilidae

Genus: *Sicamugil* Fowler

Genus: *Rhinomugil* Gill

Order: Beloniformes

Family: Belonidae

Genus: *Xenentodon* Regan

Order: Cyprinodontiformes

Family: Aplocheilidae

Genus: *Aplocheilus* McClelland

Order: Synbranchiformes

Family: Synbranchidae

Genus: *Monopterus* Lacepede

Family: Mastacembelidae

Genus: *Macrognathus* Lacepede

Genus: *Mastacembelus* Scopoli

Order: Perciformes

Family: Chandidae

Genus: *Chanda* Hamilton-Buchanan

Genus: *Parambassis* Bleeker

Family: Sciaenidae

Genus: *Johnius* Bloch

Family: Nandidae

Genus: *Badis* Bleeker

Genus: *Nandus* Valenciennes

Family: Cichlidae

Genus: *Oreochromis* Gunther

Family: Gobiidae

Genus: *Glossogobius* Gill

Family: Anabantidae

Genus: *Anabas* Cuvier

Family: Belontidae

Genus: *Ctenops* McClelland

Genus: *Colisa* Cuvier

Genus: *Trichogaster* Bloch and Schneider

Family: Channidae

Genus: *Channa* Scopoli

Order: Tetraodontiformes

Family: Tetraodontidae

Genus: *Tetraodon* Linnaeus

Factors Affecting Diversity of Species

Species richness in a region is governed by a number of factors which operate at different spatial and temporal scales. These could be large scale factors (*e.g.*, speciation, dispersal, etc,) or small scale factors (*e.g.*, competition, disturbance, predation, etc.). These biotic as well as abiotic factors act together in regulating the local species richness. Initially, more local deterministic biotic interactions were assumed to dominate community structure and composition and not much regard was given to larger scale processes like dispersal, speciation, or geological changes (Ricklefs, 1987). There has been much debate on the importance of both the local and regional processes in the regulation of local diversity (Angermeier and Winston, 1998). Stream fishes have been used extensively to examine the relative influences of local and regional factors on local species diversity. There have been a number of studies in order to understand these patterns, *e.g.*, works of Moyle and Herbold (1987) in the streams of eastern North America; works of Hugeuny and Paugy (1995) in the rivers of West Africa; and, so on. Such works seem to suggest strong regional influences on stream fish biodiversity. Conversely, studies in lentic systems (Jackson and Harvey, 1989) seem to suggest little regional control in lakes. Further studies by Angermeier and Winston (1998) had indicated that neither could be said to have a stronger role in predicting local species diversity. Concomitantly, regional diversity is said to be more influenced by biogeographic processes like speciation and dispersal, energy constraints, and also landscape diversity (Angermeier and Winston, 1998). Thus, more recent works seem to emphasise the importance of scale in determining species diversity (Bhat, 2002).

Patterns in the Distribution of Fish Species

Studies of patterns in the distribution of species diversity have shown some interesting trends that so far have been found to be consistent in most ecoregions. Though a number of these studies have been restricted to temperate regions, recent studies have extended to the tropics including Africa, South America and also South East Asia. One very clear geographical pattern that could be seen is the greater number of species found in the tropics.

Ichthyogeography

Zoogeography is that branch of science which deals with the distribution of animals including the history of their occurrence on the earth. The ichthyofauna of India could be broadly classified under two groups on the basis of their distribution, *viz.*,

1. Freshwater (FW)
2. Marine

   The third category, *viz.*,

3. The Estuarine forms could be either of the two in origin.

Ichthyogeography in General

The land masses of the world may conveniently be divided into the following realms (Darlington, 1957) keeping in view the distribution of the Primary FW fishes, particularly, those of the Order Cypriniformes.

Megagaea

1. The Palaearctic: which includes, among others, Portion of Asia like Japan; north of the Himalayas and the Yangtze Kiang watershed; Iran and portion of Africa north of Sahara.
2. The Nearctic: which includes the North America among other portions.
3. The Ethiopian Region, which includes, among others, Africa south of Sahara including Madagascar.
4. The Oriental Region, which includes, among others, India, Sri Lanka, Thailand, Southern China and Malayan

Archipelago.

Neogaea

5. The Neotropical Region, which includes, among others, America from Mexico southwards

Notogaea

6. The Australian Region, which includes mainly Australia, Tasmania, New Zealand and the Pacific Islands

Ichthyogeography of India and North-East India

In order to explain the occurrence and assemblage of FW fish fauna of India in relation to contiguous lands, various views have been put forward. A close affinity between the fish fauna of Indian and African regions had been noted by Gunther (1880), who had pointed out that, one or more representatives of the 17 out of 26 families or groups found in India also exist in Africa. Further, many African species are not generically different from those of India. Gunther had included Asia, south of the Himalayas and the Yangtze Kiang and the Islands to the west of Wallace line under the Indian region. Day (1885) had differed from Gunther and had recorded 87 genera of FW fishes in India (including Burma and Ceylon within it). Out of these, only 14 were represented in Africa. He further had observed that, out of the 369 species belonging to these genera, the distribution of only 14 were known to extend to Africa. Day had been supported by Blanford (1901) had studied the occurrence and distribution of the vertebrates in general. Blanford had concluded that the elements in Indo-Malayan fauna are generally heavily represented in the Eastern Himalayas, which diminishes gradually westwards and tends to cease in Kashmir and further west. Hora (1937) had also supported this view and advocated a South-eastern Asian origin of the FW fish fauna of India emanating, most probably, from the Yunnan. Further, according to him, this ichthyofauna had spread westwards in successive waves of migration, first to India and later to Africa, at a time, when the two land masses were said to have been connected. Still further, Hora (1937) had critically studied 41 Indian fish genera and had concluded that, not only the Indian fish fauna had marked Malayan affinity but were also related to the fauna of Thailand, Southern China and Cochin China. Menon (1955) had supported Hora's hypothesis after a detailed analysis of 89 genera of Indian FW fishes and had stated that, the fish fauna of India was related to the Malayan region; and, that, it hardly had contained any typical Ethiopean element. A close kinship with forms in countries towards the east of India, *viz.*, Myanmar (earlier Burma), Malaya, Sumatra, Java, etc.could be revealed through a careful examination of the Indian endemic genera (Jhingran, 1991). Incidentally, Menon(1955) had recorded the following fish genera as endemic and exclusive to India: *Parapsilorhynchus* Hora, *Lepidophygopsis* Raj, *Balitora* Gray, *Bhavania* Gray, *Somileptus* Swainson, *Horabagrus* Jayaram, *Hara* Blyth, *Sisor* Hamilton-Buchanan, *Ctenops* McClelland, *Ailia* Gray, *Silonia* Swainson, etc.

It is sometimes believed that, the Assam land route between peninsular India and the North became submerged during the Middle Eocene owing to the transgression of the Bay of Bengal (Menon and Prashad, 1952). From Eocene onwards extending upto the early Pliocene, the Bay of Bengal had extended northwards to China and Tibet. This had, perhaps, separated India from the rest of Asia; and, therefore, the monsoon-bearing winds crossed to the North over this Gulf; thus, making the Yunnan region; which, at that time, was, perhaps, a low land situated nearer to the equator having temperate climate (Hora, 1953). These conditions, possibly, favoured the evolution of a diverse and specialized fish fauna in the Yunnan region. The same, however, did not spread to India until the Pliocene; when, with the major upheaval of the Himalayas, the arm of the sea became shallow and transformed into brackishwater lakes and lagoons. The land-link between India and the Far East was, thus, once again formed. West tropical marshy conditions corresponding to those of the Siwaliks seem to have prevailed along the entire length of the southern face of the Himalayas extending eastwards to areas beyond China and Westwards to Baluchistan during the Pliocene synchonising with the upheaval of the Himalayas. During this period, there had been a dispersal of fishes inhabiting the marshes, as is evidenced from the fossil relics of forms like *Clarias*, *Heterobranchus*, *Channa*, *Rita*, *Bagarius*, *Silurus*, etc. Fishes, which are typical of the smaller torrential streams, also followed a more or less parallel history of emergence in the early Pleistocene. The typical Himalayan ichthyofauna, like, *Nemacheilus*, *Garra*, *Tor*, etc., perhaps, had originated somewhere in south of China and dispersed into the regions lying along the Alpine-Himalayan system and its southern loops ramifying into Europe, Western Asia and other associated mountains (Hora, 1951 b). A second major upheaval of the Himalayas had occurred towards the end of the Pliocene or early Pleistocene which had, perhaps, raised the Siwalik sediments into dry land. Concomitantly, it is assumed that, the Siwalik fore-deep (Krishnan, 1934) had gradually disappeared leaving a shallow depression, referred to as the 'Pleistocene fore-deep'; which, meanwhile, along with the formation of 'Assam Plateau', began draining northern part of India, beginning from Assam, into the Arabian Sea (Hora, 1953). It is, further, believed that, at certain point of time, there might have existed a contiguous route which had linked the Asian Plateau in the east with the headwaters of the Euphrates and the Tigris rivers along which the fishes might have got widely dispersed (Menon, 1955).

Most of the present-day fish fauna of India, however, had descended from the Pleistocene migrants. Migration along the Himalayas westwards might have commenced during the early Pleistocene and might have continued uninterrupted till the late Pleistocene, when the dismemberment of the westward-flowing drainage at the foot of the Himalayas had occurred. However, migration southwards to the peninsula might have occurred entirely during the Pleistocene. It is assumed that, from the beginning of the Pleistocene, the sea level had undergone repeated fluctuations which were supposed to have been caused by the formation of large ice-caps on the higher latitudes during the Glacial period and their subsequent melting during the inter-Glacial periods. The former, obviously, might have lowered the sea level and the latter had led to its rise. However, during the last Glacial period, the sea level had increased, probably, due to changes in climate. Hora (1951 c) had advocated that, during the Glacial periods, (Pluvial periods in the southern latitudes), the Garo-Rajmahal gap might have become a few hundred metres higher than the sea level; and, that, colder climate might have favoured increased precipitation with less evaporation and greater run off in streams like Narmada, Tapti along the Satpura-Vindhya. These conditions might have favoured the spread of torrential fishes from the region of Assam Himalayas to Peninsular India across the Garo-Rajmahal Bridge.

An Account of the Ichthyospecies

Zoogeographically, FW fish have been classified differently by different workers. Although the classification made by Myers (1949) have been proved to be the most useful, and widely accepted one, the FW fish of marine origin had been further classified as 'peripheral FW forms' by Nichols (1928) and Darlington (1957) which has also been accepted by many recent fish geographers. Incidentally, the ichthyofauna of this region, by and large, have been found to belong to the following categories (Kar, 1990):

(A) Primary FW Fish

Genera-wise break-up of the species under this group include, among others, *Notopterus, Chitala, Labeo, Cirrhinus, Catla, Cyprinus, Puntius, Rasbora, Aspidoparia, Amblypharyngodon, Barilius, Danio, Esomus, Salmostoma, Botia,Lepidocephalus, Nemacheilus, Somileptus, Rita, Mystus, Wallago,Ompok, Ailia, Eutropiichthys, Clupisoma, Silonia, Pangasius, Gagata, Glyptothorax, Clarias, Heteropneustes, Chaca, Badis, Nandus, Anabas, Colisa, Mastacembelus, Macrognathus, Tetraodon.*

(B) Peripheral FW Fish

Genera-wise break up under this group include, among others, *Gudusia, Hilsa (Tenualosa), Pisodonophis, Chanda, Xenentodon, Aplocheilus, Amphipnous, Sicamugil, Rhinomugil, Glossogobius.*

In addition to the above, on the basis of Indian and Extra-Indian fish distribution (Motwani, *et al.*, 1962), the following ichthyospecies of this region could significantly be incorporated under the following two groups:

(a) Widely Distributed Species

Genera-wise break up under this group include, among others, *Esomus, Puntius, Rasbora, Ompok, Wallago, Clarias, Xenentodon, Channa, Glossogobius, Anabas, Macrognathus* and *Mastacembelus*. Incidentally, these ichthyospecies, in addition to this region, also occur in many other parts of India, Pakistan, Bangladesh, Sri Lanka and Malaya.

(b) Species of Northern India

Species under this group include, among others, *Rasbora elanga, Botia Dario, Lepidocephalus guntea,Glyptothorax telchitta, Chanda baculis, Rhinomugil corsula, Sicamugil cascasia* and *Tetraodon cutcutia.*

In addition to the foregoing analyses, ecomorphologically (Dey, 1973), the fishes of this region could further be categorized into four distinct groups which are as follows:

(a) True Hill-stream or Rheophilic Forms

Fishes with strong body musculature and adapted to torrential abode, *e.g.*, *Garra*, *Psilorhynchus*, *Balitora*, *Glyptothorax*, etc.

(b) Semi-torrential Forms

Fishes with less body modifications as compared to rheophilic forms, *Botia*, *Lepidocephalus*, *Nemacheilus*, *Schistura*, *Somileptus*, *Gagata*, etc.

(c) Migratory Forms

Well-built fish having the power of overcoming adverse ecological conditions, *Hilsa* (*Tenualosa*), *Barilius*, *Channa*, *Badis*, etc.

(d) Plainwater Forms

Fishes having minimum body modifications and insignificant migratory habit., *Pisodonophis*, *Gudusia*, *Notopterus*, *Chitala*, *Amblypharyngodon*, *Aspidoparia*, *Catla*, *Cirrhinus*, *Cyprinus*, *Danio*, *Esomus*, *Labeo*, *Puntius*, *Rasbora*, *Salmostoma*, *Mystus*, *Aorichthys*, *Ompok*, *Wallago*, *Rita*, *Clupisoma*, *Eutropiichthys*, *Silonia*, *Pangasius*, *Clarias*, *Heteropneustes*, *Chaca*, *Xenentodon*, *Aplocheilus*, *Amphipnous*, *Chanda*, *Nandus*, *Sicamugil*, *Rhinomugil*, *Anabas*, *Colisa*, *Glossogobius*, *Macrognathus*, *Mastacembelus* and *Tetraodon*.

Chapter 8

Biodiversity of the Fishes of Barak Drainage (Assam), Mizoram and Tripura: Present Status

(A) Lotic System

As stated earlier, the NE region of India is one of the richest spots of ichthyodiversity in the world. Pilot survey conducted in 22 rivers in Barak drainage, Mizoram and Tripura, during the period 1999 to 2003, revealed the occurrence of 133 species of fishes. A systematic list of the fish diversity in these 22 rivers is presented below in Table 2 indicating the ichthyospecies assemblage and their distribution pattern (Plate 13).

Further, detailed Fish Inventory prepared for six rivers, *viz.*, Barak, Jatinga, Sonai and Dhaleswari (in Assam), Tuirial (in Mizoram) and Gomati (in Tripura) revealed the occurrence of 103 species of fishes which is presented in Table 3.

Information about the habit and habitat of these fish species is scarce and are based mainly on the present study. Data about the reproductive period, sites of breeding, feeding, etc., is scanty. According to our field observations, there is difference in habitat preference within the species in many families. Young fishes are generally found to prefer benthic zone of the rivers while the adults tend to live in all the different niches (Smith, 1994). The *Garra* spp

Table 2: Ichthyospecies Assemblage and Distribution Pattern in Barak, Mizoram and Tripura Drainages

Sl.No.	Ichthyospecies	Ba	Ji	Ma	Ja	So	Ru	Dh	Bl	Tu	Se	Tl	Tv	Mt	Ko	Ka	Mn	Kh	Go	Fe
1.	*Notopterns notopterus* (Pallas)	+	+	+			+	+				+	+			+				
2.	*Chitala chitala* (Ham-Buch)	+																	+	
3.	*Pisodonophis boro* (Ham-Buch)	+																		
4.	*Hilsa* (*Tenualosa*) *ilisha* (Ham-Buch)	+			+			+											+	
5.	*Gudusia chapra* (Ham-Buch)	+			+			+								+				
6.	*Securicula gora* (Ham-Buch)	+			+	+		+			+	+	+			+		+	+	
7.	*Salmostoma bacaila* (Ham-Buch)	+			+	+		+	+			+	+			+		+	+	
8.	*Salmostoma phulo* (Ham-Buch)												+							
9.	*Salmostoma sardinella* (Val.)													+						
10.	*Aspidoparia morar* (Ham-Buch)	+			+	+		+			+				+		+	+	+	
12.	*Barilius barila* (Ham-Buch)	+			+	+											+			
13.	*Barilius barna* (Ham-Buch)	+		+	+					+		+							+	
14.	*Barilius barnoides* (Vinciguerra)					+				+			+	+				+		
15.	*Barilius bendelish* (Ham-Buch)				+	+				+		+						+		
16.	*Barilius dogarsinghi* Hora	+			+	+														
17.	*Barilius dimorphicus* Tilak & Hussain																		+	
18.	*Barilius shacra* (Ham-Buch)					+			+	+	+		+		+		+		+	

Contd...

Table 2–Contd...

Sl.No.	Ichthyospecies	Ba	Ji	Ma	Ja	So	Ru	Dh	Bl	Tu	Se	Tl	Tv	Mt	Ko	Ka	Mn	Kh	Go	Fe
19.	*Barilius tileo* (Ham-Buch)				+	+		+	+	+	+		+							
20.	*Barilius vagra vagra* (Ham-Buch)				+	+		+	+	+		+		+		+			+	
21.	*Chela laubuca* (Ham-Buch)	+	+			+		+			+				+				+	
22.	*Esomus danricus* (Ham-Buch)	+						+				+		+					+	
23.	*Danio aequipinnatus* (McClelland)	+			+	+			+	+				+						
24.	*Danio annandalei* (Chaudhuri)																	+		
25.	*Danio devario* (Ham-Buch)				+															
26.	*Danio naganensis* (Choudhuri)				+	+				+				+	+	+	+	+		
27.	*Amblypharyngodon mola* (Ham-Buch)	+			+	+		+											+	
28.	*Tor masal* (Ham-Buch)					+							+		+					
29.	*Tor progenius* (McClelland)				+															
30.	*Tor tor* (Ham-Buch)													+						
31.	*Neolissochilus hexagonolopis* (McClelland)	+			+	+		+	+	+	+			+	+		+		+	
32.	*Neolissochilus blythii* (Day)													+						
33.	*Osteobrama cotio cotio* (Ham-Buch)	+	+					+	+						+	+				
34.	*Chagunius chagunio* (Ham-Buch)												+							
35.	*Puntius chola* (Ham-Buch)	+			+						+						+	+	+	

Contd...

Table 2–Contd...

Sl.No.	Ichthyospecies	Ba	Ji	Ma	Ja	So	Ru	Dh	Bl	Tu	Se	Tl	Tv	Mt	Ko	Ka	Mn	Kh	Go	Fe
36.	*Puntius clavatus clavatus* (McClelland)												+							
37.	*Puntius conchonius*	+	+		+	+		+		+		+	+	+	+	+	+	+	+	
38.	*Puntius sarana orphoides* (Valenciennes)									+										
39.	*Puntius sarana sarana* (Ham-Buch)				+															
40.	*Puntius sarana spilurus* (Gunther)									+				+						
41.	*Puntius sophore* (Ham-Buch)	+				+											+		+	
42.	*Puntius ticto* (Ham-Buch)	+				+		+		+										
43.	*Semiplotus semiplotus* (McClelland)													+						
44.	*Cirrhinus cirrhosus* (Bloch)																+			
45.	*Cirrhinus mrigala* (Ham-Buch)	+	+		+	+		+			+						+		+	
46.	*Cirrhinus reba* (Ham-Buch)		+		+	+		+			+						+		+	
47.	*Catla catla* (Ham-Buch)	+				+														
48.	*Labeo calbasu* (Ham-Buch)	+			+			+			+									
49.	*Labeo gonius* (Ham-Buch)		+		+		+													
50.	*Labeo boga* (Ham-Buch)																	+		
51.	*Labeo pangusia* (Ham-Buch)				+															
52.	*Labeo rohita* (Ham-Buch)	+					+													

Contd...

Table 2–Contd...

Sl.No.	Ichthyospecies	Ba	Ji	Ma	Ja	So	Ru	Dh	Bl	Tu	Se	Tl	Tv	Mt	Ko	Ka	Mn	Kh	Go	Fe
53.	*Crossocheilus burmanicus* (Hora)			+						+					+					
54.	*Crossocheilus latius latius* (Ham-Buch)				+					+				+	+		+		+	
55.	*Garra annadalei* (Hora)									+					+					
56.	*Gara gotyla gotyla* (Gray)				+	+				+			+	+	+					
57.	*Garra lamta* (Ham-Buch)													+						
58.	*Gara lissorhynchus* (McClelland)									+										
59.	*Garra nasuta* (McClelland)				+															
60.	*Psilorhynchus balitora* (Ham-Buch)				+					+		+					+		+	
61.	*Psilorhynchus gracilis* (Rainboth)									+										
62.	*Homaloptera bilineata* (Blyth)								+											
63.	*Balitora brucei* (Gray)			+						+				+			+			
64.	*Acanthocobitis botia* (Ham-Buch)	+	+	+	+	+		+		+		+			+		+		+	
65.	*Schistura multifasciatus* (Day)	+															+		+	
66.	*Schistura arunachalensis* (Menon)									+										
67.	*Schistura scaturigina* (McClelland)								+	+				+			+		+	
68.	*Schistua rupecula* (Hora)									+										
69.	*Botia dario* (Ham-Buch)	+	+	+	+	+	+	+	+			+								

Contd...

Table 2–Contd...

Sl.No.	Ichthyospecies	Ba	Ji	Ma	Ja	So	Ru	Dh	Bl	Tu	Se	Tl	Tv	Mt	Ko	Ka	Mn	Kh	Go	Fe
70.	*Botia rostrata* (Gunther)				+							+								
71.	*Acantophthalmus pangia* (Ham-Buch)									+		+								
72.	*Somileptus gongota* (Ham-Buch)																+		+	
73.	*Lepidocephalus annandalei* (Chaudhuri)					+						+								
74.	*Lepidocephalus guntea* (Ham-Buch)	+	+			+			+								+		+	
75.	*Rita rita* (Ham-Buch)	+	+					+											+	
73.	*Batasio batasio* (Ham-Buch)														+					
77.	*Batasio tengana* (Ham-Buch)		+																	
78.	*Mystus bleekeri* (Day)		+				+	+											+	
79.	*Mystus cavasius* (Ham-Buch)	+		+	+	+	+									+			+	
80.	*Mystus tengara* (Ham-Buch)					+													+	
81.	*Mystus vittatus* (Bloch)	+			+	+	+				+					+	+		+	
82.	*Aorichthys seenghala* (Sykes)	+			+		+			+					+	+		+	+	
83.	*Aorichthys aor* (Ham-Buch)												+		+	+				
84.	*Ompok bimaculatus* (Bloch)	+		+	+	+													+	
85.	*Ompok pabo* (Ham-Buch)		+																	
86.	*Wallago attu* (Bloch and Schneider)	+	+																	

Contd...

Table 2–Contd...

Sl.No.	*Ichthyospecies*	*Ba*	*Ji*	*Ma*	*Ja*	*So*	*Ru*	*Dh*	*Bl*	*Tu*	*Se*	*Tl*	*Tv*	*Mt*	*Ko*	*Ka*	*Mn*	*Kh*	*Go*	*Fe*
87.	*Ailia coila* (Ham-Buch)	+	+		+	+		+		+						+		+	+	
88.	*Clupisoma garua* (Ham-Buch)	+				+						+						+		
89.	*Eutropiichthys murius* (Ham-Buch)	+					+	+		+			+						+	
90.	*Eutropiichthys vacha* (Ham-Buch)	+	+	+	+	+	+	+		+					+	+			+	
91.	*Amblyceps mangois* (Ham-Buch)		+																	
92.	*Bagarius bagarius* (Ham-Buch)	+			+															
93.	*Gagata cenia* (Ham-Buch)																		+	
94.	*Gagata gagata* (Ham-Buch)					+													+	
95.	*Nangra nangra* (Ham-Buch)	+	+	+		+	+	+		+						+			+	
96.	*Nangra robusta* (Mirza & Awan)																		+	
97.	*Nangra viridescens* (Ham-Buch)				+			+	+	+			+							
98.	*Erethlstes pussilus* (Muller & Troschel)	+								+						+			+	
99.	*Glyptothorax cavia* (Ham-Buch)													+						
100.	*Glyptothorax striatus* (McClelland)			+	+															
101.	*Glyptothorax sinense* (Regan)												+							
102.	*Glyptothorax telchitta* (Ham-Buch)	+			+	+		+		+			+	+						
103.	*Glyptothorax trilineatus* (Blyth)												+							

Contd...

Table 2–Contd...

Sl.No.	Ichthyospecies	Ba	Ji	Ma	Ja	So	Ru	Dh	Bl	Tu	Se	Tl	Tv	Mt	Ko	Ka	Mn	Kh	Go	Fe
104.	*Exostoma berdmorei* (Blyth)																		+	
105.	*Sisor rhabdophorus* (Ham-Buch)						+													
106.	*Heteropneustes fossilis* (Bloch)							+		+										
107.	*Chaca chaca* (Ham-Buch)	+				+														
108.	*Olyra longicaudata* (McClelland)				+															
109.	*Sicamugil cascasia* (Ham-Buch)					+		+	+											
110.	*Rhinomugil corsula* (Ham-Buch)	+																+	+	
111.	*Xenentodon cancila* (Ham-Buch)	+			+	+	+			+			+	+	+	+			+	
112.	*Aplocheilus panchax* (Ham-Buch)	+				+													+	
113.	*Macrognathus aral* (Bloch Schneider)	+				+														
114.	*Macrognathus pancalus* (Ham-Buch)	+	+	+	+	+											+		+	
115.	*Mastacembelus armatus* (Lacepede)	+	+		+	+		+		+		+	+	+	+		+		+	
116.	*Chanda nama* Ham-Buch	+	+	+	+	+	+	+			+	+			+	+	+		+	
117.	*Parambassis baculis* (Ham-Buch)	+																	+	
118.	*Parambassis ranga* (Ham-Buch)	+	+			+	+									+			+	
119.	*Parambassis tenasserimensis* (Roberts)											+		+						
120.	*Johnius coitor* (Ham-Buch)	+				+		+								+				

Contd...

Table 2–Contd...

Sl.No.	*Ichthyospecies*	*Ba*	*Ji*	*Ma*	*Ja*	*So*	*Ru*	*Dh*	*Bl*	*Tu*	*Se*	*Tl*	*Tv*	*Mt*	*Ko*	*Ka*	*Mn*	*Kh*	*Go*	*Fe*
121.	*Badis badis* (Ham-Buch)	+	+		+	+				+		+	+			+			+	
122.	*Nandus nandus* (Ham-Buch)	+																		
123.	*Glossogobius giuris giuris* (Ham-Buch)	+	+	+	+	+	+	+	+		+	+			+	+	+	+	+	
124.	*Anabas testudineus* (Bloch)																		+	
125.	*Colisa fasciatus* (Schneider)								+											
126.	*Colisa lalia* (Ham-Buch)	+																		
127.	*Colisa sota* (Ham-Buch)	+																		
128.	*Channa barca* (Ham-Buch)																			
129.	*Channa marulius* (Ham-Buch)																			
130.	*Channa orientalis* (Schneider)	+											+	+	+					
131.	*Channa punctatus* (Bloch)	+	+	+	+	+			+								+		+	
132.	*Channa striata* (Bloch)		+		+															
133.	*Tetraodon cutcutia* (Ham-Buch)																			

Abbreviations: Ba: Barak; Ji: Jiri; Ma: Madhura; Jn: Jatinga; So: Sonai; Ru: Rukni; Dh: Dhaleswari; Bl: Baleswar; Tu: Tuirial; Se: Serlui; Tl: Tlawng;.Tv: Tuivai; Mt: Mat; Ko: Kolodyne; Ka: Karnafuli; Mn: Manu; Kh: Khowai; Go: Gomati; Fe: Feni and the balitorids have been generally found in the fast flowing hill streams. Sections of the rivers not having much aquatic macrophytes revealed abundance of fishes. The food of the adults generally consist of herbivorous items with occasional carnivorous components. The gut contents revealed the occurrence of mainly phytoplankton food with zooplankton encountered only occasionally.

Table 3: Detailed Fish Inventory of the Six Rivers in Barak Drainage, Mizoram and Tripura

Sl.No.	*Ichthyospecies*	*Ba*	*Ja*	*So*	*Dh*	*Tu*	*Go*
1.	*Notopterus notopterus*	+	–	–	+	–	–
2.	*Pisodonophis boro*	+	–	–	–	–	–
3.	*Hilsa (Tenualosa) ilisha*	–	+	–	–	–	+
4.	*Gudusia chapra*	+	+	–	–	–	–
5.	*Securicula gora*	–	+	–	+	–	+
6.	*Salmostoma bacaila*	+	+	+	+	–	+
7.	*Aspidoparia morar*	+	+	+	+	–	+
8.	*Barilius barila*	+	+	+	–	–	–
9.	*Barilius barna*	+	+	–	–	+	+
10.	*Barilius barnoides*	–	–	+	–	+	–
11.	*Barilius bendelisis*	–	+	+	–	+	–
12.	*Barilius dimorphicus*	–	–	–	–	–	+
13.	*Barilius dogarsinghi*	+	+	+	–	–	–
14.	*Barilius shacra*	–	–	+	–	+	–
15.	*Barilius teleo*	–	+	–	+	+	–
16.	*Barilius vagra*	–	+	–	+	+	+
17.	*Chela laubuca*	+	–	+	+	–	+
18.	*Esomus danricus*	+	–	–	+	–	+
19.	*Danio aequipinnatus*	+	+	+	–	+	–
20.	*Danio devario*	–	+	–	–	–	–
21.	*Danio naganensis*	–	+	+	–	+	–
22.	*Amblypharyngodon mola*	+	+	+	+	–	+
23.	*Tor mosal*	–	–	+	–	–	–
24.	*Tor progenius*	–	+	–	–	–	–
25.	*Neolissochilus heaxagonolopis*	+	+	+	+	+	+

Contd...

Table 3–Contd...

Sl.No.	Ichthyospecies	Ba	Ja	So	Dh	Tu	Go
26.	*Osteobrama cotio*	–	–	–	+	–	–
27.	*Puntius chola*	+	+	–	–	–	+
28.	*Puntius conchonius*	+	+	+	+	+	+
29.	*Puntius sarana orphoides*	–	–	–	–	+	–
30.	*Puntius sarana sarana*	–	+	–	–	–	–
31.	*Puntius sophore*	+	–	+	–	–	–
32.	*Puntius ticto*	+	–	+	+	+	–
33.	*Cirrhinus mrigala*	+	+	+	+	–	+
34.	*Cirrhinus reba*	+	+	–	+	–	–
35.	*Catla catla*	–	–	+	–	–	–
36.	*Labeo calbasu*	–	+	+	–	+	–
37.	*Labeo gonius*	–	+	–	–	–	–
38.	*Labeo pangusia*	–	+	–	–	–	–
39.	*Labeo rohita*	+	–	–	–	–	–
40.	*Crossocheilus burmanicus*	–	–	–	–	+	–
41.	*Crossocheilus latius latius*	–	+	–	–	+	+
42.	*Garra annandalei*	–	–	–	–	+	–
43.	*Garra gotyla gotyla*	–	+	+	–	+	–
44.	*Garra lissorhynchus*	–	–	–	–	+	–
45.	*Garra nasuta*	–	+	–	–	–	–
46.	*Garra notata*	–	–	–	–	+	–
47.	*Psilorhynchus balitora*	–	+	–	–	+	+
48.	*Balitora brucei*	–	–	–	–	+	–
49.	*Acanthocobitis botia*	+	+	+	+	+	+
50.	*Schistura multifasciatus*	+	–	–	–	–	+
51.	*Schistura rupecula*	–	–	–	–	+	–
52.	*Schistura scaturigina*	–	–	–	–	+	–
53.	*Schistura vincigurrae*	–	–	–	–	+	+
54.	*Botia dario*	+	+	+	+	–	–
55.	*Botia rostrata*	–	+	–	–	–	–
56.	*Acantophthalmus pangia*	–	–	–	–	+	–

Contd...

Table 3–Contd...

Sl.No.	Ichthyospecies	Ba	Ja	So	Dh	Tu	Go
57.	*Somileptus gongota*	–	–	–	–	–	+
58.	*Lepidocephalus annandalei*	–	–	+	–	–	–
59.	*Lepidocephalus guntea*	+	–	+	–	–	–
60.	*Rita rita*	+	–	–	+	–	+
61.	*Mystus bleekeri*	–	–	–	+	–	–
62.	*Mystus cavasius*	+	+	+	–	–	+
63.	*Mystus tengara*	–	–	+	–	–	+
64.	*Mystus vittatus*	+	+	+	–	–	+
65.	*Aorichthys seenghala*	+	+	–	+	+	+
66.	*Wallagao attu*	+	–	–	–	–	–
67.	*Ompock bimaculatus*	–	+	+	–	–	+
68.	*Ailia coila*	+	+	+	+	–	+
69.	*Clupisoma garua*	+	–	+	–	–	–
70.	*Eutropiichthys murius*	+	–	–	+	–	+
71.	*Eutropiichthys vacha*	+	+	+	+	–	+
72.	*Amblyceps mangois*	–	+	–	–	–	–
73.	*Bagarius bagarius*	+	+	–	–	–	–
74.	*Gagata cenia*	–	–	–	–	–	+
75.	*Gagata gagata*	–	–	+	–	–	–
76.	*Nangra nangra*	+	–	+	+	–	+
77.	*Nangra robusta*	–	–	–	–	–	+
78.	*Nangra viridescens*	–	+	–	+	–	–
79.	*Erethistes pussilus*	–	–	–	+	+	+
80.	*Glyptothorax striatus*	–	+	–	–	–	–
81.	*Glyptothorax telchitta telchitta*	+	+	–	+	–	–
82.	*Exostoma berdmorei*	–	–	–	–	–	+
83.	*Chaca chaca*	+	–	+	–	–	–
84.	*Sicamugil cascasia*	–	–	+	+	–	–
85.	*Rhinomugil corsula*	–	–	+	+	–	–
86.	*Xenentodon cancila*	+	+	+	–	+	+
87.	*Aplocheilus panchax*	+	–	+	–	–	+

Contd...

Table 3–Contd...

Sl.No.	*Ichthyospecies*	*Ba*	*Ja*	*So*	*Dh*	*Tu*	*Go*
88.	*Magrognathus aral*	+	–	+	–	–	–
89.	*Macrognathus pancalus*	+	+	+	–	–	+
90.	*Mastacembelus armatus*	+	+	+	+	+	+
91.	*Chanda nama*	+	+	+	+	–	+
92.	*Parambassis baculis*	–	–	–	–	–	+
93.	*Parambassis ranga*	+	+	+	–	–	+
94.	*Johnius coitor*	+	–	+	+	–	–
95.	*Bodis badis*	+	+	+	–	+	+
96.	*Nandus nandus*	+	–	–	–	–	–
97.	*Glossogobius giuris*	+	+	+	+	–	+
98.	*Anabas testudineus*	–	–	–	–	–	+
99.	*Colisa lalia*	+	–	–	–	–	–
100.	*Colisa sota*	+	–	–	–	–	–
101.	*Channa orientalis*	+	–	–	–	–	–
102.	*Channa punctatus*	+	+	+	–	–	+
103.	*Channa striata*	–	+	–	–	–	–

NB: +: Present; –: Absent.

Ba: Barak River; Ja: Jatinga River; So: Sonai River; Dh: Dhaleswari River; Tu: Tuirial River; Go: Gomati River.

Of the fish species recorded, the cyprinids were the most abundant group. Altogether, 754 number of cyprinid fishes have been collected from six rivers. The contribution of cyprinids to the total fish collection depicts its highest contribution in river Tuirial and lowest in Dhaleswari (Table 4). Further, the orderwise relative abundance of the fishes in the six extensively studied rivers have been shown in Table 5. Among the cyprinids, *Salmostoma bacaila, Aspidoparia morar, Barilius vagra vagra, Amblypharyngodon mola, Neolissochilus hexagonolopis* and *Puntius conchonius* were recorded in almost all the study sites (Table 3). Further, of the total fish species recorded, five species, *viz., Puntius sarana sarana, Ompok bimaculatus, Ailia coila, Eutropiichthys vacha* and *Bagarius bagarius* could be categorized as 'vulnerable' (Molur and Walker, 1998). Still further, species like *Gudusia chapra, Tor mosal, Labeo gonius, Crossocheilus latius*

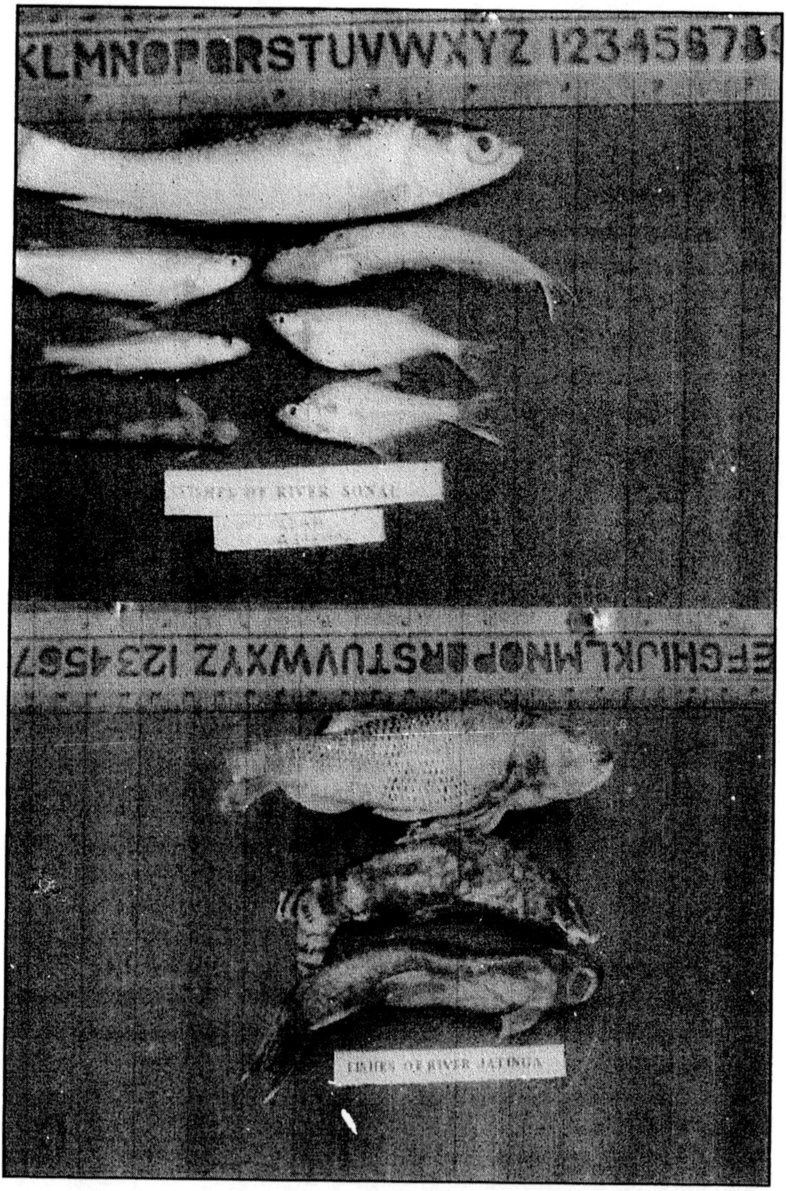

Plate 13: Fish Biodiversity (in general) in the North-East India

Contd...

Plate 13–Contd...

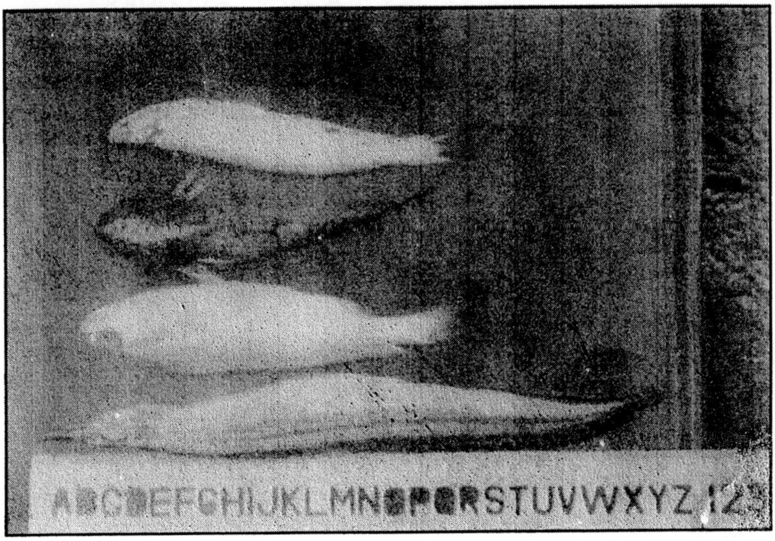

latius, Balitora brucei, Mystus tengara, Xenentodon cancila, Mastacembelus armatus and Nandus nandus could be typified as 'Indeterminate' (Ibid, 1998). Status of certain fish species in some of the rivers in NE India have been depicted in Table 6.

Table 4: Total Fish Collected, species richness, per cent contribution of cyprinids and Shannon-Weaver (SW) Index of species diversity in six extensively studied rivers

Name of River	*Habitat Volume (m^3)*	*Total No. of Fish Collected*	*Species Richness*	*% Contribution of Cyprinids*	*SW Index of Species Diversity*
Barak river	117000	239	52	42.58	1.4582
Jatinga river	7200	213	51	62.91	2.1894
Sonai river	7000	211	47	75.82	1.4016
Dhaleswari river	6400	141	33	6.82	1.1217
Tuirial river	4500	211	30	92.41	1.2755
Gomati river	4189.5	310	47	69.03	1.3670

Table 5: Order-wise Relative Abundance of Fishes in the Six Studied Rivers

Fish Order	*Relative Abundance (%) in Rivers*					
	Barak	*Jatinga*	*Sonai*	*Dhaleswari*	*Tuirial*	*Gomoti*
Osteglossiformes	1.25	–	–	0.7	–	–
Anguilliformes	1.25	–	–	–	–	–
Clupeiformes	0.41	1.4	–	–	–	0.32
Cypriniformes	42.58	62.91	75.82	6.82	92.41	69.03
Siluriformes	17.15	10.32	13.27	21.98	1.42	11.29
Beloniformes	0.41	1.4	0.47	–	0.47	1.61
Cyprinodontiformes	0.41	–	–	–	–	0.64
Synbranchiformes	2.92	1.87	1.42	3.54	2.84	1.61
Perciformes	33.89	22.06	7.1	4.96	2.84	16.77
Mugiliformes	–	–	1.42	2.83	–	0.32

Many fish species of North-East India are quite less abundant today because of many reasons, notably, habitat degradation, disease, etc. A list of less abundant fishes is given below (Plate 14) in Table 6:

Higher species richness (52 and 51) have been recorded from the rivers Barak and Jatinga respectively while lower values (33 and 30) have been recorded in rivers Dhaleswari and Tuirial respectively. Species diversity Index was found to be higher (H = 2.1894) in river Jatinga and in river Tuirial (H= 1.2755) (Shannon and Weaver, 1949), the least value being recorded in river Dhaleswari (H = 1.1217) (Table 4).

The study (Kar, 2003 a)further revealed that, almost all the cyprinid species are generally distributed in pools, pool edges, backwater pools, riffles and riffle edges. Runs and cascades have been found to be least preferred habitats by the cyprinids. The edges of the run habitats have been found to be inhabited mainly by *Chanda nama*, *Puntius conchonius* and *Esomus danricus* in river Barak while the cascade habitats are generally colonized by *Laaeo pangusia* and *Garra lissorhynchus* in river Jatinga. Dammed pools, backwater pools and deep pool edges with bedrock substratum are generally the highly preferred habitats for *Botia dario, Danio dangila, D. naganensis, Badis badis, Puntius conchonius, Tor mosal, Salmostoma bacaila* and *Cirrhinus reba* as revealed in rivers Sonai and Jatinga. *Barilius*

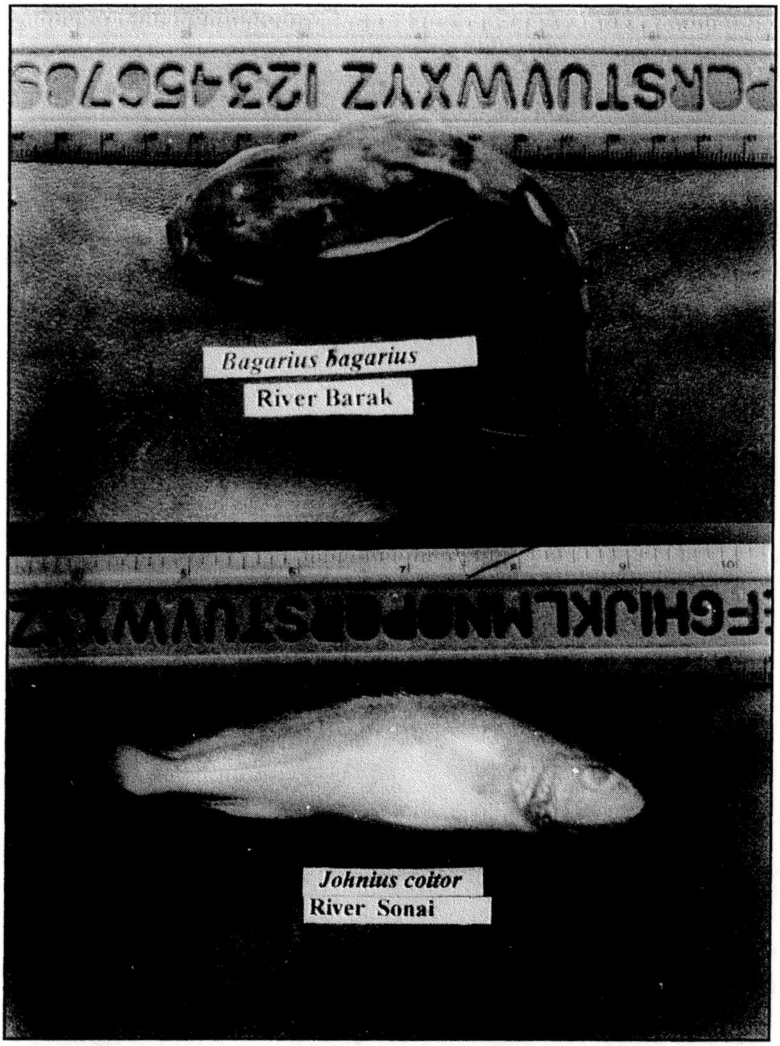

Plate 14: Some Less Abundant Fishes of the Region

Contd...

Plate 14–Contd...

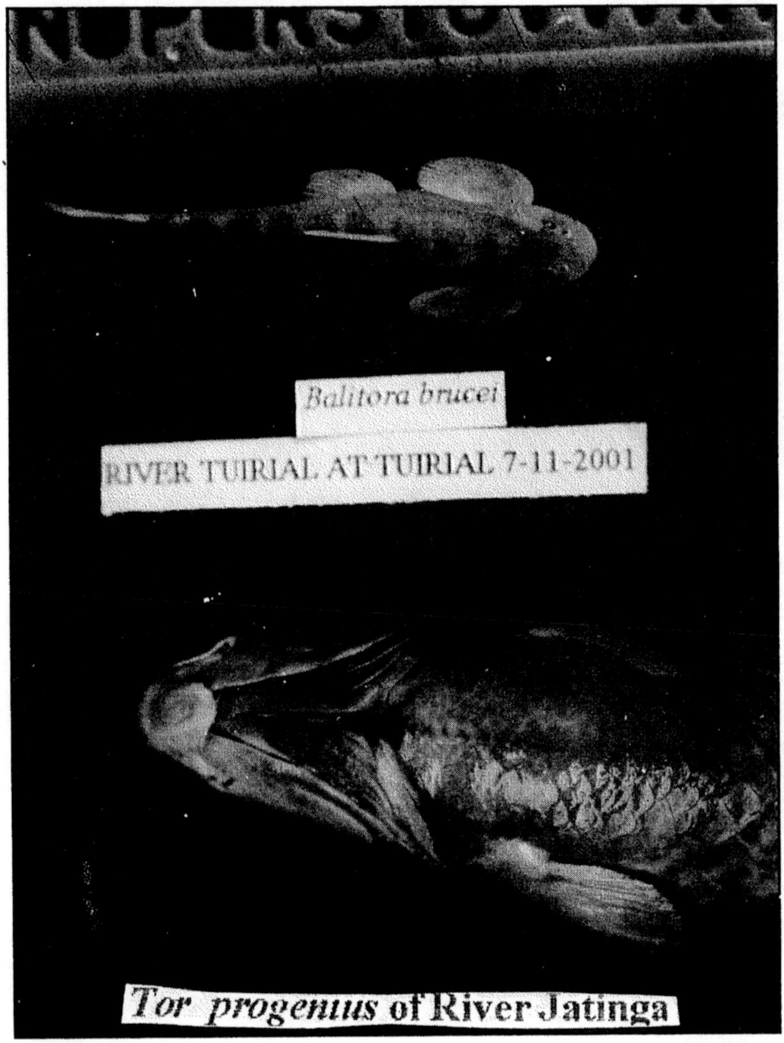

bendelisis, B. dogarsinghi, Schistura spp. are generally abundant in the riffle-type of micro-habitats in river Tuirial where the substratum have been found to be mainly dominated by cobbles. Among the cyprinids, *Tor progenius* and *Neolissochilus hexagonolopis* are generally confined to large deep pools in river Jatinga. Nevertheless, species like, *Crossocheilus latius latius, Psilorhynchus balitora* and *Balitora brucei* have been recorded generally from the cascade to riffle regions in the upper gradient zones of rheophilic streams.

Table 6: Status of Some of the Fish Species in the Studied Rivers (Based on informations collected from January 2000 to March 2002)

Sl.No.	*Ichthyospecies*	*River in which Recorded*	*Status*
1.	*Chitala chitala*	Barak	Critically endangered
2.	*Danio annandalei*	Khowai	Endangered
3.	*Bengana elanga*	Sonai	Endangered
4.	*Tor mosal*	Sonai	Vulnerable
5.	*Tor progenius*	Jatinga	Endangered
6.	*Puntius sarana orphoides*	Tuirial	Critically endangered
7.	*Puntius sarana sarana*	Jatinga	Critically endangered
8.	*Puntius sarana spilurus*	Tuirial	Vulnerable
9.	*Labeo pangusia*	Jatinga	Endangered
10.	*Garra nasuta*	Jatinga	Endangered
11.	*Psilorhynchus gracilis*	Tuirial	Endangered
12.	*Balitora brucei*	Tuirial, Mat	Endangered
13.	*Schistura multifasciatus*	Barak, Gomati	Vulnerable
14.	*Schistura scaturigina*	Tuirial	Endangered
15.	*Botia rostrata*	Jatinga	Vulnerable
16.	*Mystus tengara*	Sonai, Gomati	Vulnerable
17.	*Bagarius bagarius*	Barak, Jatinga	Endangered
18.	*Erethistes pussilus*	Barak, Gomati	Endangered
19.	*Chaca chaca*	Barak, Sonai	Vulnerable
20.	*Rhinomugil corsula*	Barak, Gomati	Vulnerable
21.	*Parambassis tenasserimensis*	Tlawng, Kolodyne	Vulnerable
22.	*Johnius coitor*	Dhaleswari	Endangered

Macro- and micro-habitats combine to create the total habitat available for organisms. Macro-habitats control the general pattern of species distribution and abundance which governs the flow of energy through the system. In our study, the riparian cover, substrate type and habitat volume were found to be associated with fish density. Habitat volume is generally considered as an important attribute in determining species diversity as observed in rivers Jatinga and Barak. Low habitat is generally linked with low species diversity as observed in rivers Dhaleswari and Tuirial. These findings support that habitat volume is a major determinant in fish assemblage. Cyprinids are generally and largely recorded from pool habitats. Conversely, riffle and cascade-dwelling species include fishes like *Psilorhynchus balitora, Balitora brucei, Glyptothorax telchitta,* etc. The habitat structure of the rivers Tuirial and Jatinga have been found to be quite suitable for both surface dwelling and typical bottom dwelling species, *e.g., Danio aequipinnatus, Salmostoma bacaila, Puntius conchonius* and *Garra lissorhynchus, G.gotyla gotyla, G. nasuta, Psilorhynchus balitora, Balitora brucei, Glyptothorax striatus, Tor mosal* and *T. progenius.* However, all the habitat components were not available in rivers Sonai and Dhaleswari, with concomitant less fish cover. Lesser abundance of cyprinids in river Barak could be due to comparative abundance of carnivorous fishes like *Notopterus notopterus, Aorichthys seenghala, Wallago attu* and *Bagarius bagarius.*

In addition to the above, some specific information regarding the studied rivers, as mentioned above, are given below:

In river Barak, population of Large Growing Fishes (LGF), *e.g., Labeo rohita, Cirrhinus mrigala,* etc., appeared to have declined due to overfishing, siltation of the river bed and lack of fish cover. Even the population of *Bagarius bagarius, Aorichthys seenghala* have dwindled as compared to earlier days. The semi-rheophilic forms like *Neolissochilus hexagonolopis, Danio aequipinnatus, Nangra nangra, Botia dario, Barilius barila* were found to occur in the little upstream region of the river where the water current in the river is faster. The rheophilic forms, *e.g., Glyptothorax telchitta, Nemacheilus multifasciatus, Garra gotyla gotyla* were found to occur in the more upstream regions having boulders and cobbles in the substratum. Nevertheless, these strongly rheophilic fishes are sometimes swept into the downstream region of the river (Kar, 2005 a).

In view of the fact that, river Sonai has both rheophilic upstream portion and plainwater downstream portion (within the Barak valley region of Assam) with concomitant fish cover provided by undercut banks and overhanging vegetations, fish species like, *Cirrhinus mrigala, Catla catla, Salmostoma bacaila, Puntius conchonius, P. ticto, Mystus vittatus, Mystus cavasius, Mystus tengara, Ompok bimaculatus, Eutropiichthys vacha, Ailia coila, Sicamugil cascasia, Rhinomugil corsula, Glossogobius giuris*, etc, were found mostly in the plainwater downstream region with mostly run and sheet characteristics which is quite colonisable by them. Conversely, species like *Neolissochilus hexagonolopis, Barilius barila, B. shacra, Botia dario, Bengana elanga, Chaca chaca, Lepidocephalus guntea, Danio aequipinnatus, Chela laubuca*, etc., were found in the little upstream region with mostly 'run' characteristics which is said to be quite suitable as their habitat. Further, rheophilic species like *Garra gotyla gotyla, Gagata gagata*,etc., have been recorded further upstream region with faster flow of water (Kar, 2003 a). It has been found that Jorkhal and Nagakhal, from the surrounding catchment, join river Sonai as two rivulets. These rivulets were found to have higher water transparency and faster flow of water as compared to the main river Sonai. Gravels and cobbles formed the principal component of their substratum. Such a condition is said to have paved the way for some fishes to migrate locally from the main river Sonai to these rivulets, as has been revealed by the occurrence of the fry of genera, like *Balitora, Barilius, Tor*, etc., in these rivulets. Attainment of a depth of >6.0 m at some spots having bedrock as substratum and overhanging vegetation cover was also found to be one of the specialities of river Sonai during the monsoon season. Such spots revealed high fish catch (Kar, 2005 a).

In river Jatinga, fishes like *Danio aequipinnatus, Botia dario, Barilius vagra, B. tileo*, etc., take shelter mostly in the interspace between the logs and fallen trees. Further, fishes, like *Garra gotyla gotyla, Psilorhynchus balitora, Glyptothorax striata, Crossocheilus latius latius, Labeo pangusia*, and *Neolissochilus hexagonolopis* occur mostly in the rheophilic upstream region where mainly riffle-pool characteristics exist. LGF, like *Bagarius bagarius* occur mostly in the deeper pools in the upstream region of the river (Kar, 2005 a).

The critically endangered Mahseer fish, *Tor progenius* (K.C. Jayaram, Personal communication) have been recorded in the riffle-pool region of upstream of river Jatinga (Kar *et al.*, 2002 b).

In river Dhaleswari, fishes like *Amblypharyngodon mola*, *Puntius ticto*,*Securicula gora*, *Esomus danricus*, *Mystus bleekeri*, *Ailia coila*, *Eutropiichthys vacha*, *Glossogobius giuris*, *Mastacembelus armatus* are generally found in the downstream plainwater region. Conversely, fishes like *Osteobrama cotio*, *Botia dario* and *Rita rita* generally occur in the little upstream region and, probably, sometimes, get swept into the downstream region. However, species like *Barilius tileo*, *Neolissochilus hexagonolopis*, *Acanthocobitis botia* and *Glyptothorax telchitta* generally occur in the rheophilic upstream region with some of them being swept into the downstream region (Kar, 2005 a).

In river Tuirial, fishes like *Puntius conchonius*, *Danio aequipinnatus* and *Mastacembelus armatus* are generally found to occur in small pools while species like *Barilius vagra*, *B. tileo* and *Aorichthys seenghala* are generally found to occur in little turbulent regions. On the other hand, ichthyospecies like *Neolissochilus hexagonolopis*, *Garra annandalei*, *G. gotyla gotyla*, *Psilorhynchus gracilis*, *Crossocheilus latius latius* and *Schistura vinciguerae* generally occur in quite high turbulent regions. Significantly, in this river, *Balitora brucei* are generally found hidden under the rocks (Kar, 2005 a).

In river Gomati, species like *Cirrhinus reba*, *Puntius conchonius*, *P. sophore*, *Salmostoma bacaila*, *Amblypharyngodon mola*, *Securicula gora*, *Mystus bleekeri*, *M. tengara*, *M. vittatus*, *Ailia coila*, *Chanda nama*, *Parambassis ranga*, *Xenentodon cancila* and *Mastacembelus armatus* are generally found in the downstream plainwater regions; while, species like *Somileptes gongota*, *Gagata cenia*, *G. nangra* and *Nangra nangra* are generally recorded from the little upstream regions of the river. On the other hand, species like, *Neolissochilus hexagonolopis*, *Psilorhynchus balitora*, *Crossocheilus latius latius*, *Barilius vagra*, *B. shacra*, *Balitora brucei* and *Schistura multifasciatus* are generally found in the upstream turbulent regions of the river. Some of these are, probably, swept into the downstream regions. Significantly, *Ompok bimaculatus* have also been recorded in the downstream region (Kar 2005 a).

It may be noted here that, the truly freshwater fishes can generally pass from one isolated drainage system to another only through the slow physiographic changes of the land itself. They are most likely to preserve old distributional pattern (Darlington, 1957). However, the range of distribution of the torrential fishes is much more limited. Thus, in view of their limited power of migration, the torrential fishes afford better evidence of zoogeographical relationship (Hora, 1937).

(B) Lentic

There is a bewildering diversity of fishes in the lentic systems of this region.

An account of the principal lentic bodies of this region, with regard to their ichthyodiversiy and fish yield trend in briefly given below:

(a) Fish Diversity in Sone Beel of Assam

70 species of fishes belonging to 49 genera under 24 families and 11 orders have been recorded in Sone Beel, the biggest wetland in Assam. The ichthyospecies are listed below systematically:

1. *Pisodonophis boro*
2. *Gudusia chapra*
3. *Hilsa (Tenualosa) ilisha*
4. *Chitala chitala*
5. *Notopterus notopterus*
6. *Amblypharyngodon mola*
7. *Aspidoparia morar*
8. *Barilius bendelisis*
9. *Catla catla*
10. *Cirrhinus mrigala*
11. *C. reba*
12. *Cyprinus carpio*
13. *Danio devario*
14. *Esomus danricus*
15. *Labeo bata*
16. *L. calbasu*
17. *L. gonius*
18. *L. nandina*
19. *L. rohita*
20. *Puntius chola*
21. *P. conchonius*

22. *P. sarana sarana*
23. *P. ticto*
24. *Rasbora daniconius*
25. *B. elanga*
26. *Salmostoma bacaila*
27. *Botia dario*
28. *Lepidocephalus guntea*
29. *Acanthocobitis botia*
30. *Somileptus gongota*
31. *Mystus cavasius*
32. *Mystus corsula*
33. *M. tengara*
34. *M. vittatus*
35. *Aorichthys seenghala*
36. *Rita rita*
37. *Ompok bimaculatus*
38. *Wallago attu*
39. *Ailia coila*
40. *Clupisoma atherinoides*
41. *C. garua*
42. *Eutropiichthys vacha*
43. *Silonia silondia*
44. *Pangasius pangasius*
45. *Gagata nangra*
46. *Glyptothorax telchitta*
47. *Clarias batrachus*
48. *Heteropneustes fossilis*
49. *Chaca chaca*
50. *Xenentodon cancila*
51. *Aplocheilus panchax*
52. *Channa gachua*
53. *C. marulius*

54. *C. punctatus*
55. *C. striata*
56. *Amphipnous cuchia*
57. *Parambassis baculis*
58. *Chanda nama*
59. *Badis badis*
60. *Nandus nandus*
61. *Rhinomugil corsula*
62. *Sicamugil cascasia*
63. *Glossogobius giuris*
64. *Anabas testudineus*
66. *Colisa fasciatus*
67. *Macrognathus aral*
68. *M. pancalus*
69. *Mastacembelus armatus*
70. *Tetraodon cutcutia*

Of the 70 ichthyospecies of Sone Beel, 59 species under 39 genera belong to the Primary FW group while 11 species under 10 genera belong to the category of Peripheral FW group (Nichols, 1928; Darlington, 1957). On the other hand, on the basis of Indian and Extra-Indian Territorial Distribution (Motwani *et al.*, 1962; Kar, 1990), 28 ichthyospecies of Sone could significantly be incorporated under two groups, *viz.*, (a)Widely distributed species and (b) Species of Northern India. Further, among the other species, one species, *viz.*, *Glyptothorax telchitta* was found to be a true hill stream form; while, five species, *viz.*, *Botia dario*, *Lepidocephalus guntea*, *Acanthocobitis botia*, *Somileptes gongota* and *Gagata nangra* were recorded as Semi-torrential forms (Dey, 1973). Thirtynine fish species were found to belong to the Plainwater group (Dey and Kar, 1990).

(b) Fish Diversity in Sat Beel of Assam

14 species of fishes belonging to 12 genera, 6 families and 3 orders have been recorded in Sat Beel which are listed below (Kar, 1998):

1. *Amblypharyngodon mola*
2. *Puntius chola*

3. *P. ticto*
4. *Labeo calbasu*
5. *Lepidocephalus guntea*
6. *Cyprinus carpio*
7. *Mystus vittatus*
8. *M. bleekeri*
9. *Aorichthys seenghala*
10. *Wallago attu*
11. *Ompok bimaculatus*
12. *Clupisoma garua*
13. *Channa punctatus*
14. *Macrognathus pancalus*

(c) Fish Diversity in Chatla Haor of Assam

57 species of fishes, belonging to 28 genera, 17 families and 9 orders, have been recorded in Chatla Haor. They are listed below (Kar and Barbhuiya, 2000 b):

1. *Gudusia chapra*
2. *Hilsa (Tenualosa) ilisha*
3. *Chitala chitala*
4. *Notopterus notopterus*
5. *Amblypharyngodon mola*
6. *Catla catla*
7. *Cirrhinus mrigala*
8. *C. reba*
9. *Cyprinus carpio*
10. *Danio devario*
11. *Esomus danricus*
12. *Ctenopharyngodon mola*
13. *Hypophthalmichthys molitrix*
14. *Labeo calbasu*
15. *L. gonius*
16. *L. rohita*

17. *Puntius chola*
18. *P. conchonius*
19. *P. ticto*
20. *P. sarana sarana*
21. *Rasbora daniconius*
22. *Salmostoma bacaila*
23. *Botia dario*
24. *Lepidocephalus guntea*
25. *Acanthocobitis botia*
26. *Mystus cavasius*
27. *M. corsula*
28. *M. tengara*
29. *M. vittatus*
30. *Aorichthys seenghala*
31. *Ompok bimaculatus*
32. *Wallago attu*
33. *Ailia coila*
34. *Clupisoma atherinoides*
35. *C. garua*
36. *Eutropiichthys vacha*
37. *E. marulia*
38. *Clarias batrachus*
39. *Heteropneustes fossilis*
40. *Xenentodon cancila*
41. *Aplocheilus panchax*
42. *Channa gachua*
43. *C. punctatus*
44. *C. striata*
45. *Amphipnous cuchia*
46. *Parambassis baculis*
47. *P. ranga*
48. *Chanda nama*

49. *Badis badis*
50. *Nandus nandus*
51. *Glossogobius giuris*
52. *Anabas testudineus*
53. *Colisa fasciatus*
54. *C. lalia*
55. *Mastacembelus armatus*
56. *Macrognathus pancalus*
57. *Macrognathus aral*

Zoogeographically, the ichthyospecies of Chatla Haor contains 79.62 per cent of Primary FW fish while the rest (20.38 per cent) belong to the Peripheral class (Nichols, 1928; Darlington, 1957; Kar, 1990). Further, on the basis of Indian and Extra-Indian territorial fish distributional pattern (Motwani *et al.*, 1962), ichthyospecies of Chatla Haor did contain fishes belonging to the groups called 'widely distributed species (notably, *Puntius, Ompok, Channa, Anabas*) and species of Northern India (notably, *Botia dario, Lepidocephalus guntea*, etc.). Ecomorphologically (Dey, 1973), fish species of Chatla Haor contain only the 'Semi-torrential' forms and the 'Plainwater' forms (notably. *A.mola, C.catla, C. carpio, Puntius spp., Mystus spp*).

(d) Fish Diversity in Puneer Haor of Assam

The Puneer Haor contains altogether 24 species of fishes belonging to 22 genera, 15 families and 8 orders, which are listed below (Laskar, *et al.*, 2002):

1. *Notopterus notopterus*
2. *Amblypharyngodon mola*
3. *Securicula gora*
4. *Salmostoma bacaila*
5. *Cirrhinus reba*
6. *Labeo calbasu*
7. *Puntius ticto*
8. *Puntius conchonius*
9. *Mystus vittatus*
10. *Mystus bleekeri*

11. *Rita rita*
12. *Ompok bimaculatus*
13. *Clarias batrachus*
14. *Heteropneustes fossilis*
15. *Aplocheilus panchax*
16. *Xenentodon cancila*
17. *Mastacembelus armatus*
18. *Parambassis ranga*
19. *Nandus nandus*
20. *Oreochromis mossambica*
21. *Anabas testudineus*
22. *Colisa fasciatus*
23. *Colisa sota*
24. *Channa punctatus*

(e) Fish Diversity in the Anuas of Assam

In Rupairbala Anua

24 species of fishes belonging to 21 genera, 15 families and 9 orders have been recorded in this Anua which are listed below (Kar, *et al.*, 2000 b):

1. *Puntius conchonius*
2. *P. ticto*
3. *Aplocheilus panchax*
4. *Botia dario*
5. *Gudusia chapra*
6. *Glossogobius giuris*
7. *Lepidocephalus guntea*
8. *Labeo gonius*
9. *Oreochromis mossambica*
10. *Anabas testudineus*
11. *Chanda nama*
12. *Xenentodon cancila*
13. *Amblypharyngodon mola*

14. *Notopterus notopterus*
15. *Badis badis*
16. *Mystus cavasius*
17. *Parambassis ranga*
18. *P. baculis*
19. *Securicula gora*
20. *Ompok bimaculatus*
21. *Colisa fasciatus*
22. *C. lalia*
23. *Macrognathus pancalus*
24. *Tetraodon cutcutia*

In Baskandi Anua

13 species of fishes belonging to 10 genera, 6 families and 4 orders have been recorded in Baskandi Anua which are listed below (Kar, *et al.*, 2000 b; Dhar, 2004):

1. *Channa striata*
2. *Glossogobius giuris*
3. *Channa punctatus*
4. *Channa marulius*
5. *Macrognathus aral*
6. *M.pancalus*
7. *Salmostoma bacaila*
8. *Badis badis*
9. *Chanda ranga*
10. *Puntius ticto*
11. *Rasbora daniconius*
12. *Amblypharyngodon mola*
13. *Puntius conchonius*

In Fulbari Anua

In course of a pilot survey, 7 species of fishes belonging to 6 genera, 6 families and 6 orders have been recorded from Fulbari Anua which are listed below (Kar *et al.*,2000 b):

1. *Gudusia chapra*
2. *Puntius ticto*
3. *Mystus cavasius*
4. *M. bleekeri*
5. *Macrognathus pancalus*
6. *Parambassis ranga*
7. *Channa punctatus*

In Sibnarayanpur Anua

22 species of fishes belonging to 21 genera, 10 families and 5 orders have been recorded from Sibnarayanpur Anua which are listed below (Kar, *et al.*, 2000 b):

1. *Gudusia chapra*
2. *Securicula gora*
3. *Salmostoma bacaila*
4. *Cirrhinus mrigala*
5. *Chela laubuca*
6. *Esomus danricus*
7. *Labeo rohita*
8. *Labeo gonius*
9. *Puntius ticto*
10. *P. conchonius*
11. *Catla catla*
12. *Osteobrama cotio*
13. *Rita rita*
14. *Clupisoma garua*
15. *Eutropiichthys vacha*
16. *Clarias batrachus*
17. *Xenentodon cancila*
18. *Monopterus cuchia*
19. *Macrognathus pancalus*
20. *Mastacembelus armatus*
21. *Chanda nama*
22. *Badis badis*

(f) Fish Diversity in some of the Reserve Forests (RF) in Barak Valley Region of Assam

Principal ichthyofauna in River (stream) Jalengachhara in Innerline RF, as recorded during a pilot survey in 1996-98, is listed below (Kar, 1998):

1. *Badis badis*
2. *Amblypharyngodon mola*
3. *Rasbora daniconius*
4. *Anabas testudineus*
5. *Channa orientalis*
6. *Chanda baculis*
7. *Danio devario*
8. *Clarias batrachus*

Major fish fauna in Bhuban Hill section of Sonai RF, as recorded during a pilot survey in 1996-98 (Kar, 1998) is listed below. The fishes belong mostly to the Nagakhal stream in the area:

1. *Amblypharyngodon mola*
2. *Rasbora daniconius*
3. *Lepidocephalus guntea*
4. *Mystus bleekeri*
5. *Glossogobius giuris*
6. *Parambassis baculis*
7. *Amphipnous cuchia*
8. *Puntius sophore*
9. *Puntius* sp.
10. *Channa orientalis*
11. *Barilius bola*
12. *Colisa sota*

(g) A Glimpse of the Fish Fauna of North Cachar (NC) Hills and Lumding Region of Assam

A consolidated Inventory of major fish fauna found in NC Hills and Lumding region of Assam is given below. The list has been prepared through preliminary surveys conducted in the above mentioned regions during 1996 to 1998 (Kar, 1998):

In NC Hills, the fishes have been collected mostly from Diyung river and few from Haflong Lake. 13 species of fishes belonging to 11 genera, 6 families and 4 orders are listed below:

1. *Eutropiichthys vacha*
2. *Botia dario*
3. *Puntius conchonius*
4. *Amblypharyngodon mola*
5. *Notopterus notopterus*
6. *Garra gotyla*
7. *Xenentodon cancila*
8. *Neolissochilus hexagonolopis*
9. *Barilius dogarsinghi*
10. *Barilius barila*
11. *Barilius bendelisis*
12. *Osteobrama cotio*
13. *Erethistes pussilus*

In Lumding area, the fishes have been collected mostly from small semi-torrential streams. 11 species of fishes belonging to 10 genera, 5 families and 3 orders are listed below:

1. *Osteobrama cotio*
2. *Mystus vittatus*
3. *Glossogobius giuris*
4. *Oreochromis mossambica*
5. *Puntius conchonius*
6. *Tor* sp.
7. *Danio naganensis*
8. *Barilius barila*
9. *Barilius evezardi*
10. *Cirrhinus mrigala*
11. *Nemacheilus corica*

(h) List of Principal Fishes in Tumdil of Mizoram

The principal fish fauna found in Tumdil Lake in Mizoram are the following as revealed from a pilot survey conducted during 1997 (Kar, 1998):

1. *Cirrhinus reba*
2. *Puntius conchonius*
3. *Danio* sp.
4. *Aplocheilus panchax*

(i) Fish Diversity in Kurti Beel of Tripura

42 species of fishes have been recorded in Kurti Beel of Tripura which are listed below (Kar, 1998):

1. *Gudusia chapra*
2. *Notopterus notopterus*
3. *Catla catla*
4. *Cirrhinus mrigala*
5. *Labeo calbasu*
6. *L. gonius*
7. *L. rohita*
8. *Cyprinus carpio*
9. *Ctenopharyngodon idellus*
10. *Hypophthalmichthys molitrix*
11. *Rasbora daniconius*
12. *Puntius conchonius*
13. *P. sophore*
14. *Esomus danricus*
15. *Chagunius chagunio*
16. *Amblypharyngodon mola*
17. *Salmostoma bacaila*
18. *Lepidocephalus guntea*
19. *Botia dario*
20. *Mystus vittatus*
21. *M. cavasius*

22. *Ompok bimaculatus*
23. *Wallago attu*
24. *Clupisoma garua*
25. *Eutropiichthys vacha*
26. *Channa orientalis*
27. *C. marulius*
28. *C. punctatus*
29. *C. striata*
30. *Heteropneustes fossilis*
31. *Clarias batrachus*
32. *Anabas testudineus*
33. *Badis badis*
34. *Nandus nandus*
35. *Chanda nama*
36. *Parambassis baculis*
37. *Xenentodon cancila*
38. *Colisa fasciatus*
39. *C. lalia*
40. *Macrognathus aral*
41. *M. pancalus*
42. *Mastacembelus armatus*

Chapter 9

An Account of the Mahseer Fishes of the Region

'Mahseers' form an important group of game and food fish in India. Several species of mahseers inhabit different rivers throughout India, Pakistan, Myanmar, Sri Lanka, etc. (Desai, 1994). Most significant of them are the *Tor tor, T. mosal* and *T. putitora* which occur in the Sub-Himalayan rivers of North India including Narmada and Tapti basins. *Tor khudree and T. mussulah* are called the Deccan mahseer, which occur in the rivers Godavari, Krishna, Kaveri and their tributaries in peninsular India and also in Sri Lanka (Kar *et al.*, 2002 a, b, c; Jayaram, 2003). There are also sub-species of the genus *Tor, viz., Tor mosal mahanadicus, T.khudree malbaricus* and *T. khudree longispinus.* Further, the anglers consider two other fish species as Mahseer, *viz., Neolissochilus hexagonolopis* and *N. blythii,* which are popularly called 'Chocolate Mahseers'. Mirza (1967) reported a new species of Mahseer called *Tor zhobensis* sp. nov. from the river Zhob in Pakistan. Mahseers provide one of the principal fisheries in the Siwalik Himalayas and in the lower reaches of the uplands of Deccan plateau. An estimated elevation of 2000 m above MSL in the Himalayas is said to be the upper limit of migration of the Mahseers during the SW monsoon (Jun-Oct) (Cordington, 1939). Some authors suggest that the occurrence and distribution of the Mahseers is generally controlled by the prevailing water temperature in the streams rather than the altitude (Jhingran and Sehgal, 1978).

However, Mahseers can generally tolerate a temperature range of 6-36°C. All these species, recognized as 'Mahseers' or 'large-scale carp of India', have been so named because of big size of their head (Hora, 1939).

There have been many reports on the Mahseer fishes of India, notably, Hora (1939), Mirza (1967), Pisolkar (1994), Menon (1994), Singh *et al.* (1994), Yadava and Chandra (1994), Kotwal (1994), Dubey (1994), Sehgal (1994), Sandhu *et al.* (1994), Pandey *et al.* (1994), Pathani (1994), Joshi and Kapoor (1994), Sinha (1994), Gupta and Khan (1994), Nautiyal (1994), Soni (1994), Agarwala (1994), Ogale (1994), Narayani and Tamot (2002), Indulkar and Ogale (2002), Kar (2000, 2003a, 2004), Kar *et al.* (2002 a, b, c).

Mahseers are known to occur in the rivers of Assam, Mizoram and Tripura (Figure 2).

In Barak valley region of Assam, Mahseers have been recorded from rivers Jatinga, Sonai and Dhaleswari (Figure 3). The river Jatinga (24° 57′ N to 92° 45′E, Altitude 29.57 m MSL, Gradient 2°), is a tributary of River Barak, which originates near Jatinga village in North Cachar Hills district of Assam and joins river Barak in the Cachar district of Assam. It is a hill stream of 4th order (stream order is hierarchical ordering of streams based on the degree of branching) (Armontrout, 1990) being formed by the union of two 3rd order streams (the latter was formed by the union of two similar streams belonging to the preceding orders). A glimpse into the habitat parameters of this river is essential here in order to show the suitability of this river habitat for colonization by the Mahseers and other rheophilic fishes. River Jatinga has alluvial valley segment and pool-riffle and braided type of reach. The micro-habitat type is dominated by riffle and run with occasional occurrence of trench pools. The river is less entrenched because of V-shaped valley segment and the substrate type is dominated by gravels and cobbles with the occurrence of some boulders and bedrocks; thus, providing suitable feeding and breeding habitat for Mahseer fishes, like *Tor tor* (Ham-Buch), *Tor mosal* (Ham-Buch), *Tor progenius* (McClelland), *Neolissochilus hexagonolopis*, etc. The riparian type is generally mesoriparian (the soil/substrate is dry seasonally) while the riparian vegetation type is mostly dominated by deciduous trees with some amount of shrubs and grasses. The fish covers, which enable the fish to hide, are of various types ranging from undercut bedrock to bottom-free big boulders and the pools in many places. Limnological studies revealed average values of water temperature (24°C), turbidity (16.8 NTU), pH (7.31), Dissolved oxygen (8.53 mg/lit), Free carbon di-oxide (23.33 mg/lit), Total alkalinity (40.0 mg/lit) and conductivity (126 μ mhos/cm). Out of the 52 species of fishes recorded in river Jatinga during January 1999 to March 2002, the following species of Mahseers have been found to occur:

Tor mosal (Ham-Buch), *Tor tor* (Ham-Buch), *Tor progenius* (McClelland) and *Neolissochilus hexagonolopis* (McClelland). Of these, *Tor progenius* (McClelland), recording a population of only 3 specimens from a single location at village Ditokchhara along the course of the river, during the period 1999-2002, has been assigned the status of 'critically endangered' (Menon, 1994; Kar *et al.,* 2002 b; Jayaram, 2003) (Plate 14).

In addition to river Jatinga, Mahseer fishes have been recorded from others rivers in Barak valley, such as, *Neolissochilus hexagonolopis* in river Barak and in river Dhaleswari, *Tor mosal* in river Sonai, etc.

In Mizoram, *Neolissochilus hexagonolopis* has been recorded in river Tuirial. However, a detailed study of the Mahseer fishes has been conducted in River Mat in Mizoram, a river joining the Kolodyne-Irrawady drainage system of the Indo-Myanmar region. The study, conducted during 2000-2002 (Kar *et al.,* 2002 c) is briefly described below:

In river Mat, only three species of Mahseer fishes have been recorded, *viz., Tor tor* (Tor Mahseer), *Neolissochilus hexagonolpis* and *N. blythii* (Chocolate Mahseer). Records of their occurrence along with physico-chemical characteristics of the river water at the site of collection during the period Jan 2000 to May 2002 have been presented in the Table 1.

Information pertaining to various habitat parameters of the river Mat for the same period of study have been shown in the Table 2.

A hillstream of 2nd order, river Mat has colluvial (landslide from adjacent hill slopes delivering sediments and organic matter) valley segment and pool-riffle type of reach. This reach type is most commonly associated with small to mid-size streams and is a quite prevalent type of reach in the rheophilic zones. On the other hand, runs are generally laminar flow of water with sandy substratum.

Table 1: Record of Mahseer Fishes in Mat River in Mizoram along with Physico-chemical Characteristics of River Water at the Site of Collection for the Period from January 2000-May 2002

Name of the Mahseer	Total No. Collected	Length (cm)	Weight (g)	Water Characteristics						
				Temp. (°C)	pH	DO (mg/l)	FCO_2 (mg/l)	TA (mg/l)	Turbidity (NTU)	Conductivity (µ mhos/cm)
Tor tor	5	14	85							
Neolissochilus hexagonolopis	11	17	6	26	8.62	6.0	2.5	12.2	6	161
Neolissochilus blythii	6	12	70							

Pools, riffles and runs are generally found to dominate the micro-habitat type with frequent occurrence of trench pools. This river is said to be more entrenched (entrenchment ratio being 0.566) based on V-shaped valley segment. The substrate type has been found to be dominated by gravels and cobbles with frequently-occurring quite large number of boulders and some bedrocks. This kind of habitat with concomitant physico-chemical characteristics of water is said to be a suitable feeding and breeding ground for *Tor* spp and *Neolissochilus* spp. The riparian type is generally Mesoriparian while the riparian vegetation type is generally deciduous trees with some amount of shrubs and grasses. Different types of fish covers have been recorded in river Mat ranging from undercut bedrock to bottom-free big boulders and pools. Signs of erosion are visible. Riparian land use pattern mainly involves human habitation and some amount of agriculture through 'jhooming'. Field observations revealed differences in habitat preferences among the fishes of different size groups. Nevertheless, the Mahseers in river Mat seem to prefer deep pools as one of their favourable habitats in view of depth providing strong cover to them. However, they also make use of covers like bottom-free big boulders, undercut banks, etc. Riparian vegetation did not seem to provide much cover to these fishes.

Table 2: Record of Habitat Parameters of River Mat of Mizoram for the Period from January 2000-May 2002

Habitat Parameters	*Observation*
Geographical position	N 220 23' 48.5" E 920 57' 59.5"
Altitude:	152.43 m
Stream order	Second order
Microhabitat type	Pool, riffle and run
Cover type	Overhanging vegetation, undercut bank
Substrate type	Dominated by gravels, cobbles, boulders and occasionally bedrocks
Average wetted width	40 m
Average wetted depth	1.25 m
Riparian vegetation type	Shrubs and Trees
Riparian land use pattern	Human habitation and jhuming
Sign of erosion	Visible

In Tripura, *Neolissochilus hexagonolopis* has been recorded in river Gomati.

Notwithstanding the above, the introduction of a completely new factor within recent times has readily altered the situation. This factor is the unprecedented overwhelming dominance of a single species, man, in the evolutionary history. It is, thus, essential to differentiate between natural extinction on the one hand and extinction due to human agency on the other, when considering the extinction of a fish species. Natural selections, are, however, productive in the evolutionary process, in that, they contribute to the continuance of life; but, man-induced extinctions are negative without any succession and without contributing to biological diversity (Kar, 2003 b).

Probable Causes of Decline of Mahseer Fishery

Concerns for depletion of Mahseers have been voiced from different parts of India since few decades. Raj (1945), Pathani (1981) reported about decline of Mahseers from UP; while Sehgal (1970) reported about the status of Mahseers in Jammu province. Nautiyal and Lal (1981) projected about declining trend of Gangetic and Himalayan population of Mahseers. Mahseer population of Mizoram rivers is also on decline (Kar *et al.*, 2002, a, c). Probable causes of depletion are given below:

1. Drawbacks in general planning and developmental policies, like construction of dams, barrages and roads along the river course.
2. Introduction of exotic species causing threat to the indigenous fish germplasm.
3. Improper management practices in the waterbodies.
4. Lack of infrastructural facilities for *in situ* conservation.
5. Lack of public awareness with regard to destruction of forest in the catchment area
6. Extraction of sand and stone from the river bed.

Suggestions for the Protection of Mahseers

Habitat loss and environmental degradation has seriously affected the fish and wildlife resources in India. Knowledge of available resources and biological characters of a species serves as the baseline information for taking-up further studies and steps on resource management and conservation. Although some works have been done on the game fishes of India (Menon, 1988); scattered information are available on the Mahseer fishes of NE India. And, virtually, no information is available on the conservation of Mahseer fishes in Mizoram. As such, it is strongly suggested to declare certain potential sites in river Mat, having favourable habitat characteristics, as 'aquatic sanctuaries' in order to protect and conserve the germplasm of these Mahseer fishes.

Concomitant to above, in river Jatinga in Assam, practically no effort has been made to protect *Tor progenius* from extinction. Although the riverine habitat of Jatinga, with the above-mentioned characteristics, seems to be quite suitable for *Tor progenius*, harmful fishing methods, like the use of banned nets and explosives, are causing a huge toll of the fish, including the juveniles and also degrading the breeding grounds. Harvest of cobbles and boulders from the river substratum affect the habitat of these fishes; thereby, adds to their deteriorating trend. This has been posing serious threat to its survival and conservation. Efforts, like proposing creation of 'aquatic sanctuaries', are being made to protect *Tor progenius* (McClelland) from extinction.

Chapter 10

An Account of the Fish Yield from Water Bodies in Barak Drainage of Assam, in Mizoram and in Tripura

A conservative survey conducted by North Eastern Council (NEC) (Govt of India) resulted in the following informations: 172 fish species belonging to 73 genera and 27 families have so far been recorded from NE Hill region. Of these, 33 representatives are quite common to Indo-Gangetic fish stock; 19 representatives are common with the fauna of the Eastern Himalayas and 11 representatives are common with the Myanmarese fauna. 23 fish species are said to be endemic in this region (NEC, 1991).

Satewise Data on Fish Production

Assam

Total fish production in Assam during 1986-87 was 52,477 metric tonnes (mt); while, during 1987-88, it was estimated at 60,990 mt. The resourcewise contribution is as follows:

Resources	*Production (mt)*	*% of Total Production*
Rivers	24300	46.30
Beels	21000	40.02
Reservoirs	110	0.21
Swamps/lowlying areas	1000	1.91
Ponds and Tanks	6067	11.56

Mizoram

The total estimated fish production in Mizoram from various sources during 1987-88 was 2400 mt. The contribution from various sources is given below:

Resource	Production (mt)
Ponds and Tanks	1920
Rivers	450
Paddy-cum-Fish culture	30

Tripura

The fish production in the province increased from 10,978 mt in 1985-86 to 14,012 mt during 1987-88. Fish production during 1988-89 from various waterbodies in the province, as shown in the State Plan Document, is presented below:

Resource	Total Production (mt)	Yield Rate
Rivers	770.0	641.67/km
Reservoirs	337.5	75.0 kg/ha
Ponds/Mini barrages/ Govt.Co-op farms	15592.5	1519.0 kg/ha

An account of the fish yield from various water bodies in each of the above-mentioned provinces is briefly given below:

Fish Yield from Some of the Waterbodies in Assam

Lentic

Sone Beel

The data for annual fish yield were extrapolated from daily catch statistics recorded at all the landing stations of the Beel (FAO, 1974) while the trend and cyclic variations were constructed by applying 12 month's moving average method (Coxton and Cowden, 1950).

Of the 70 ichthyospecies, 84.2 per cent belong to the Primary FW group while the rest are of peripheral class. The annual fish yield from the Beel was determined as 358.21 mt during 1979-80 and 312.16 mt during 1980-81 having an annual average yield of 355.18 mt. Per hectare fish yield was determined as 103.5 and 90.26 kg respectively during the investigating years (Kar, 1990). The present data show a declining trend revea'ing an annual average yield of *c* 315.0 mt (Kar, 2003 a). Inconsistent trend of natural recruitment in such water bodies (Dey, 1981) coupled with higher catches in the turbid water might be the causative factor for higher yield during the first year. The forage fish containing mainly the cyprinids and chandids which constitute the bulk landing, could have gained ascendance over the predatory catfish to add further reasons to such propositions. The fish yield from Sone Beel appeared to be low as compared to 250 kg/ha obtained by Vaas and Schurman (1949) in some tropical FWs of Java. Further, Jhingran and Tripathi (1969) had recorded the range of fish yield from the Indian reservoirs to be as low as 6-7 kg/ha/year.

The constant deposition of 'silt' expose the fishes in the shallower areas to shortage of living space (Welcomme, 1979). This has impact on the production through an eventual diminution of the water spread area (Kar, 1990). Moreover, the lake soil with pH 5.0 to 5.9, organic carbon 0.25 to 1.74 per cent, available phosphorus 0.15 to 1.93 per cent; and, water with pH 6.0 to 7.9 and dissolved oxygen 2,4 to 5,9 mg/lit were not very suitable for fish production (Banerjea, 1967; Michael, 1969). Also, dense strands of aquatic macrophytes, often known to bind up nutrient materials, thus, could result in reduction of natural food availability for the fish and lead to poor fish yield (Bennet, 1962). Further, as an attempt to increase fish yield (Fernando and Furtado, 1975) by introduction of *Cyprinus carpio* into the Beel, there have been adverse effect on the autochthonous ichthyodenizens of the Beel which could well account for overall poor yield from Sone Beel (Dey and Kar, 1990).

Of the 70 fish species, only *Puntius chola, Parambassis baculis, Chanda nama* and *Gudusia chapra* contained Annual Average Yield (ARY) more than 10 per cent. Amongst them, *Puntius chola,* the minnow, contributed to the bulk of the fish landing from the Beel registering an ARY of 21.39 and 27.82 per cent during two successive years; thus, portraying its significant contribution to the overall fish yield from the Beel. It is significant to note here that, the Indian major carps (IMC) of the Beel consists of five distinct species depicting ARY (as a group) of 21.39 and 27.82 per cent respectively during two successive years. Of them, *Labeo rohita* alone portrayed an ARY of 0.23 and 0.24 per cent during two successive years, which, incidentally, was the highest amongst the species of the group. Non-judicious and irrational harvest of the juvenile IMCs in greater proportions have caused imbalance in the IMC fishery of the wetland culminating in overall poor contribution of this group in the total fish yield of the lake. Concomitant to IMCs, the Indian shad, *Hilsa(Tenualosa) ilisha*, with an ARY of 0.04 per cent (which is admittedly low) revealed a single run in Sone Beel during the monsoon against two runs in other water bodies of India. Hilsa is generally more prevalent in the NE, SE and SW zones of the Beel, probably, due to fluviatile nature of these zones. Indeed, the day-in and day-out fishing operations could spare very little scope for the recruited shads to make their return journey. *Cyprinus carpio* was the only exotic carp recorded in the Beel with an average ARY of 0.155 per cent. Among the other carps and minnows, *Puntius sarana sarana*, *Rasbora daniconius* and *Salmostoma bacaila* depicted ARY of 0.085, 0.63 and 0.93 per cent respectively. Among the cobitids, which significantly includes the 'ornamental fishes', *Botia dario* accounted for 0.485 per cent of the average ARY; while, *Lepidocephalus guntea* dominated the cobitids with an ARY of 0.97 per cent.

The 'C'-class fish (Swingle, 1950) of the Beel usually contain ARY of 1.75 and 4.83 per cent collectively. However, of all the predatory fishes, *Wallago attu* alone contributed to an average ARY of 2.94 per cent as, revealed from the study (Kar, 1990). Nevertheless, *Mystus vittatus* dominated the predatory fishes with an average ARY of 7.75 per cent. Fishes like *Aorichthys seenghala, Mystus cavasius, Ompok bimaculatus* and *Eutropiichthys vacha* portrayed high commercial value and depicted average ARY of 0.075, 0.03,0.77 and 0.09 per cent respectively. The live catfishes include *Clarias batrachus* and *Heteropneustes fossilis* which exhibit average ARY of 0.11 and 1.81 per cent respectively. The bottom-dwelling murrels, which usually contribute significantly to the total fish yield of the lake, contain species like *Channa gachua, C.punctata* and *C. striata* with average ARY of 0.27, 7.0 and 0.24 per cent respectively. *Channa marulius* is generally a rare species in Sone Beel with an ARY of only 0.015 per cent.

Among the perches, *Anabas testudineus* depicted an ARY of 1.89 per cent as revealed from the study (Kar, 1990) and it fetched high market value. The perches also include two ornamental species of fishes, *Colisa fasciatus* and *C.lalia* with ARY of 1.28 and 0.245 per cent respectively. The chandids generally constitute the bulk of the trash fishes with two species, *viz.*, *Parambassis baculis* and *Chanda nama* with average ARY of 7.41 and 6.5 per cent respectively being usually portrayed. Interestingly, *Glyptothorax telchitta* was found to be the only true rheophilic fish which is supposed to have been swept into the Beel from higher gradient zones of the inlet and depicted an ARY value of 0.10 per cent. There are different kinds of eels in Sone Beel. However, *Mastacembelus armatus*, *Macrognathus pancalus* and *Macrognathus aral* have been found to be the most abundant with ARY of 0.78, 5.57 and 1.07 per cent respectively (Kar, 1990).

Average ARY of different Fish Groups in Sone Beel is briefly tabulated below:

Fish Group	*Average ARY (%)*
Featherbacks	0.27
Clupeids	8.30
Carps and Minnows	29.93
Loaches	1.68
Bagrids	7.94
Silurids	3.72
Schilbeids	0.88
Pangasiids	0.0011
Sisorids	0.015
Live cat fish	1.92
Chacids	0.025
Needle fish	0.28
Killi fish	0.03
Murrels	7.53
Glass fish	13.91
Perches	7.45
Mullets	0.045
Gobies	6.31
Eels	7.43
Puffer fish	0.94

Chatla Haor

57 species of fishes, belonging to 28 genera, 17 families and 9 orders, have been recorded in Chatla Haor. Species diversity by Shannon-Weaver Index revealed high diversity of the fish species of this wetland (H=–6.15) (Kar and Barbhuiya, 2000 b).

Of the 57 species, as revealed from our studies (Kar and Barbhuiya, 2000 b), Cypriniformes was found to constitute 32.3 per cent of the total fish population followed by Channiformes (22.8 per cent), Siluriformes (14.02 per cent), Clupeiformes (10.52 per cent), and Osteoglossiformes (0.00029 per cent). Family Notopteridae showed the least abundance (0.00064 per cent). Interestingly, among the clupeids, occurrence of *Hilsa (Tenualosa) ilisha* (0.000041 per cent) is a remarkable feature in the distribution of the species in freshwater. Further, occurrence of advanced fry stages of hilsa (45.5 to 128.0 mm) is an exceptional feature in the zoogeography and biology of the fish. On the other hand, yield of *Gudusia chapra* revealed an alarming trend of decline in this wetland. Furthermore, occurrence of juveniles of Indian major carps (IMC) in Chatla indicate the possibility of this wetland serving as a breeding ground for these large growing fishes (Kar and Dey, 2002).

Other Lentic Systems in Barak Valley Region of Assam

The 'Anuas'are potential oxbow wetlands having, in general, rich population of *Gudusia chapra* and *Salmostoma bacaila* both of which together constitute *c* 45 per cent of the total fish yield from these oxbow wetlands. On the other hand, a comparative study of a floodplain wetland (Puneer Haor) and an Oxbow wetland (Sibnarayanpur Anua) revealed Orderwise distribution of ichthyospecies as follows: Osteoglossiformes, 1 and nil species; Clupeiformes, nil and 1; Cypriniformes, 7 and 11; Siluriformes, 6 and 4; Synbranchiformes, 1 and 3; Beloniformes, 1 and 1; Cyprinodontiformes, 1 and nil; and, Perciformes, 7 and 2 (Kar *et al.*, 2002 d). In wetlands, like Hathichhara Beel, Doloo Beel, Sumeri Beel, Karkari Beel, Jonamara Beel, Sushka Beel, Tapang Beel, Angang Beel, Sagar Beel, Rani-Meghna Beel, Deochhara Beel, etc., the small trash fishes constitute *c* 83 per cent of the total yield from the wetland. However, in Rata Beel, which is situated adjacent to Sone Beel along its northern boundary, the trash fishes compose *c* 73 per cent of the total yield from the wetland (Kar, 2004).

Lotic

River Barak in its Assam region used to contain a rich diversity and population of ichthyospecies like *Chitala chitala, Hilsa (Tenualosa) ilisha, Labeo rohita, Catla catla, Salmostoma bacaila, Aorichthys seenghala, Wallago attu, Ompok bimaculatus, Ailia coila, Eutropiichthys vacha, Clupisoma garua, Nangra nangra,* etc.about 50 years ago from now. *Mystus cavasius* and *Bagarius bagarius* also had a moderate population. These fishes used to constitute *c* 42 per cent of the total fish yield from the river. Today, they are few and far between. Their population has declined to an alarmingly low level of *c* 9 per cent of the total fish yield (Kar, 2004). Situation had almost been the same in other major rivers of Barak drainage, notably, Jatinga, Sonai, Dhaleswari, Jiri, Chiri, Madhura, Baleswar, Singla, Longai, Kushiara, Surma, etc. registering a declining trend in fish diversity and yield (Kar, 2004).

Fish Yield from Some of the Waterbodies in Mizoram

Lentic

Fish yield from the three major lakes in Mizoram, *viz.*, Tum Dil, Rung Dil and Polok Dil, have not been very high. The trash fishes generally dominate the catch registering *c* 77 per cent of the total yield from the lakes (Kar, 2004).

Lotic

There are 7 major rivers in Mizoram, *viz.*, Tuirial, Tlawng, Tuirini, Tuivai, Mat, Kolodyne and Karnafuli. As indicated earlier, all these rivers had a rich diversity and population of fishes about five decades ago containing quite a rich variety and number of LGF. There had been alarmingly declining trend of ichthyospecies in almost all these rivers. The trash fishes today account for *c* 77 per cent of the total fish yield from the rivers in Mizoram. The rivers are said to be sometimes stocked with IMC (Kar.2003 a; Kar, 2004).

Fish Yield from Some of the Waterbodies in Tripura

Lentic

The tanks constructed by the ancient Monarchs ('Maharaja') of Tripura constitute the principal lentic systems in addition to the natural Beels. The important tanks are Rudra Sagar, Fatik Sagar and Amar Sagar. They have a fairly big waterspread area; but, today, the trash fishes contribute significantly to the overall yield from the lentic bodies upto the extent of *c* 74 per cent on average. In addition to these, the Dombaru Reservoir formed by damming the river Gomati contains rich population of trash fishes contributing to *c* 77 per cent of the total yield from the reservoir (Kar, 2004).

Lotic

The principal rivers, *viz.*, Monu, Khowai, Gomati, Feni and Howrah. contain a rich diversity of fishes. Trash fishes constitute *c* 84 per cent of the fish yield from these rivers on an average. The rivers are said to be sometimes stocked with IMC (Kar, 2004).

Chapter 11

An Account of the Indian Shad in the Barak Drainage of Assam

Hilsa (*Tenualosa*) *ilisha* (*Hamilton-Buchanan*) is undoubtedly one of the important edible fishes in the Indian Sub-continent. It is essentially an estuarine fish of the 'Herring' group. It is called the 'Indian shad' and it periodically ascends the freshwaters. Hilsa, locally known as 'ilish', forms an important fishery in the Brahmaputra and Barak drainages of Assam. Hilsa is known to be distributed along the Indian coasts and also to ascend the large rivers to breed (Day, 1889). However, distribution of hilsa ranges from the Persian Gulf in the West to the Bay of Bengal in the East and it is also known to ascend the large rivers in the Middle East. India, Bangladesh and Myanmar to breed (Hora, 1940; Dey and Kar, 1989b; Kar and Dey, 1982, a, b; Yadava *et al.*,1989).

There have been many studies done on the hilsa in the waterbodies of India. Some of the significant works are cited below:

Ahmad (1954); Anon (2000); Bhanot (1973); Bensam (1987); Chacko (1954); Chacko and Ganapati (1949); Choudhury *et al.* (1990); Day (1873); De (1996, 1999); De and Datta (1990); De and Sinha (1997); De *et al.*.(1998, 1999); Dey (1981); Dey and Kar (1989 b); Doha and Hye (1971); Dutta Munshi (1960); Dutt (1966); Dunn (1992); Ghosh and De (1999); Hora (1938, 1940, 1941); Jayaram (1981, 1999); Jenkins (1910); Jones and Sujansingani (1951); Jhingran (1957); Kar

(1990, 2001); Kar and Dey (1982 a,b, 2002); Kulkarni (1950); Malhotra (1969); Mitra and Devasundaram (1954); Motwani *et al.*. (1957); Menon (1974, 1999); Nair (1939, 1954); Pillay (1959); Prasad *et al.* (1940); Pal (1963); Quddus (1982); Rajyalakshmi (1973); Rao and Pathak (1972); Swarup (1961); Shafi *et al.* (1978); Southwell and Prasad (1918); Sinha *et al.* (1998); Sinha and De (1996, 1999); Yadava *et al.* (1989). Further, more specifically, occurrence of hilsa in Chilka lake had been reported by some workers, notably, Choudhuri(1916), Jones and Sujansingani(1951), Mitra and Devasundaram(1954), Ramakrishnaiah(1972). Hora (1941) had reported on the bionomics of the Indian shad in Bengal waters; while, Chacko and Ganapati (1949) studied biology of hilsa from river Godavari.

Notwithstanding the above, investigations on hilsa from Assam waters, particularly from the seasonal and perennial wetlands (locally called 'Haor' and 'Beel' respectively) had not been many; and, are restricted mainly to the works of Pillay and Ghosh (1958), Pillay *et al.* (1962), Rao and Pathak (1972), Dey (1981), Dey and Kar (1989 b), Yadava *et al.* (1989) and Choudhury *et al.* (1990). Kar and Dey (1982 a,b), for the first time, reported the occurrence of *Hilsa* (*Tenualosa*) *ilisha* in Sone Beel of Assam.

The data for annual yield of hilsa were extrapolated from daily catch statistics recorded at all the landing stations of the Beel (FAO, 1974) while the trend and cyclic variations were constructed by 12 month's moving average method (Coxton and Cowden, 1950).

The hilsa fishery of Sone Beel is not a detached and isolated one as the crop is dependent on the stocks in the adjoining rivers. The present investigation reveals a single run of hilsa. The fish enters the lake in shoals on the advent of monsoon unlike in other lentic (Jones and Sujansingani, 1951) and lotic (Pillay, 1958) bodies of India. The total landing of hilsa from Sone Beel was 163.5 kg during 1979-80 and 220.7 kg during 1980-81, the trend, thereby, showed per hectare yield of 0.055 kg during the investigating period. Concomitantly, the Monthly Relative Yield (MRY) (per cent) recorded during the first year was 0.04 (Apr), 0.028 (Jul), 0.11 (Aug), 0.01 (Sep) and 0.004 per cent (Oct). In the following year, the MRY values were 0.07 (May), 0.09 (Jun), 0.06 (Jul), 0.04 (Aug) and 0.37 per cent (Sep). The Annual Relative Yield (ARY) was 0.04 and 0.07 per cent during the first and second years respectively (Dey and Kar, 1989 b). Recent studies revealed further decline in yield of hilsa to the tune of 129 kg during 1999-2000 and 117.2 kg during 2000-01 (Kar, 2001). Further, the hilsa entering Sone Beel generally vary in size from 23 -60 cm.

Hilsa was found to be comparatively more prevalent in the SE, SW and NE regions of the Beel (Figure 4), probably, due to fluviatile nature of these areas. Unusual turbidity of water may not always be considered as a criterion for increased rate of natural recruitment of shad (Leach, 1925) as Sone Beel exhibited a good population of the fish within a turbidity range of 20.5–83.6 TU. Indeed, wind velocity and its direction coupled with ingress of water through the perennial channels (Figure 4) could be the influential factors for higher rate of recruitment of this anadromous fish generally during the months of July, August and September. However, temperature was not found to have any direct impact on the migratory behaviour of the shad. Nevertheless, their occurrence in Sone Beel only during the period from April to October indicated that the fish preferred to live in the Beel during the hot season when its water is warm.

Field investigations, further, revealed that, with the onset of rains, shoals of hilsa enter the Beel, generally, through the outlet (Figure 4) when they are caught in appreciable number without any discrimination of size. Subsequently, intensive fishing in the Beel does not generally spare most of the stragglers to perform the return journey.

Chapter 12

Occurrence of Advanced Fry of *Hilsa* (*Tenualosa*) *ilisha* in the Wetlands of Barak Valley Region of Assam

The anadromous fish hilsa, portrays a single run in the wetlands of Barak valley region of Assam (Kar and Dey,1982 a,b). Occurrence of advanced fry of hilsa in the Beels and Haors of Assam had not been reported by any of the earlier workers. The following is a brief account of the occurrence of advanced fry of hilsa in the Chatla Haor (Plate 2) wetland situated *c* 10 km away from Silchar city.

Advanced fry of hilsa (Plate 15) has been collected from Chatla Haor through experimental fishing operations from August 1995 to July 1998 with the help of 8910 m^2 encircling gear (locally called 'Mahajal') having mesh size of 1.0 mm^2. The fish specimens were preserved in 10 per cent formalin and were identified through morphometric measurements and meristic counts after Day (1878, 1889), Jayaram (1981, 1999); Menon (1974, 1999). The physico-chemical characteristics of water have been estimated after APHA (1995).

Data pertaining to morphometric and meristic measurements and physico-chemical parameters of water have been presented in Table 1 and Table 2 respectively.

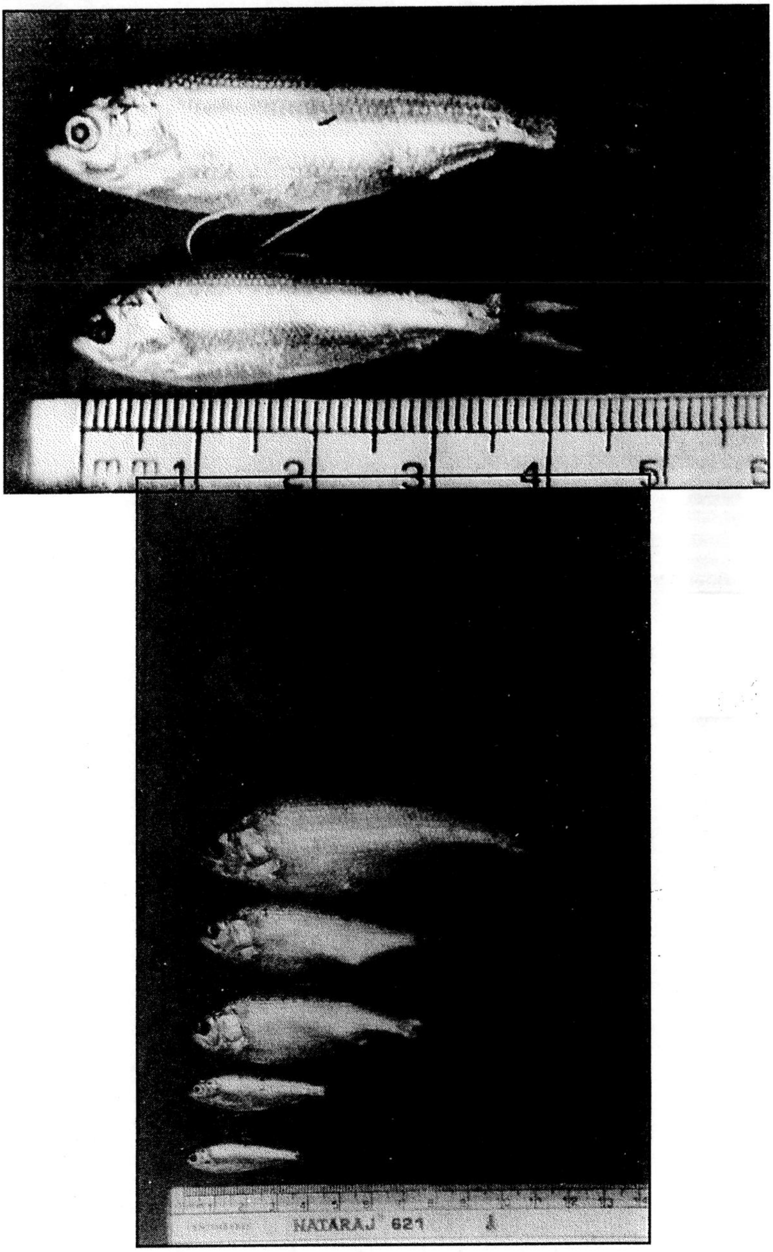

Plate 15: Advanced Fry of Hilsa in Chatla Haor in Assam

Table 1: Morphometric Measurements and Meristic Counts of *Hilsa (Tenualosa) ilisha* Collected from Chatla Haor in Assam During 1995-98

Parameter	*Measurement*
Total length (mm)	45.5–220.0
Standard length (mm)	37.0–176.0
Pre-dorsal length (mm)	16.5–85.0
Caudal peduncle depth (mm)	3.5–7.0
Caudal fin length (mm)	44.0
Pre-pelvic scutes	15
Post-pelvic scutes	14
Pectoral fin rays	15
Pelvic fin rays	7
Dorsal fin rays	17
Anal fin rays	20
Branchiostegal rays	8
Lateral line scales	45

Table 2: Physico-chemical Characteristics of the Water of Chatla Haor Wetland During the Period 1995-98 (average values) (study months: May to October)

Parameter	*Average value*
Temperature (°C)	33.3
Turbidity (NTU)	83.27
pH	6.09
Dissolved oxygen (mg/lit)	5.75
Free carbon di-oxide (mg/lit)	6.59
Total Alkalinity (mg/lit)	83.39
Conductivity (micromhos/cm)	142.91

The knowledge of minimal size of fish at its first maturity is of great significance which may be utilized for the development of the fishery (De, 1991). Jenkins (1910) recorded a single young hilsa in Monghyr market and tried to reveal the possibility of hilsa breeding ground in the river Ganges in the vicinity of Monghyr town (Hora,

1940). Spawning activity of hilsa brings it to the fore, cleavage of opinion with regard to its spawning habit, duration of spawning season and the location of spawning grounds. There could be several spawning grounds in Hooghly and Ganga rivers which could change from year to year (De, 1991).

Southwell and Prashad (1918) had reported the occurrence of hilsa in the waterbodies in Bengal, Bihar and Orissa. Finlow (1920) had opined that the probable spawning grounds of hilsa could be occurring in Eastern Bengal.

Contrary to this, Chatla Haor is a seasonal floodplain wetland with no significant fluviatile condition existing in it. However, advanced fry of hilsa ranging from 45.5 to 220.0 mm have been recorded in Chatla Haor. Further, the advanced fry collected from Chatla Haor conformed to the diagnostic characters of hilsa (Table 1). Further, mature hilsa swarms have also been recorded in this Haor between June and September. Study, elsewhere (De *et al.*, 1994), revealed that, upstream migration of hilsa mainly occurs during the monsoon season.

Hora (1940) had collected hilsa from shallow waters while Bensam (1987) had collected post-larvae from a depth of *c* 3 m. Availability of both mature and juvenile hilsa in Chatla Haor, only during the period from June to October every year, suggests a single run of the fish in this lentic habitat; thus, conforming earlier studies conducted by the present author in Sone Beel (Kar and Dey,1982 a, b; Dey and Kar, 1989 b). Also, the physico-chemcial characteristics of Chatla Haor (Table 2), having similarity with that of Sone Beel (Dey and Kar, 1987), suggests that, hilsa can live even in lentic condition which is quite different from the characteristics of lotic system where two runs of hilsa have been reported by many workers. Nevertheless, it is assumed that, flood and seasonal maturity of the migrants seemed to induce the spawning hilsa in the sea to undertake upward journey into the rivers (Hora, 1940). Occurrence of mature hilsa, as far as near the Assam-Manipur border in the upstream of river Barak, supports the previous observation (Kar, 2001). Our study is also in conformity with the occurrence trend of hilsa juveniles in the Palta Waterworks in West Bengal (Hora, 1940), who had opined that, hilsa could enter the settling tanks of the Waterworks through the water intake pipes either in the egg stage or as very young larvae. It is interesting to note here that, fishermen of Chatla Haor (most of them are immigrant 'Kaibarta' from the earstwhile East Pakistan) use the term 'Jatka' to denote the advanced fry of hilsa (Finlow, 1920). Studies elsewhere (De *et al.*, 1989) portrayed that the hilsa juveniles generally start their downstream migration after attaining a size of 80-110 mm while upstream migration has been found to be associated mainly with the state of sexual maturity supported by factors, like rainfall, temperature, water current, etc.

Notwithstanding the above, hilsa forms a lucrative fishery also in the Brahmaputra river system. Study revealed that, hilsa occupied, in order of abundance, the foremost position in the overall fish landing records from the river Brahmaputra at Dhubri (5427 to 12045 kg/year), moved to second position at Guwahati (1980 to 8648 kg/year at Uzan Bazar and 3796 to 15,967 kg/year at Fancy Bazar) and ranked third at Tezpur (1635 to 3282 kg/year). A sharp decline in the commercial catches of hilsa have been registered (trend equation: $X_1 = 448.1 - 979.3\,t + 72.5\,t^2$) from the river Brahmaputra in the recent past (Yadava *et al.*, 1989).

Chapter 13

An Account of Indian Major Carps in the Beels of Assam

The Indian major carps (IMC) constitute an important aspect of fishery in Brahmaputra and Barak drainages of Assam. They are distributed throughout the Indian Freshwaters and ascend the large rivers and floodplains to breed (Jayaram, 1999). Their occurrence in the Ganga river system was reported, among others, by Jhingran and Chakraborty (1958); in the East coast system by Chacko *et al.* (1954);and, in the West coast system by a number of different workers (Anon, 1964). However, investigations on IMC from Assam waters were not many (Motwani *et al.*, 1982; Dey, 1981; Kar, 1990). The present account portrays the yield pattern and abundance trend of the IMC (*Labeo calbasu, L. gonius, L. rohita, Cirrhinus mrigala* and *Catla catla*) in Sone Beel, the biggest Beel in Assam.

The data on annual fish yield of IMC have been extrapolated from daily catch statistics recorded at all the landing stations of the Beel (FAO, 1974) while the trend and cyclic variations have been constructed by 12 month's moving average method (Coxton and Cowden, 1950).

The IMC fishery of the Beel is not an isolated one as its crop is dependent on the stocks present in the adjoining rivers. The total landing of IMC from Sone Beel was 3,184.4 kg (0.86 per cent of the total fish yield) during 1979-80 and 1958.65 kg (0.61 per cent of the total fish yield) during 1980-81; the trend, thereby depicted average per hectare yield of 0.74 kg during investigating period. The species-wise average Monthly Relative Yield (MRY) recorded during 1979-80 was 0.06, 0.014, 0.25, 0.37 and 0.22 per cent respectively for *C. catla*, *C.mrigala*, *L. calbasu*, *L. gonius* and *L. rohita;* while, those recorded during 1980-81 was 0.07, 0.13, 0.26, 0.076 and 0.36 per cent respectively. Concomitantly, the species-wise Annual Relative Yield (ARY) successively was 0.015, 0.02, 0.21, 0.39 and 0..23 per cent during 1979-80 and 0.01, 0.13, 0.16, 0.07 and 0.24 per cent during 1980-81 (Kar and Dey, 2000). However, there has been a declining trend of IMC yield from Sone Beel reflecting an annual yield of 1257.2 kg during 1999 and 1162.7 kg during 2000 (Kar, 2004). The IMCs entering Sone Beel generally vary in size from 19.0-66.8 cm for *C. catla*, 12.9- 65.0 cm for *C. mrigala*, 15.0-54.3 cm for *L. calbasu*, 6.8–50.0 cm for *L. gonius* and 12.0–78.5 cm for *L. rohita.*

The Indian major carps have been found to be more prevalent in the SE, SW and NE regions of the Beel, probably, due to fluviatile nature of these areas. The inflow-stimulated upstream migration during the spawning ascent generally results in a moderate population of the group during the monsoon months; thus, indicating that physiological urge of the fish to breed in such shallower belts adjoining the rivers could be a prerequisite for the stock. Unusual turbidity of water may not always be considered as a criterion for increased rate of natural recruitment of IMC, as Sone Beel exhibited considerable population of the fishes within a turbidity range of 20.5-83.6 TU (Dey and Kar, 1987). Indeed, direction, coupled with ingress of water through the perennial channels could be the influential factor for higher rate of recruitment of these species of fishes during monsoon. Temperature was not found to have any direct impact on the movement of IMCs. Their occurrence in Sone Beel, almost throughout the whole year, without much impact being caused by temperature, lends support to this view. Field investigations, further, revealed that, with the onset of rains, shoals of IMC enter the Beel through the outlet called river Kachua (Fig.4). During this time, they are caught in an appreciable number without much discrimination in size. Such irrational catches may sometimes cause imbalance in the IMC fishery of the Beel leading to depletion of the IMC population in the wetland. This also results in poor contribution of this important group to the overall fish yield of Sone Beel.

Chapter 14

Interrelationship and Dynamics of Fish Population in the Beels of Assam

Fishes in a waterbody may be herbivorous, carnivorous, omnivorous or exclusively planktophagous and they maintain some kind of relationship among themselves. Swingle (1950) evolved the techniques for determining the interrelationships and dynamics of balanced and unbalanced fish populations through periodic determination of changes in selected populations. He considered items like F/C and Y/C ratios; and A_T, A_F, I_F and S_F values. The present chapter portrays an analysis of the interrelationship and dynamics among the 70 ichthyospecies (Dey and Kar, 1990) in 3,458.12 ha Sone Beel situated in the Barak valley region of Assam (Kar and Dey, 1993).

As stated earlier, the data for annual fish yield were extrapolated from daily catch statistics recorded (FAO, 1974) at all the landing stations of Sone Beel. The trend of cyclic variations were constructed by 12 months' moving average method (Coxton and Cowden, 1950). The interrelationship and dynamics of fish population were analysed after Swingle (1950).

Among the 70 fish species recorded in Sone Beel, 23 species were found to belong to C-class and the rest to the F-class constituting respectively 13.16 and 86.84 per cent of the total yield respectively (Dey and Kar, 1990). The F/C values, 3.79 and 4.70, obtained during successive sessions, appeared to be the most desirable in the balanced range (Swingle, 1950). It explains that occasional exploitation of the C-class at a little more vigorous rate might have allowed the movement of F-fishes from the group of small to intermediate stage and finally to large-sized fishes. This could be the reason for increase of population of F-class as a whole. The Y/C values, 7.43 and 9.68 successively, showed unbalanced nature and indicated their incapability of producing yearly crops of harvestable fish because of overcrowding with forage fishes. The A_T values and 33.42 per cent overlap the ranges of balanced and unbalanced populations and the population is said to be less efficient, in that, relatively large percentage of their weights are of small and intermediate fishes.

Analysis of A_F, I_F and S_F values respectively portray as 2.32, 20.7, 76.75 per cent and 1.64, 17.63, 80.71 per cent during two successive years of studies. A_F reveals unbalanced status because bulk of the F-class fishes consist of small and intermediate sizes. The I_F values overlap in the present case because of movement of individuals from this group to the large fish group due to increased food available to the former. Further, as a result of increased growth and predation in the S-group, replacements move up into the I-group. As a result of harvest of part of F-species and accompanying increase in food availability for A_F group, more rapid growth and increased reproduction results. Since the S_F fishes subsequently increase in number, and because they have available larger percentage of food supply, the weight of this group tends to increase.

The foregoing account, thus, portrays an overall mixed status of the interrelationship and dynamics of fish population of Sone Beel in respect to the various parameters.

Chapter 15

Fish Catching Devices in Barak Drainage (Assam), in Mizoram and in Tripura

Fish catching devices, popularly called 'Fishing gears and crafts', are the implements to harvest fishes from a water body. These exhibit bewildering diversity globally, nationally, regionally and locally. Variations also exist according to fish type, fishermen type and season type.

The efficacies of the gears have been determined on the basis of fish catch (kg/person/gear/hour: CPGH) (Dey, 1981); while, wherever necessary, per cent of fishing communities operating a particular kind of gear has been evaluated from field survey. Attempts have been made to calculate Mechanisation Index (MI) (Gadgil and Reddy, 1989), Hanging Co-efficient (HC), and Niche' Width (NW) (Smith, 1986) from the field data (Kar and Dey, 1991, 1993, 1996; Kar *et al.*, 1999). ANOVA tests, wherever done, indicated that, there exists highly significant differences between the different kinds of scooping gears with regard to their CPGH (Kar and Dey, 1996).

An account of the fishing gears and crafts, in vogue, in Barak drainage (Assam), in Mizoram and in Tripura is briefly given below:

In Barak Drainage (Assam)

The province of Assam is gifted with myriads (Kar, 1984) of riverine and tectonic wetlands and floodplains, locally called 'Beels' and 'Haors' respectively, which cover a total water spread area of 0.12×10^6 ha (Dey, 1981). The fish catching devices play important role in the commercial exploitation of the wetland and riverine fisheries of the province. Various types of fishing gears are operated in lentic and lotic systems in this region. They are briefly described below:

In Lentic Systems

Differences are found in the fishing gears operated in various lentic bodies.

In Sone Beel

Sone Beel is the biggest wetland in Assam. 26 different types fishing gears under 7 classes (FAO, 1963; Brandt, 1980) have been recorded in Sone Beel (Kar, 1990; Kar and Dey, 1991, 1993, 1996). Also 4 types of crafts under 2 categories have been found in operation with the gears. The smaller nets are generally knit by the local fishermen according to their preferences and methods of fishing. However the size of the boats generally vary according to the nature of fishing and the money available for her construction. Sone Beel is one of very few wetlands in India in which 26 different types of fishing gears have been recorded in a single wetland (Plates 16-19).

The various types of gears and crafts along with their salient features are shown in Table

Fabrication, *Modus Operandi* and Catch Record of the Gears

Hook and Line gears

Lar Barshi [Plate 16(2)]

Metallic hooks of various sizes are tied to a coconut coir, *c* 400-600 m in length, by *c* 45 cm long nylon strings at an interval of about a metre. The dry weight of Lar barshi generally varies from 5-6 kg. Both the ends of the nylon line are fixed to the Beel bottom with two bamboo poles to allow the baited hooks hang nearer the Beel bottom.

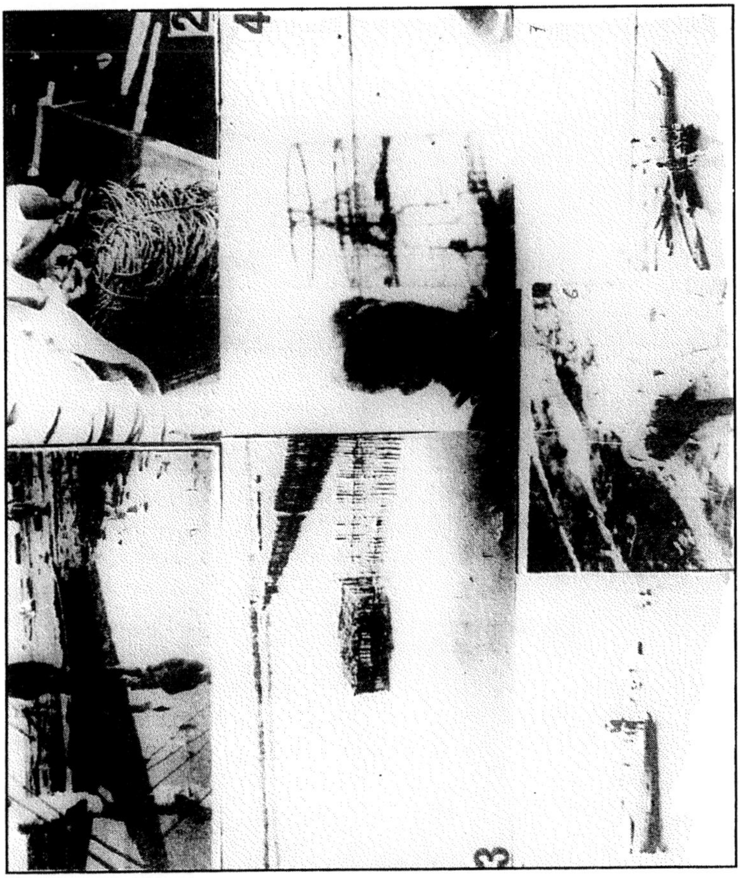

Plate 16: Fish Catching Devices in Sone Beel in Assam

It is lifted from a canoe by a person after 4-8 hours of its placement in the lake and hooked fish, if any, are collected into the boat.

Table 1: Diversity of Fish Catching Devices in Sone Beel in Assam

Category	*Gear (Local Name)*	*Size (m)*	*Mesh Size (mm)*
I. Hook and Line gears	1. Lar barshi	L. 400–600	—
	2. Kupa barshi	L. 450–500	—
	3. Tanga barshi	L.350–450	—
II. Traps	4. Dori	H. 0.45–1.35	Bamboo slits 1 mm apart
		W. 0.30–0.60	
	5. Gui	H.0.90–2.00	Bamboo slits 10 mm apart
		W. 0.75–1.00	
	6. Paran	H. 0.20–0.40	Bamboo slits 7 mm apart
		W. 0.10–0.14	
	7. Khati bundh	H. 2.00–3.00	Dense bamboo mat
		L/unit 4.50	
III. Trawls	8. Chhatjal	L. 3.50–4.50	Upper part: 4.0
		W. in front	Lower part: 1.0
		2.00–3.00	
	9. Pelain	L. 2.50–3.00	1.0–5.0
		W. in front	
		1.00–2.00	
IV. Scooping gears	10. Dheki jal	L. 8.00–12.00	Wider part: 50.0
		W. in front	Narrower part: 6.0
		7.0–9.00	
	11. Darar jal	L. 4.50–5.00	5.0
		W. 4.00–5.00	
	12. Kuchrung jal	L. 1.25–1.50	5.0
		Dia: 1.00–1.15	
		Cir: 3.00–3.50	
V. Entangling gears	13. Patan jal:		
	Big variety	L. 300–400	46.0
		H. 4.00–4.50	
	Medium variety	L. 300–390	19.0
		H. 3.00–3.50	

Contd...

Table 1–Contd...

Category	*Gear (Local Name)*	*Size (m)*	*Mesh Size (mm)*
	Small variety	L. 300–400	15.0
		H. 1.50–2.00	
	14. Haran jal	L. 4.50–6.50	Upper part: 8.0
		W. of widest portion	Lowet part: 6.0
		3.50–4.50	
VI. Encircling gears	15. Maha jal	L. 700–750	Episurface: 5.0
		H. 3.75–6.00	Underwater: 1.0
	16. Dal jal	L. 80–90	1.0
		H. 2.00–5.00	
	17. Chat jal	L. 110–120	1.0–8.0
		H. 2.50–3.50	
	18. Ghuran jal	L. 30.00–40.50	1.0–6.0
		H. 1.50–2.50	
	19. Jhaki jal	L.4.50–5.40	6.0
		Cir.11.25–14.00	
		Dia. 4.00– 4.50	
	20. Tana jal	L. 45.00–67.50	46.0
		H. 3.60– 4.50	
	21. Rekh jal	L. 5.00–5.50	35.0–38.0
		Dia. 4.50–5.50	
		Cir. 11.50–15.00	
VII. Miscellaneous	22. Arar jal	L. 3.50–4.50	4.0–5.0
		H. 2.50–2.75	
	23. Fal jal	L. 100–2000	1.0
		H. 3.00–4.00	
	24. Impoundment: with bailing	L. 0.50–5.00	—
	vessel ('Kuin')	W. 0.20–0.45	
	Without bailing Vessel ('Debli')	—	—
	'Dhagamara'	—	—

Contd...

Crafts and their Salient Features

Craft	*Size (m)*	*Construction*	*Propulsion*	*Crew*
Plank–built boat		Wooden planks nailed together to form a rigid frame		
Flat bottom:				
Large	L. 25.0–35.0		Manual	20–25
	W. 8.0– 10.0			
	H. 1.5– 2.0			
Small	L. 10.0–20.0		Manual	3–6
	W. 2.0– 4.0			
	H. 0.75– 1.0			
Round bottom:	L.10.0–20.0		Manual	3–6
	W. 2.0– 4.0			
	H. 0.75– 1.0			
Dug-out boat	L. 5.0–10.0	Hollowing-out single	Manual	1–3
(Canoe)	W. 1.0–2.0	log of wood.		
	H. 0.5–0.75			

This type of gear is generally found in operation throughout the year in the depth range of 1-5 m. Different species of both small (<30 cm) and big (>30 cm) fishes are caught with such hooks.

Kupa Barshi [Plate 16(1)]

This gear consists of a hook, a nylon line of 30-50 cm in length and a bamboo stick of 1.0-1.5 m in length, 4.5-8.0 cm in width. The line is attached to the hook on one end while the other end is tied by the middle of the stick. Earthworms or small forage fish are used as baits. A single fisherman operates *c* 80-100 Kupa barshis at a time. During operation, the rod remains afloat in water and the baited hook is hung from it. These are hauled after a few hours and the fish caught in the baited hooks are collected.

Usually, Kupa barshi in large numbers are found in operation in the shallow weed-infested areas of the Beel generally during the dry season. Mainly *Wallago attu* is captured with this gear.

Tanga Barshi [Plate 16(6)]

Baited hooks of different sizes are hung from a coconut coir, *c* 10-12 m in length, by *c* 30 cm long nylon strings at an interval of 1.0-1.5 m. Both the ends of the nylon line are fixed to the lake bottom with two bamboo poles to allow the baited hooks hang *c* 0.5-1.0 m below the surface of water. The line may be lengthened by extending the nylon string, if needed. It is lifted from a canoe by a person after 4-8 hours of its placement in the Beel and hooked fish, if any, are collected into the boat.

This type of gear is generally found in operation from November to April in the depth range of 0.5-1.25 m. *Wallago attu* is usually the chief catch with this gear.

Traps

Dori [Plate 16(7)]

It is fabricated in U-shaped form by knitting bamboo slits (1mm thickness) together with the help of cane string. The bamboo mat is strengthened with intermittent thick bamboo slits. On the inner side of the trap, a V-shaped door with a narrow slit (1.5-2.0 mm wide) is made for the fish to enter. The Dori is placed in the depth range of 0.45-1.8 m on both sides of the barricade extending 50-500 m into the interior of the Beel. The fish, being obstructed by the barricade, enter the Dori through its door and take shelter in its corners. After 8-10 hours of its placement, the catches are removed by the fishermen, inverting the trap on his boat.

It is generally found in operation throughout the year for catching small fish of various types, notably, *Rasbora daniconius, Clupisoma atherinoides* and *Salmostoma bacaila*.

Gui [Plate 16(3)]

A bigger-sized Dori is called a 'Gui. It is fabricated exactly like that of Dori, except that the thickness of the bamboo slits vary from 4-5 cm and are stitched wider apart.

Gui is operated principally for entrapping the big fish.

Paran [Plate 16(4)]

It is fabricated in a conical form by knitting bamboo slits (1.0 mm thick) together with the help of cane string. The bamboo slits are strengthed with thicker bamboo slits placed in between. On the inner side of the trap, a V-shaped door with a narrow slit (1.5-2.0 mm wide) is kept. The slits at the upper free end are tied with a string in such a way that the fish are taken out with bare hand simply by loosening it. The Paran is placed irregularly in the depth range of 0.30-1.25 m on each side of a barricade, extending 50-500 m from the shore line. Due to the miniature size of the cage, it is tied to the barricade by the upper string. The fish enters the trap through the slit after being obstructed by any means. The fisherman takes out the catches after 8-10 hours of its placement.

This type of trap is generally operated in the Beel between October to March. The main catches in the trap are *Puntius chola, Mystus vittatus, Clupisoma atherinoides* and *Parambassis baculis.*

Khati Bundh [Plate 17(2)]

It is fabricated in the form of a mat by knitting bamboo slits (height 2.5-8.0 m) with coconut coirs without leaving any gap in between the bamboo slits. The SL of a Khati is usually 4.5 m. The Khati is inserted into the Beel bottom for obstructing the movement of both big and small varieties of fishes throughout the year and the fish are caught by the traps placed along its sides.

Trawls

Among the trawls of Sone Beel, operation of 'Chhatjal' had been initiated by the fishermen belonging to the 'Maimal' community since very ancient times (Kar, 1990). Pelain, on the other hand, is mainly a domestic gear used by all the four fishing communities of the Beel, *viz.*, Kaibarta, Patni, Maimal and Namasudra. An account of the two types of trawls are given below:

Chhat Jal [Plate 17(7)]

This type of trawl is fabricated by tying two bamboo poles at their bases in the form of a 'V'. One of the poles is slightly longer which acts as a handle. The handle is tied to the opposite pole by a *c* 1.0 m long nylon string with which the Chhat jal is operated. The net, which contains an upper big and a lower fine-meshed portion, is tied to the wide extremities of the gear by a nylon string passing through it. In operation, the gear is trawled to a fishing site by two persons on a canoe. The fish allured at the site are caught by the trawl and brought out of water. Interestingly, a *Saccharum* sp (locally called 'Bhutang') is usually used for alluring the fish.

Plate 17: Fish Catching Devices in Sone Beel in Assam

Chhat jal is found in operation generally from May to October, mainly by the Maimal community, in the depth range of 3.5–5.0 m in order to extensively harvest *Chanda nama*.

Pelain [Plate 17(3)]

It is fabricated in the form of a triangle by tying three bamboo poles with nylon strings. One of the poles is longer to act as a handle for the trawl. The net is tied to the vertices of the triangular frame. A single man, tying a collecting basket to his waist, trawls the gear from the shore to knee-deep water. The fish caught in this operation are kept in the collecting basket.

Pelain is generally seen in operation at a depth range of 0.5-1.0 m, throughout the year, mainly for catching the trash fishes.

Field studies revealed that, Chhat jal generally accounts for *c* 7.5 per cent of the trash fish catches, of which, *c* 75.8 per cent of the fish catch usually consist of *Chanda nama*. This gear is totally ineffective to LGF. Pelain is mainly a domestic gear and its catches generally account for 9.3 per cent of the trash fishes. Fish catch composition by Pelain usually consist of *Puntius* spp, *Mystus* spp, *Glossogobius giuris, Amblypharyngodon mola, Chanda nama,* etc. This gear is also ineffective to LGF.

Investment-return analysis of the trawls has been presented in the following table. It may be noted here that Chhat jal costs more than Pelain in its fabrication and also depicts higher return in terms of CPGH. Conversely, the Pelain costs lesser; and, though it usually portrays lower CPGH, it is a common type of fishing gear used by many fishermen, as reflected by its higher value of NW. It generally also depicts higher values of MI and HC.

Gear Type	*Fabrication Cost (Rs.)*	*Catch (g/haul)*	*Return (Rs/haul)*	*CPGH (kg)*	*MI*	*NW (%)*	*HC*
Chhat jal	2000.00	300.00	16.00	0.129±0.0062	1.66	7.69	0.23
Pelain	700.00	100.00	10.00	0.104±0.0060	4.66	16.61	0.80

CPGH: Catch/Person/Gear/Hour; MI: Mechanisation Index; NW: Niche' Width; HC: Hanging Co-efficient.

In Sone Beel, fishing gears exhibit seasonality in their operation. Chhat jal is the monsoon variety. Pelain has no seasonality and is operated throughout the year. Studies on gear-specificity to fishermen type, revealed that, all the three classes of fishermen, *viz.*, occasional, part-time and professional (Kar, 1990; Dey and Kar, 1989 c) generally use Pelain both on domestic as well as commercial scale; while, the Chhat jal is operated only by the Maimal community, which constitutes *c* 5.54 per cent of the professional class in Sone Beel (Kar, 1990). Studies on gear-specificity to fish types revealed that Pelain is effective to all types of trash fishes; while, Chhatjal is used specifically for hauling *Chanda nama*. Efficacy of the trawls, analysed through CPGH experiments, revealed that Chhat jal is more efficaceous than Pelain. Pelain, though had a lower CPGH, is generally used by a larger section of the fishermen belonging to all the four fishing communities as revealed from its high value of NW.

Scooping Gears

In Sone Beel, operation of scooping gears have been initiated by the Patni and Maimal communities of fishermen since very ancient times for exploiting the fish resources of the Beel (Kar, 1990). The fishermen belonging to the Kaibarta community had migrated into Sone Beel during the 1950s. They did not adopt this traditional method of fishing.

An account of the three types of scooping gears in Sone Beel is given below:

Dheki Jal [Plate 18(3)]

It is a fixed type of triangular net made from cotton/nylon. It consists of two long bamboo poles tied together to form an acute angle.The net is fastened on the two poles of the bamboo while at the free end of the net, it is supported by a rope. Although the shape of the net is triangular, it possesses a large bag behind. The net is kept hanging on a rigid frame made of bamboos. The net is fitted to the bamboo frame in such a manner that it is usually put under water by pulley design. To lift the net, extra pressure on the base of the triangular corner of the net is exerted as weight by the operator. The net, in general, is lifted at an interval of *c* 15-20 minutes or even earlier depending on detection of a shoal of fish on the net.

Dheki jal or Chinese dip net is usually operated from the last part of August to mid-January in the connecting channels of Sone Beel or in the shallow areas of the Beel for catching different varieties of big and small fishes. Sometimes, a bamboo barricade is also constructed from the bank to the net for preventing the escape of the fishes which pass through the water.

Dorar Jal [Plate 19(4)]

It is fabricated by fixing an erect bamboo pole to the sail-hole of a flat bottom boat. Two other bamboo poles, incurved downwards, are attached to the apex of this pole in the form of a chiasma. The gear assumes the shape of an umbrella when the net is fitted to the four corners of the two sticks. Four persons and three boats are employed to operate the gear. Two of them operate the gear while the other two engage themselves in chasing the fish towards the gear. The Dorar jal is lowered into the water by its weight and the fish chased are caught and scooped out of water by drawing the net along the operator's back.

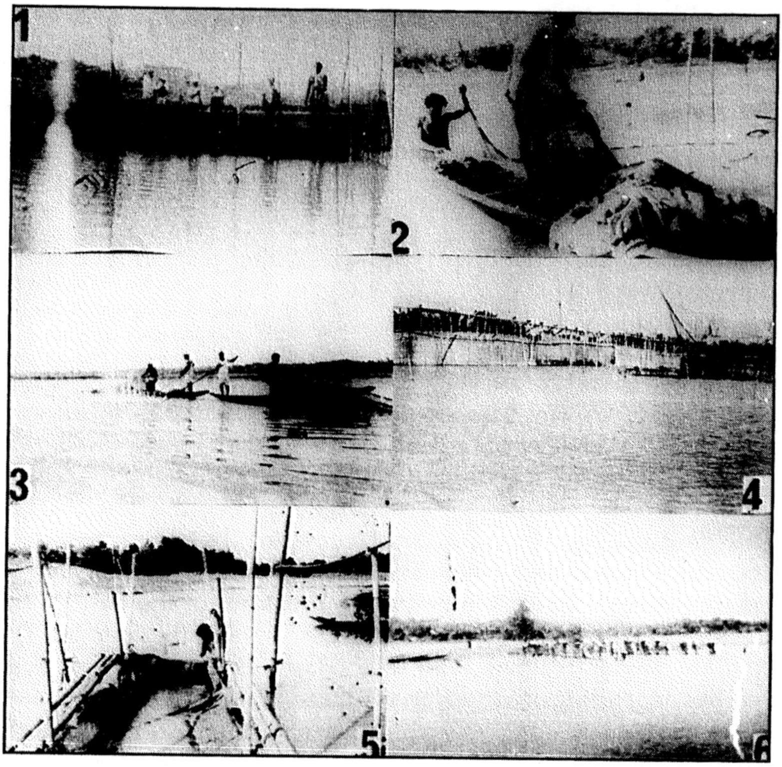

Plate 18: Fish Catching Devices in Sone Beel in Assam

In Sone Beel, the Dorar jal is generally seen in operation from September to November for catching different types of small fishes.

Kuchrung Jal [Plate 17(1)]

It is fabricated in the form of a circular bamboo frame with a long handle (*c* 2.0) passing through it. The net is fitted along the circumference of the circle. The Kuchrung jal is operated by two persons on a canoe. The person sitting in front operates the gear while the boat is rowed by the fisherman sitting at the rear end. The fish are scooped out of water as the boat is rowed.

This gear is seen in operation generally throughout the year in a depth range of 1.5 -2.5 m mainly for catching *Macrobrachium* sp.

Plate 19: Fish Catching Devices in Sone Beel in Assam

Kar (1990) and Kar and Dey (1996) reported that Dhekijal accounted for 2.4 per cent of the IMCs and 15.3 per cent of the trash fish catches. However, Dorar jal hauled only 18.2 per cent of the trash fishes of the Beel and it was found to be quite ineffective to the LGF. Kuchrung jal is effective only to *Macrobrachium* sp. and it accounted for 66.3 per cent of its haul of *Macrobrachium* sp. from Sone Beel. This gear is also totally ineffective to LGF.

Investment-return analysis of the scooping gears, as given below, show that, among the three scooping gears, Dheki jal costs the highest in its fabrication; but, accounted for poor return in terms of CPGH, thus, portraying economically less viable. Conversely, Kuchrung jal costs the least, but found to have high CPGH and more viability in trms of its economics. This is also reflected by the highest per cent of fishermen operating this gear.

Cost-benefit Analysis and Efficacy of the Scooping Gears of Sone Beel (Figures are average values)

Gear Type	*Fabrication Cost (Rs.)*	*Catch (g/haul)*	*Return (Rs/haul)*	*CPGH (kg)*	*MI*	*NW (%)*	*HC*
Dheki jal	10,000.00	400	12	0.028±0.0082	0.10	2.30	0.50
Dorar jal	5,000.00	250	8	0.033±0.0013	0.26	5.51	0.50
Kuchrung jal	2,000.00	300	6	0.084±0.0088	2.66	44.12	0.60

CPGH: Catch/Person/Gear/Hour; MI: Mechanisation Index; NW: Niche' Width; HC: Hanging Co-efficient.

In Sone Beel, the scooping gears also exhibit seasonality in their operation. Dorar jal is the monsoon variety. Kuchrung jal has no seasonality and is operated throughout the year. Studies on gear-specificity to fishermen revealed that, both the Occasional and Part-time classes of fishermen (Dey and Kar, 1989 c) use mainly the Dheki jal and the Dorar jal; while, the Professional class of fishermen use all the three types of scooping gears with a tendency towards extensive operation of the Kuchrung jal being restricted to the Patni Community of fishermen who constitute *c* 6.91 per cent of the Professional class in Sone Beel (Kar and Dey, 1991; 1996). Nevertheless, the Maimals, constituting *c* 5.54 per cent of the Professional class (Kar and Dey, 1993), are quite efficient in operating the Dheki jal. Gear-specificity to fish types revealed that, Dheki jal hauls different types of both big and small fishes, *e.g.*, *Labeo rohita*, *Salmostoma bacaila*, *Puntius* spp, etc.; while, Dorar jal catches mainly the small fishes, like, *Puntius* spp., *Mystus* spp, *Amblypharygodon mola*, etc. The Kuchrung jal is specifically meant for catching *Macrobrachium* sp and is operated only by the Patni community of fishermen.

Efficacy of the scooping gears analysed through CPGH experiments revealed that Kuchrung jal is the most efficacious of all the scooping gears. It is followed by Dorar jal and Dheki jal. ANOVA reveals highly significant differences between the different types of scooping gears with regard to their CPGH. Efficiency of Kuchrung jal is also evident from its high HC value. Analysis of MI portrayed that Kuchrung jal possesses the highest MI, which; thus, makes it a popular gear as is reflected by its highest value of NW among the scooping gears.

Entangling Gears

These are popularly called 'gill nets' and locally called 'Patan jal' and 'Haran jal'. In Sone Beel, these were introduced by the Kaibarta community of fishermen during the 1950s (Dey and Kar, 1989 c). Nevertheless, today, fishermen belonging to other communities of fishermen do also use the gill nets in Sone Beel (Kar and Dey, 1991).

The principal types of gill nets, seen in operation in Sone Beel, are briefly described below:

Patan Jal [Plate 19(5,6,8)

This type of gill net is a widely-used gear in Sone Beel designed to capture fishes of particular sizes, notably, *Hilsa ilisha, Wallago attu, Channa striata, Labeo* spp and *Puntius* spp. The nets vary in mesh size in accordance with the size of the fish sought to capture. The big variety of the net is usually operated to capture mainly *Wallago attu, Hilsa (Tenualosa) ilisha* and *Labeo* spp while the small category of Patan jal is operated principally to haul *Puntius* spp. The medium variety of Patan jal is generally used to catch medium-sized fishes belonging to these species. Although, in general, nets with smaller mesh size are generally made of heavy twine; thin, rigid, nylon twine is generally found to be more effective in entangling the fishes. The Patanjal usually consists of a single wall of webbing kept vertical in the water with sink at the bottom and float above. The fishes are entangled at their operculum while trying to pass through the mesh in vain. The Patan jal are generally set in the Beel at dusk and hauled (lifted) at dawn; and, they are operated throughout the year at a depth range of 2-4 m.

Haran Jal [Plate 19(2)]

This gill net is fabricated in the form of a funnel by tying two longitudinal nets at their extremities. This kind of gear is provided with a head rope tied with float above and a foot rope fitted with iron sinks below to enable maintain its shape properly underneath. The gear is operated by two persons from a round bottom boat at a depth range of 3-5 m and 300-800 m away from the shore. After lowering the gear into the water, it is brought to episurface by drawing coconut coirs tied to it. The fish which enter the gear through its funnel-shaped mouth, get entangled into the meshes. This type of gear is generally found in operation in Sone Beel from May to December. Haran jal is found to be effective in catching *Aorichthys seenghala* and *Mastacembelus armatus* in addition to different types of small fishes.

Impact of Gill Nets on the Fishery of Sone Beel

Dey and Kar (1990) reported that catch by Patan jal consist, on average, of 0.01 per cent of *Labeo* spp; 0.083 per cent of *Wallago attu*; 0.012 per cent of *Hilsa* (*Tenualosa*) *ilisha*; 0.23 per cent of *Channa striata* and 16.9 per cent of *Puntius* spp. Conversely, Haran jal hauls usually include 0.004 per cent of *Aorichthys seenghala* and 0.018 per cent of *Mastacembelus armatus* in addition to 18.4 per cent of trash fishes. This indicates that, judicious use of gill nets could help in harvesting good quantity of trash fishes from the Beel. Moreover, exploitation of the predatory cat fishes, which are quite vulnerable to gill nets, could give scope to the IMCs to breed and thrive in large numbers.

Patan jal generally costs more than Haran jal in fabrication; but, generally assumes poorer return in terms of CPGH. Hence, it could be considered as economically less viable. In fabrication of a Patan jal of 300 m length and 19.0 mm mesh size, 1.0-1.5 kg (@ Rs.400.00/kg) of knotless nylon webbing (locally called 'parachute' quality) is generally required. On the other hand, only 500-600 g of the same kind of nylon webbing is usually required for fabricating a Haran jal. As such, Haran jal is generally owned by individual fishermen in view of less cost in its fabrication. On the average, a Patan jal or a Haran jal, generally, lasts for five years, if operated for approximately 8 hours a day.

In Sone Beel, Haran jal is generally seen to be operated during the entire monsoon period and part of the winter; while, Patan jal is seen to be hauled throughout the year. Studies on gear-specificity to fishermen (with regard to gill nets) reveal that both the kinds of gill nets are usually operated only by the 'Professional' class of fishermen (Anon, 1962; Kar and Dey, 1991). Indeed, operation of Haran jal is generally restricted exclusively to the Kaibarta community of fishermen, which constitutes *c* 55 per cent of the professional class (Dey and Kar, 1989 c). Nevertheless, the Maimals, which generally constitute *c* 5.54 per cent of the Professional fisherfolk in Sone Beel, also, today, operate Patan jal in Sone Beel after its introduction by the Kaibartas.

Cost-Return Analysis and Efficacy of Gill Nets

Gear Type	*Fabrication Cost (Rs.)*	*Average Catch (g/haul)*	*Average Return (Rs/haul)*	*Average CPGH (kg)*	*% of Fishermen Using the Gear*
Patan jal:					
L. 300 m	2,000.00	600	30.00	0.0076	7.0
Mesh: 46 mm					
L. 300 m	2,250.00	900	41.00	0.0018	5.0
Mesh: 19.0 mm					
L. 300 m	2,500.00	700	28.00	0.002	8.0
Mesh: 15.0 mm					
Haran jal	800.00	400	16.00	0.020	11.0
L. 4.5 m					
Mesh:					
Upper: 8.0 mm					
Lower: 6.0 mm					

L: Length; CPGH: Catch/Person/Gear/Hour.

Efficacies of the gill nets, analysed through CPGH experiments, revealed that, Haran jal reflected higher CPGH (0.020 ± 0.0077) than Patan jal (0.00189 ± 0.001) on the average.

Imposition and practice of 'closed season' could prove rewarding in protecting the juvenile IMCs of the Beel. Further, organizing credit facilities; marketing of product and purchase of domestic as well as production requirements through co-operatives; encouragement of culture of IMCs in the confined 'Bundhs' at dead storage of the Beel; and, above all, proper rehabilitation of the poor fishermen, could go a long way in augmenting fish production in the Beel vis-à-vis uplifting the socio-economic condition of the poor fishermen.

Encircling Gears

The encircling gears in Sone Beel have been introduced by the Kaibarta community of fishermen during the early 1950s (Kar, 1990). The fishermen belonging to the other communities, *viz.*, the Patnis, the Maimals and the Namasudras, had no knowledge of the mechanization of the gears before initiation by the Kaibartas (Dey and Kar, 1989 c).

The seven different types of encircling gears, seen in operation in Sone Beel, are briefly described below:

Maha Jal [Plate 18(6)]

It is the largest type of encircling gear observed in Sone Beel. Its height ranges from 6.3-7.2 m in the middle and 1.5-3.5 m at the outer edges. Its length, however, ranges from 700-1000 m. Usually, it consists of two distinct parts, *viz.*, a fine-meshed underwater part and a broad-meshed episurface part, both the portions being joined together in operation. The underwater part, however, forms the main part of the Maha jal. The episurface part has a bigger mesh (5.0 mm) than the undersurface (1.0 mm). The nets are made up of cotton/ nylon yarn. However, indigenous bamboo floats, metal sinks and weights and bamboo poles form the essential components of a Maha jal. The floats are tied to separate monofilament rope attached to the head rope, *c* 1.5 m apart. 100-150 m of hauling ropes are fixed for dragging the net ashore. During operation, the underwater portion of the net is kept vertically erect through weights and floats while the episurface part of the net is maintained through poles. The net is extended above water to a good height to prevent fishes from escaping during operation. In Sone Beel, Maha jal is seen to be operated from May to October at a depth range of 2.0-5.0 m. 30-40 men, carrying equal portions of the net on two massive flat bottom boats (*c* 4-5 m dia), proceed towards the centre of the Beel, and drop the net in a semi-circular fashion as the boat returns towards the shore. On being dragged by both its head and foot ropes, the catches confined to the fine-meshed middle portion, are emptied into a canoe for sale. The catches include *Hilsa (Tenualosa) Ilisha, Gudusia chapra*, IMCs, Exotic carps, *Wallago attu, Salmostoma bacaila*, etc. It is to be noted here seriously, that, a overwhelmingly enormous quantity of (mostly fry) *Chanda nama* are caught regularly by the Maha jal; thus, posing a serious threat to their survival.

Dal Jal [Plate 18(1)]

This type of fishing, locally called 'Jheng' or 'Kheo', incidentally, forms one of the major methods of fishing at Dead Storage Level (DSL) of Sone Beel. Dal jal is used, at DSL, to encircle certain fishing areas of the Beel, which is generally selected previously and made ready through accumulation of weeds, branches of trees, bamboos, etc., in order to lure fish and take shelter therein. Dal jal is generally 80-90 m in length, 2-6 m in height and has a mesh size of 1.0 mm. However, the length and height of the gear varies according to the area of the site. This type of gear is generally operated by 8-10 fishermen from from November to March at a depth range of 0.5-2.0 m. The principal types of catches include the IMCs, exotic carps, *Puntius* spp, *Wallago attu, Chanda* sp, etc.

Chat Jal [Plate 18(3)]

This type of gear is usually fabricated by joining 10-15 pieces of nylon nets together, each 8-10 m in length and 5.0-5.5 m in height. Its total length generally ranges from 80-150 m while the mesh size varies from 1-8 mm. The foot rope is made heavier than the head rope for fast sinking. Spindle-shaped wooden buoys (locally called 'ula') are tied to the head rope, generally, at an interval of 4-5 cm for enabling the gear to remain afloat. In addition, two nylon strings, each 2-3 m long, are tied to both head and foot ropes to facilitate manipulation. This type of gear is operated by four persons on two round bottom (RB) boats. In operation, two persons, each in two RB boats (dia. *c* 2.4 m), carrying equal portions of the gear, row parallel for some distance, become face to face on a deep portion of the Beel; and, start dropping the gear in an encircling manner at a depth range of 1.0-5.0 m. Among the 4 operators, 2 persons lift the net by manipulating the 2-3 m long nylon strings which are tied to both the head and foot ropes. The other two persons make peculiar knocking sound as well as sprinkle water with the help of the oars in order to agitate the water and allure the fishes to enter the net when they are caught.

In Sone Beel, Chat jal was seen to be operated generally from May to December for hauling specifically *Gudusia chapra* and *Salmostoma bacaila*. It is important to note here that, there was no engine-driven ferrying boat in Sone Beel till around 1988. After their initiation, it is said that, fishes like *G. chapra* and *S. bacaila*, which are said to be mainly surface dwellers, left the Beel in large numbers, being disturbed by the outboard engines. Having happening this, the Chat jal operators failed to get enough catches in their gears. Due to this situation, there have been a sharp decline in the operation of Chat jal in Sone Beel since 1990.

Ghuran Jal [Plate 18(2)]

It is fabricated almost in the same manner as 'Chat jal, except that, the foot rope is provided with iron balls in order to enable it sink effectively. This type of gear varies in length from 30-40 m, height from 5.0-5.5 m and mesh size from 1.0-6.0 mm. Ghuran jal is operated by three persons from a RB boat. In operation, one person holds one end of the head rope on the bank; or, after rowing to a shallow area of *c* 1.0 m deep, one person gets down from the boat holding one end of the head rope of the gear. The boat is rowed gradually to a deeper water of *c* 1.5 m, when the second man, who holds the other extremity of the net, also gets down from the boat. By this time, the net has been cast from the coracle in a semi-circular manner. While the area is being encircled and the net is gradually being dragged towards the shore, the second man gets into the boat to lift the net, the first man keeps the net under his feet, and the third man creates a mild current therein by revolving the boat slowly. The fish moving along the current are caught by the gear and brought above water.

In Sone Beel, this kind of gear has been seen in operation from April to December at a depth range of 0.50-1.50 m for capturing mainly carps and minnows.

Jhaki Jal [Plate 19(1)]

This is locally called 'Jhaki jal' or 'Ural jal' and is one of the commonest types of fishing gears used in Assam and Bengal. It is popularly called 'cast net' in English. This cotton or nylon-woven light circular, somewhat bell-shaped net has a length of 4.5-5.4 m, an inside diamenter of 4.0-4.5 m, circumference of 11.2-14.0 m and is weighted by 4-6 kg of iron balls around its perimeter; thus, making the gear weigh 7-8 kg. Further, the circumference of the gear is recurved inward to form a pouch around its inner edge to prevent fish from swimming out, when the gear is lifted out of water after operation. In operation, the fisherman folds the net, so that, it could be thrown over the water to land horizontally. When cast over a school of fish, the weighted edges sink rapidly. The net is retrieved by a string attached near the centre.

In Sone Beel, this gear is seen to be operated throughout the year at a depth range of 0.5-4.0 m. The catches include, among others, the IMCs. *Puntius* spp, *Mystus* spp, *Channa* spp, *Wallago attu*, etc.

Rekh Jal [Plate 17(5)]

This gear is fabricated in the same manner as Jhaki jal except that the meshes are much bigger in size. The dimension of Rekh jal are: L 5.0-5.5m, Circumference 11.5-15.0 m, Dia 4.5-5.0 m, mesh size 35-38 mm. Its operation is also almost similar to that of Jhaki jal. In operation, a portion of the Beel is pre-selected and four such nets, carried on four canoes, are cast at a time from four corners of the site for maximising catch.

In Sone Beel, Rekh jal is seen in operation, generally from November to March, for catching fishes of some particular size and type, notably, *Wallago attu*, *Cirrhinus mrigala*, *Catla catla*, etc.

Tana Jal [Plate 19(7)]

This type of encircling gear is fabricated by fixing wooden floats above and sinkers below the net line. The floats enable the top line remain afloat while the weights help the bottom line to sink. The dimension of Tana jal are as follows: L 45.0-67.5 m, H 5.0-5.5 m, mesh size 45-46 mm. The gear is operated by three persons on two round bottom boats. In operation, a small portion of the Beel is encircled with the net. The first person operates the boat, the second lowers the net and the third arranges to encircle the area to capture fish.

In Sone Beel, Tana jal is generally seen to be operated during the period September-February at a depth range of 0.50-5.0 m usually for catching *Wallago attu*.

Kar (1984) and Dey and Kar (1990) reported that Maha jal efficiently accounted for 17.63 per cent of the IMC and 63.10 per cent of the trash fish catches from Sone Beel; Chat jal and Ghuran jal hauled 35.40 per cent and 30.33 per cent of the IMC and 18.40 per cent and 8.20 per cent of the trash fishes respectively. On the other hand, Tana jal accounted for 2 per cent of the big cat fishes; while, Dorar jal catches included 12 per cent of the IMC and the exotic carps and 16 per cent of the trash fishes. Jhaki jal found to have favoured the IMC by harvesting a big bulk (29.9 per cent) of the big cat fishes and 10 per cent of the trash fishes of Sone Beel. Rekh jal hauls included 7 per cent of the big cat fishes. Ghuran jal and Chat jal are effective gears to to haul the IMCs. On the other hand, Tana jal and Rekh jal are totally ineffective to trash fishes because of big size of their mesh.

Investment-Return analysis of the encircling gears is tabulated below:

Investment-Return Analysis of the Encircling Gears

Gear Type	*Fabrication Cost (Rs)*	*Average Catch (kg/haul)*	*Average Return (Rs/haul)*	*Mahajan*		
				Per Head Share (%) (Rs)	*Per Head Investment (Rs)*	*Per Head Return/ Haul (Rs)*
Maha jal (L 750m)	62,000.00	70.0	750.00	25.00	15,500.00	187.50
Dal jal (L 80 m)	3,000.00	70.0	2,450.00	20.00	600.00	490.00
Chat jal (L 110m)	5,000.00	0.5	10.00	25.00	1,250.00	2.50
Ghuran jal (L 30 m)	2,000.00	0.5	8.00	33.00	666.00	2.66
Jhaki jal (D 4.0 m)	800.00	0.1	1.50	50.00	400.00	0.75
Tana jal (L 55 m)	2,000.00	0.5	15.00	25.00	500.00	3.75
Rekh jal (D 4.0 m)	700.00	0.1	3.00	50.00	350.00	1.50

The foregoing analysis reveals that, Maha jal costs the highest in its fabrication but accounted for poor return in terms of CPGH. It, thus, portrays less viable economically.

As has been mentioned earlier, in Sone Beel, gears exhibit seasonality in their operation. Maha jal and Tana jal are the gears operated exclusively during the rainy season while Dal jal and Rekh jal are operated mainly during the winter. Chat jal and Ghuran jal are seen to be operated during both the seasons. Jhaki jal has no seasonality and is used throughout the year

Studies on gear-specificity to fishermen portrayed that, both Occasional and Part-time classes of fishermen (Kar, 1990; Dey and Kar, 1989 c) use mainly the Jhaki jal; while, the Professional class of fishermen use all the types of encircling gears with tendency to operate Maha jal, Chat jal and Ghuran jal intensively. Indeed, operation of Maha jal, Chat jal, Ghuran jal and Tana jal are restricted exclusively to the Kaibartas, who constitute *c* 55 per cent of the Professional class (Kar and Dey, 1991). Neverthess, the Maimals, who constitute *c* 5.54 per cent of the Professional class (Kar and Dey, 1996), are very efficient in operating the Rekh jal.

Efficacy of the Encircling gears and other relevant informations are tabulated below:

Efficacy of the Encircling Gears

Gear Type	*Average CPGH*	*Average MI*	*Average HC*	*Niche' Width (%)*
Maha jal	0.000084	6.67	0.60	1.39
Dol jal	0.00015	3.34	1.00	20.97
Chat jal	0.00054	4.00	0.40	34.96
Ghuran jal	0.0083	5.00	0.60	41.95
Jhaki jal	0.024	5.34	0.75	61.53
Tana jal	0.00074	1.50	0.67	0.69
Rekh jal	0.016	1.34	0.67	15.38

Analysis of efficacy of the encircling gears through CPGH, as shown in the above table, reveals that Jhaki jal seems to be by far the most efficacious encircling gear. The table also reflects highly significant differences between the different kinds of encircling gears with regard to their CPGH. Efficacy of the Jhaki jal is also reflected from its high HC value. On the other hand, analysis of MI portrayed that, Maha jal possesses the highest value of MI. Also, high value of MI of Jhaki jal makes it a popular gear and its extensive use in Sone Beel is reflected by its highest value of niche' width among the different varieties of encircling gears (Kar and Dey, 1993).

Miscellaneous Types of Gears

Arar Jal [Plate 18(5)]

In Sone Beel, fishes from deeper portions (locally called 'Kuri') are sometimes seen to be caught by making an artificial outflow and placing an 'Arar jal' against the current, in such a way, that, the water passes only through the net. Where natural outflow exists, the net is simply placed over it. The sides of the jal are raised, so that, it assumes the shape of a boat with a trough wherein the fishes accumulate. The net is manipulated by a nylon string passing across its width.

During operation, when the water from the 'Kuri' flows out, the fishes therein also try to swim out and eventually accumulate in the trough of the 'Arar jal' finding it deeper. The fishes are caught by raising the net with the help of the string as and when they accumulate in sufficient quantity.

In Sone Beel, Arar jal is seen in operation generally from September to March for catching different types of big and small fishes.

Fal Jal [Plate 17(2)]

Locally, the word 'Fal' means 'to jump'. As such, this kind of gear is used to catch mainly the fishes which have much jumping habit. It is designed in the following manner:

A nylon net is attached to the 'Fish barricade' (locally called 'Khati' or 'Khati bundh') found in the Fishing Centres at DSL of the Beel. The net is attached at different parts of the Khati forming pouches. The length and height of the Fal jal varies in accordance with the height of the'Khatis' and the area of the fishing centre. The fishes which are obstructed in the fishing centre, sometimes, try to escape by jumping over the 'Khati'. In doing so, they are trapped in the pouches of the Fal jal when they are caught by the fishermen.

In Sone Beel, Fal jal is seen to be operated generally from September to March mainly for catching different types of big fishes.

Impoundment

This is a device of trapping fishes by raising small embankments out of bottom soil of the Beel. The fishes could also be trapped by other kinds of devices of impoundment. In Sone Beel, impoundments could be of the following kinds:

With Bailing vessel (Locally Called 'Kuin') [Plate 17(4)]

Fish catching by bailing out of water is a lucrative practice at DSL of Sone Beel. The 'Kuin' is fabricated in the form of a boat by nailing three pieces of wood, in which, two pieces form the sides and the third piece forms the base. Posteriorly, the base of the Kuin is slightly bent. The 'Kuin' is supported by an 'X'-shaped bamboo structure having a horizontal bar above it. Anteriorly, it hangs from the horizontal bar by a *c* 50-75 cm long coconut coir. Posteriorly, it is tied to one of the bamboo poles at the chiasma.

During operation, the anterior end of the implement is lowered into the source of water to be emptied by a person exerting his weight. As the 'Kuin' is filled, it is released of the weight; and, at the same, it is slightly pulled out with the help of the rope with which it is tied to the horizontal bar. In this process, water is emptied to enable the fishermen to grab the impounded fish with bare hand.

In Sone Beel, this kind of fishing method is seen in operation, generally from November to March, for grabbing fishes like *Wallago attu, Cyprinus carpio* and *Channa* spp.

Without Bailing Vessel (Locally Called 'Debli') [Plate 17(6)]

In Sone Beel, this kind of fishing method, is employed only under certain circumstances. When a 'Kuri'becomes so shallow that, fishing by other methods become impossible, the fishes are caught by a method, locally called 'Debli'. In this process, the 'Kuri' is connected to a pit (*c* 2 m deep) on the land by a canal, *c* 15-20 cm deep and *c* 1.0-1.25 m long. The canal is dug slightly inclined, so that, the water from the 'Kuri'is easily drained out. A Pelain is placed at the mouth of the canal just over the pit. In addition, the canal is kept shady by placing shrubs over it. Finding safe, the fishes move towards the pit along with the current of water when they are caught with the help of the Pelain which is already placed there.

In Sone Beel, this method of fishing is seen in vogue generally from November to March for catching mainly different kinds of small fishes, notably, *Puntius* spp, *Channa* spp, *Chanda* sp, etc.

Dhagamara [Plate 16(5)]

This indigenous method of fishing is seen to be in vogue only in Sone Beel.

In this method, the fishes are chased from all sides to a 'U'-shaped fishing site containing some vegetations and debris and bound from three sides by certain ridges made out of bottom soil of the Beel. After chasing the fishes for some time, the fishing site is closed from all sides by raising another earthen ridge out of the bottom soil of the Beel and the water within this enclosure is made highly turbid by churning. Four 'Dori' (cage traps) are placed at the four corners of the fishing site. The plants and the debris are eventually cleared. Finding the site unfit for living, the fishes swim towards the corners and, thus, get entrapped by the'Doris' which are already placed there. Those fishes, which come to the water surface due to asphyxiation, are caught with the help of 'Pelain'.

In Sone Beel, this method of fishing, in which 10-20 persons take part, is generally seen to be in vogue from mid-February to mid-April for catching mainly different kinds of small fishes, notably, *Chanda* sp, *Puntius* spp, *Mystus* spp, etc.

The different types of fishing crafts (boats) found to be used in Sone Beel have been mentioned in the Table showing classification of gears and crafts.

Analysis of the Operation of Fishing Gears in Sone Beel

Of the 24 categories of fishing gears recorded in Sone Beel, all but Pelain could be regarded as belonging to commercial types. Indeed, Paran and Jhaki jal are sometimes seen in domestic use as well without involving crafts (boat). The Jhaki jal is sometimes seen to be operated along the littoral belt of the Beel, particularly, at the DSL. Further, impoundment, which often leads to rich harvest from the Beel at the DSL, does not involve any gear at all except some implements to bail out the impounded water.

Notwithstanding the above, seasonality in operation of some of the types of gears is of special significance. The monsoon varieties include the Chhat jal, Dorar jal, Haran jal, Maha jal and the Tana jal. On the other hand, Kupa barshi, Tanga barshi, Gui, Paran, Dal jal,

Rekh jal, Arar jal and Fal jal are the gears which are operated mainly during the winter. Dheki jal, Chat jal and Ghuran jal are seen in operation during both monsoon and dry seasons. Indeed, the Lar barshi, Dori, Pelain, Kuchrung jal, Patan jal and Jhaki jal has no seasonality and are found in use throughout the year.

A critical survey of the gear-specificity to fishermen types (FAO, 1962) revealed that the Occasional type of fishermen, who resort to fishing only in commemoration of an event, use the simple varieties of domestic gears like the Pelain. The Part-time class of fishermen, in addition to Pelain, also generally operate the Dori, Jhaki jal and the Lar barshi for a rich harvest in the event of their hardship. The Professional category of fishermen, conversely, use most of the types of gears seen in operation in Sone Beel. Indeed, a tendency by this group for intensive fishing with the help of gears like Maha jal, Chat jal and Ghuran jal is generally observed. Further, it is generally seen that, the Kuchrung jal, Dorar jal and Dhagamara varieties are restricted exclusively to the Patni community; while, Dori, Paran and Gui are seen to be extensively operated by the Namasudras. Interestingly, the Maimals are generally seen to be more inclined to operate Chhat jal, Rekh jal, Patan jal, Dheki jal and the Barshis. The Kaibartas, who constitute the major proportion of the Professional category of fishermen, operate almost all the types of fishing gears, in vogue, in Sone Beel. However, the operation of gears like Maha jal, Chat jal, Ghuran jal, Dal jal, Fal jal, Tana jal and Haran jal are seen to be restricted exclusively to the Kaibarta community.

Study of gear-specificity to fish types reveal that, IMCs of Sone Beel are caught mainly by the Maha jal. However, Chat jal, Ghuran jal, Dheki jal, Patan jal, Jhaki jal, Rekh jal and Fal jal are also found to be effective to this group. Of the three Ichthyospecies of Sone Beel fetching high yield, *Puntius chola* is caught in enormous quantity by Maha jal followed by Chat jal, Dheki jal, Jhaki jal and Dori. The other two species, *viz.*, *Parambassis ranga*, *Chanda nama* are caught in large number by the Maha jal, Chat jal and Chhat jal. In fact, the fry of *Chanda nama*, caught with Maha jal in enormously bulk quantity, are sold in the local fish markets as 'Bhujiya'. Chhat jal catches consist exclusively of *Chanda nama*; and, in the operation, as already mentioned earlier, a plant, locally called 'Bhutang' (*Saccharum* sp), is used to allure the fishes towards the gear. *Gudusia chapra*, which was abundantly available in Sone Beel once upon a time has been very rare since the iniation of engine-operated motor boats in Sone Beel around 1990. *Gudusia chapra* used to be caught in large number in the Maha jal.

The 'clap net' (locally called 'Sanglo jal' in Brahmaputra valley and 'Chairon jal' in Barak valley) is designed specifically to capture *Hilsa* (*Tenualosa*) *ilisha* in the rivers Barak and Brahmaputra. This gear is not seen in operation in Sone Beel. Rather, Maha jal, Chat jal and Patan jal are found to be used for catching the Indian shad in Sone Beel. Among the other fish groups, mullets are generally caught with the help of Dorar jal; while, the Haran jal is generally seen to be effective in catching, among others, the *Aorichthys seenghala* and *Mastacembelus armatus*. It also catches a large quantity of trash fishes. Among the cat fishes, *Wallago attu* is found to be caught generally with the help of Kupa barshi and Tanga barshi. The Lar barshi is generally found to be very effective in catching the bottom-dwelling murrels.

In Chatla Haor

Chatla Haor is a seasonal floodplain wetland in Cachar district of Assam. It retains water for about six months in a year from around May to around October every year.The fishing gears operated in Chatla Haor are not as diverse as in Sone Beel. The principal types of fishing implements in vogue in Chatla Haor are the following: Their nomenclature in local language, fabrication and *modus operandi* of the gears in Chatla Haor are almost the same as, in vogue, in Sone Beel; except that, the gears are smaller in dimension as compared to their counterparts operated in Sone Beel.

Lar barshi, Dori, Paran, Khati bundh, Pelain, Dheki jal, Patan jal, Maha jal, Dal jal, Ghuran jal and Jhaki jal, are the principal varieties of fishing gears and devices found to be operated in Chalta Haor.

In the Anuas

The 'Anuas' are river-formed oxbow wetlands which are peculiar mainly to the Barak Valley region of Assam. Many of these anuas were once reservoirs of mainly *Gudusia chapra* and *Salmostoma bacaila*. As such, gears, which are specifically effective in catching these two types of fishes, are generally seen in operation in the 'Anuas'.

Lotic System

In River Barak

The river Barak, forming the Barak valley region in Assam, is the second biggest river in NE India. There is a great diversity of fish catching devices in different stretches of the river, *viz.*, upstream, mid-stream and down stream, depending on the gradient(s), fish types (rheophilic/plainwater), fishermen types (tribal/non-tribal) and season types (rainy/dry). However, a brief account of the fish catching devices (named in the local language) seen in the stretch of the river Barak around Silchar city and its environs, covering a stretch of *c* 25 km, is given below:

The monsoon varieties include, among others, the Lar barshi, Lewa jal, Maha jal, Basta jal, Bara jal and the Pan jal; while, the winter varieties include, among others, the Lar barshi, Pelain, Dheki jal and the Patan jal. Jhaki jal is seen to be operated both during the monsoon and winter months. Operation of Dheki jal generally ranges from around October to April. In addition to the above, a special type of gear, popularly called 'clap net' (locally called 'Chairon jal'), is seen to be in operation in River Barak during the rainy season for catching specifically *Hilsa* (*Tenualosa*) *ilisha*. The local name of the 'clap net' varies from 'Chairon' to 'Akon' or 'Buribain', with 'Tinon' and 'Duion' in between, depending on the size of the mesh. Except Pelain, Jhaki jal and Dheki jal, *modus operandi* of the other types of gears in vogue, in river Barak, as mentioned above, are quite different from the *modus operandi* of the same types of gears seen in operation in lentic bodies like Sone Beel, etc., because of the fluviatile condition in the river. However, operation of Lar barshi and Patan jal in river Barak seem to show some amount of similarity with those operated in Sone Beel.

The upstream of river, beyond village Lakhipur in Assam, is inhabited mostly by triabals belonging to Hmar community admixed with Kuki, Naga, Meitei, etc. Fish catching devices vary in different stretches with a tendency towards using fish toxins (mostly herbal) and explosives occasionally. Use of hook and line gears, cast nets, gill nets and drag nets seem to be quite common among the tribals.

In the major tributaries of river Barak, in Assam, *viz.*, Jiri, Chiri, Madhura, Jatinga, Sonai, Rukni, Dhaleswari, etc., the fish catching devices in the downstream portion of each of these rivers is almost the same as mentioned in the case of river Barak.

In Mizoram

In Lentic Systems

In Tum Dil, Rih Dil and Palak Dil

In Mizoram, the lentic water bodies are generally under the administrative governance of both Fishery and Toursim departments of the provincial government. For the former, it is cultural ground for fish; while, for the latter, it is a source of earning revenue through tourism. As such, the lentic water bodies in Mizoram are, generally, not much open to the public for vigorous fishing. However, sport fishing or angling is usully allowed as part of promotion of tourism. Therefore, fishing with simple hook and line gear is usually seen in Tumdil. This is also, by and large, the situation in Rih Dil and Palak Dil.

In Lotic Systems

As has been indicated earlier, the tribals, in general, operate the cast nets, gill nets and the drag nets for fishing in the turbulent rivers. In Mizoram, the Professional fishermen, among the Mizos, are seen to operate generally the cast nets, gill nets and the drag nets for catching fishes in the rivers Tuirial, Tlwang, Tuirini, Tuivai, Serlui, Mat, Kolodyne, Tuichong and Karnafuli. However, the diameter of the cast nets generally vary from 1.0-1.5 m for frequent operations in the fast torrents, easy handling and manipulation, unlike the bigger ones seen in operation in the plain water systems. The size of the gill nets generally vary with the width of the rivers. The cast nets are seen in operation in rivers Tuirial, Mat, and Kolodyne generally from around 8 am to around 2 pm. The gill nets are seen in operation in rivers Tuirial, Tlwang, Mat, Kolodyne where they are set in the rivers around 8 pm and the fishes caught in the nets are collected around 7 am next morning.

Cast net catches, generally, include *Crossocheilus latius latius*, *Puntius* spp, *Semiplotus semiplotus*, *Garra* spp, *Barilius* spp, *Danio* spp and sometimes *Tor* spp, and *Neolissochilus* sp. On the other hand, the gill net catches include generally the *Tor* spp, *Neolissochilus* sp, *Puntius* spp, *Barilus* spp, *Eutropiichthys vacha*, *Aorichthys seenghala*, etc. The gill nets are sometimes held on the two banks of the rivers and dragged along its length for harvesting a big catch through sieving.

In river Karnafuli, flowing along the Mizoram-Bangladesh border, the professional fishermen fishing in the river mostly belong to the Chakma community and there are not many Mizo fishermen. The Chakmas fish almost all along the course of the river Karnafuli from its source around Marpara along the Mizoram-Tripura border upto Diglibak along the Indo (Mizoram)-Bangaledsh border. The Chakmas, in addition to using the cast nets, generally fish with the help of typical indigenous bamboo made baskets. The catches include, among others, *Puntius* spp, *Danio* spp, *Barilius* spp, *Osteobrama* sp, *Neolissochilus* sp, etc. Moreover, non-tribal fishermen, from Barak valley region of Assam, take portions of river Karnafuli on lease from the Govt.of Mizoram, and migrate, in groups, to Demagiri on the bank of river Karnafuli generally during November to February when they fish in the river with the help of Maha jal for harvesting a big haul of fish.

In river Tuirini, a comparatively smaller river, fishing is done generally with the help of cast nets. As indicated earlier, like many other tribals, the Mizos also use explosives in the rivers with the expectation of having a rich harvest. However, our observation and experience shows that the explosives are used mostly by the non-fishermen Mizos when they go for joyful fishing. Explosives have been seen in use in rivers Tuirial and Tlwang.

In Tripura

In Lentic Systems

In the 'Sagars'

In Tripura, the wetlands created by the earstwhile Monarchs (locally called 'Sagar') constitute important components of the lentic systems. The significant Sagars are the Amar Sagar, the Rudra Sagar and the Fatik Sagar.

The Amar Sagar is situated in the small Sub-divisional town of Amarpur in South Tripura district. It has an average depth of *c* 2.0 m at the FSL. The Sagar is given on lease by the Tripura Government, generally to the Fishermens' Co-operative Society functioning there. The fishermen belonging to the Co-operative Society are generally seen to operate Patan jal, Tana jal and sometimes the small size Maha jal for harvesting the fishes. The fishes caught, are generally sold out in auction.

The Rudra Sagar is situated near a village called Melagarh, in the small Sub-divisional town of Sonamura in West Tripura district. It has an average depth of *c* 1.5 m at the FSL, being tremendously silted and infested with weeds and paddy cultivations in between. As in the case of Amar Sagar, Rudra Sagar is also leased out to the Fishermens' Co-operative Society prevalent there. The fishermen operate almost the same types of gears as are operated in Amar Sagar; but, being shallower than Amar Sagar, in general, cage traps, like Dori, Paran, etc., are used here unlike in Amar Sagar. The fishes caught, are generally sold out in auction.

The situation in Fatik Sagar is almost the same as in Amar Sagar.

In Lotic Systems

In Tripura, the major rivers are the Manu, Khowai, Gomati and Feni. The river Howrah (on the bank of which, the Agartala city is situated), though, has a short stretch of flow, is quite big in its downstream region. Also, the river Deo, flowing between the towns of Dharmanagar and Kumarghat, is also quite a big river.

The fish catching devices, in vogue, in Tripura, could again be dealt with under two major aspects, *viz.*, (a) those operated by the tribals mainly in the upstream rheophilic stretch of the rivers; and, (b) those operated by the non-tribals (mostly Bengalis) in the downstream plainwater stretch of the rivers.

In river Manu, the upstream stretch, is inhabited, among others, by the Jamatias and Riangs. They use, during both rainy and dry seasons, mainly the cast nets (Dia 2.5-3.0 m), the gill nets (L100-125 m; Ht 2.5-3.0 m) and the peculiar indigenously prepared bamboo basket traps for fishing.The catches generally include, among others, the *Barilius* spp, *Danio* spp, *Puntius* spp, *Acanthocobitis* sp, *Garra* spp, *Glyptothorax* spp, etc. Its downstream stretch is inhabited mainly by the Bengalis. They fish mostly with cast nets (Dia 4.0-5.0 m) and gill nets (L 200-250 m; Ht 2.5-3.0 m). The Chinese dip net ('Dheki jal') is also sometimes seen in operation. The catches include, among others, the *Puntius* spp, *Amblypharyngodon mola*, *Danio* spp, sometimes the IMCs, *Mystus* spp, *Eutropiichthys vacha*, etc.

In river Khowai, the upstream stretch is inhabited, among others, by the tribals belonging to Riang and Hrangkhol tribes. They use, during both rainy and dry seasons, mainly the cast nets

(Dia 2.5-3.0 m), the gill nets (L 75-100 m; Ht 2.0-3.0 m) and also the peculiar indigenously prepared bamboo basket traps for fishing. The catches include, among others, the *Barilius* spp, *Danio* spp, *Garra* spp, *Puntius* spp, *Glyptothoroax* spp, etc. Its downstream stretch is inhabited mainly by the Bengalis (and, also, to some extent, by the tribals belonging to the Deb Barma community). They fish mostly with the help of cast nets (Dia 4.0-5.5 m) and gill nets (L 250-300 m; Ht 3.0-3.5 m). Also, the Chinese dip net ('Dheki jal') is sometimes seen in operation. The catches include, among others, the *Puntius* spp, *Amblypharyngodon mola, Salmostoma bacaila, Danio* spp, *Xenentodon cancila, Chanda nama*, etc.

In river Gomati, the upstream stretch, around Mandirghat, Tirthamuksh and Jatanbari, is inhabited, among others, by the tribals belonging to Deb Barmas and Jamatias. They use, during both rainy and dry seasons, mainly the indigenously prepared bamboo baskets and sometimes the cast nets (Dia 2.5-3.0 m) and gill nets (L 100-125 m; Ht 2.5-3.0 m) for fishing. The catches include, among others, the *Barilius* spp, *Danio* spp, *Garra* spp, *Puntius* spp, *Neolissochilus hexagonolopis, Schistura* spp, *Glyptothorax* spp, etc. Its downstream stretch, around Natun Bazar, Amarpur, Udaipur and Melagarh, is inhabited mainly by the Bengalis (and, also, to some extent, by the tribals belonging to the Deb Barma community). They fish mostly with the help of cast nets (Dia 4.5-5.5 m) and gill nets (L 125-150 m; Ht 2.5-3.0 m) for fishing. Also, the Chinese dip net ('Dheki jal') is sometimes seen in operation. The catches include, among others, the *Puntius* spp, *Salmostoma bacaila, Amblypharyngodon mola, Danio* spp, *Xenentodon cancila, Mystus* spp, *Eutropiichthys vacha, Aorichthys* spp, etc. Sometimes, *Bagarius bagarius* is found in the down stream markets (*e.g.*, in the Udaipur fish market), possibly, being swept away from the upstream region of the river.

In river Feni, the upstream stretch is inhabited, among others, by the tribals belonging to Deb Barma and Jamatia communities. They use, during both rainy and dry seasons, mainly the cast nets (Dia 2.0-2.5 m), gill nets (L 100-125 m; Ht 2.0-2.5 m) and the indigenously prepared bamboo basket traps for fishing. The catches include, among other, the *Schistura* spp, *Barilius* spp, *Garra* spp, *Puntius* spp, *Glyptothorax*, spp, etc. The downstream stretch, around Sabrum and Ravindra Nagar (along the Indo-Bangladesh border, the mid-stream being the International border) is inhabited mainly by the

Bengalis. They fish mostly with the help of cast nets (Dia 4.0-4.5 m) and gill nets (L 125-150 m; Ht 2.0-3.0 m) for fishing. Sometimes, the Maha jal and the Dheki jal are also seen in operation. The catches include, among others, the *Puntius* spp, *Mystus* spp, *Aorichthys* spp, *Salmostoma bacaila*, *Amblypharyngodon mola*, *Danio* spp, etc.

Chapter 16

Fisherfolk in Barak Drainage (Assam), Mizoram and Tripura

'Fisherfolk' or the 'Fishermen' are the instruments to harvest fish from any water body. They are trained, not only in various methods of fishing, but also, in fabrication of different types of fishing implements and devices according to fish type, habitat type and season type. Sometimes, a fishing implement is fabricated specifically for a particular fishing community and often they feel proud of such identification of a community specialized in operating a particular type of fishing implement or device (Kar, 2000 b).

In NE India, particularly, with regard to Barak drainage (Assam), Mizoram and Tripura, there are a large number of fishermen belonging to diverse types of fishing communities. They could be broadly classified under different categories, based on the communities, intensity of fishing, and so on. Most of the fishermen in this region belong to the Scheduled Caste community (if non-tribals) and to the Scheduled Tribe community (if tribals) as notified in the Constitution of India. The fishermen, in general, are poor and not very literate. There has not been much ambitious welfare measures for their upliftment. The co-operative movement has also been not much well organized among the fishermen, except in the province of Tripura. However, the Sone Beel Fishermens' Co-operatve Society (SBFCS) is, perhaps, one of the biggest of its kind in this region; but, not very active and progressive; although, it ought to have been (Kar, 2003 c).

In Assam

In Lentic systems

In Sone Beel

The villages around the shore of a Beel are generally inhabited by fishermen belonging to different communities. The activities of the fishermen and their socio-economic condition have been influencing the development and utilization of its fishery resources in a significant way. This aspect has gained importance in view of the comparatively so-called inferior social status of the fishermen *vis-à-vis* overfishing of the water bodies by these poverty-stricken, so-called less progressive people.

Abounding Fishermen of the Beel

Fishermen belonging to 4 principal communities have been recorded in the 39 villages around Sone Beel (Kar, 1990). These are the 'Kaibarta', the 'Patni', the 'Maimal' and the 'Namasudra'. Certain relevant information pertaining to relative composition of each community in respect of both fishermen as well as total population living in the villages around the Beel; as well as, annual average area exploited by individual fisherman of each community are tabulated below:

Fishermen Community	*Relative Abundance (%) Out of*		*Fishing Area (ha/fisherman)*
	Total Fishermen	*Total Inhabitants*	
Kaibarta	68.14	47.67	0.32
Patni	14.66	10.24	1.51
Maimal	12.29	8.59	1.81
Namasudra	4.91	3.43	4.53
		30.07 (Non-fishermen)	

Moreover, on the basis of fishing intensity, the fisherfolk of Sone Beel have been further classified, after FAO (1962), into three categories, *viz.*, Professional, Part-time and Occasional, an account of which, with relevant information (Kar, 1990), is tabulated below:

Fisherman Category	*Relative Abundance (%) Amongst*				*Total*
	Kaibarta	*Patni*	*Maimal*	*Namasudra*	
Professional	55.04	6.91	5.54	2.51	70.00
Part-time	8.00	3.00	2.60	1.40	15.00
Occasional	4.50	4.75	4.75	1.00	15.00
Total	**67.54**	**14.66**	**12.89**	**4.91**	**100.00**

A brief description of the three categories of fishermen, as mentioned above, is given below:

Occasional Fishermen

They take recourse to fishing only when an occasion arises or when the situation compels. Such category of fishermen are prevalent in Sone Beel generally during the dry days when the fishing sites are easily accessible on foot. The relative ease of access coupled with certain amount of free time between sowing and harvesting of the Beel crops (notably, paddy), makes this type of fishing attractive and often sportive to the group. This category of fishermen constitute *c* 15 per cent of the fishermen population of the Beel. Both affluent and poor section of the fishermen belong to this category: those compelled to this profession, are the poors; while, those adopt it as a sport, are the riches. Certain deeper portions of the Beel, locally called 'Kuri', are found to be traditionally or occasionally reserved for them. These are seen to be fished during certain festivals or feasts commensurating with a holiday where a large section of the Society generally participate. Most of the fishes caught in such operations usually go directly into their diet. The individual time spent is short and the gears used are usually simple and less efficacious. As the number of participants are large, it is difficult to assess the contribution of individual efforts to total catch. Migratory habit is not generally seen in this category of fishermen.

Part-time Fishermen

This non-nomadic group of fishermen constitute *c* 15 per cent of the fisherfolk population of Sone Beel. They fish only during part of a year and consider this activity as equal to or inferior to other activities of the group. The flood cycle, the biological cycle of fish (Welcomme, 1979) and the seasonal needs of agriculture seem to impose a cyclicity on such communities. As such, reservation of fishing sites is not practiced by this group. Rather, they generally prefer cultivation to fishing in shallow water. Although their interest lies in hauling the large growing fishes, the yield status is usually poor because of the lack of adequate knowledge of the distribution of fish species in the Beel. Nevertheless, the Part-time fishermen try to operate almost all the types of gears as are operated by the Professional fishermen. In addition, they have a tendency to practice certain aquacultural techniques.

Professional Fishermen

This is the largest group among the fishing communities in Sone Beel constituting *c* 70 per cent of the total fishermen population around the Beel. Both poor and rich fishermen belong to this group. They fish throughout the year with the help of diverse types of fishing gears and devices without any restriction or reservation of fishing sites. However, they have a tendency of operating some of the sophisticated and little mechanized fishing gears and they are proud of being identified as operators of certain specific gears, *e.g.*, Professional fishermen belonging only to the Kaibarta community operate the Maha jal, Chat jal, Ghuran jal and the Haran jal. Similarly, Professional fishermen belonging only to the Patni community operate the Kuchrung jal and the Dorar jal. Also, Professional fishermen belonging only to the Maimal community operate the Chhat jal and Rekh jal. Although the bamboo cage traps and the bamboo barricades, are used by almost all the categories of fishermen, the Professional fishermen belonging to the Namasudra community operate only the bamboo made cage traps of different kinds, *e.g.*, Dori, Paran, Gui, Polo, Runga, etc. However, the Namasudras also sometimes operate the Lar Barshi in Sone Beel. The catches by the professional fishermen include a rich diversity and quantity of fishes, notably, the IMCs, the exotic carps, *Puntius* spp, *Salmostoma bacaila, Danio spp, Botia dario, Mystus vittatus, Aorichthys* spp, *Wallago attu, Eutropiichthys vacha, Channa* spp, *Chanda nama. Mastacembelus armatus, Macrognathus* spp, etc. Their catches also include a large number of juveniles. The Professional category of fishermen often become nomadic. They construct temporary thatched fishing huts and lead camp life on high grounds around the Beel. They spend days together on boats and rafts while continuing fishing in the Beel as well as in the inlets and outlets of the Beel.

The Fishing Villages

The names of the fishing villages situated around Sone Beel, with the village areas and the number of fishermen inhabiting therein (Kar, 1990), are tabulated below. The position of the villages have been shown in the map of Sone Beel (Figure 4).

Sl.No.	*Name*	*Area (ha)*	*Fishermen Population*
1.	Shingua	121.41	25
2.	Debodwar	153.78	1100
3.	Fultali	142.45	500
4.	Balirbond	161.88	800
5.	Tulakona	101.17	218
6.	Sangatipur	62.72	130
7.	Dhritipur	145.69	175
8.	Shekikanisail	158.82	150
9.	Belala	100.32	275
10.	Phakhoagram	261.43	2000
11.	Nayanabin	88.62	555
12.	Sanyasipatta	55.03	351
13.	Kodali	105.22	479
14.	Digalikandi	79.32	16
15.	Birgram	76.89	100
16.	Nagendranagar	485.64	2177
17.	Bhairabnagar	41.27	10
18.	Tinghari	46.94	6
19.	Basantapur	116.55	12
20.	Kaiyakuri	50.50	8
21.	Kishnanagar	52.61	162
22.	Bhabanipur	170.30	439
23.	Saitanala	165.90	25
24.	Gallasail	110.00	32
25.	Dolu	259.00	20
26.	Mangalpur	169.97	100
27.	Shantipur	174.02	413

Contd...

Sl.No.	Name	Area (ha)	Fishermen Population
28.	Nabingram	263.05	955
29.	Kalibaribazar	169.97	388
30.	Sailanagar	157.83	600
31.	Bagantila	129.50	210
32.	Sontoshpur	259.00	645
33.	Swastipur	64.75	400
34.	Kalyanpur	119.79	1180
35.	Bargool	61.91	95
36.	Samriddhipur	65.20	15
37.	Dargarbond	54.16	550
38.	Anandapur	92.27	300
39.	Shrinagar	57.46	80

Socio-economics of the Fisherfolk Population of Sone Beel

In India, fishing is generally considered as a low category profession. However, in Sone Beel, it is the principal way of earning livelihood by the poor fishermen. The entire water spread area of the Beel is used for intensive fishing both on domestic as well as commercial scales by operating different types of fish catching devices which have been mentioned earlier (Table 1 in Chapter 15). Of these fishing gears, all but Pelain, are of commercial types. Jhaki jal and Paran are sometimes found in domestic use as well. CPGH (Catch/Person/Gear/Hour: Dey, 1981) of the principal gears operated by the fishermen of Sone Beel, and calculated on the basis of 12 months' moving average method (Coxton and Cowden, 1950), indicates that, Dori is the most efficacious and is sequentially followed by Jhaki jal > Haran jal > Ghuran jal > Patan jal > Chat jal > Lar barshi > and Maha jal. Use of fine-meshed nets involving catch of juveniles, which is seen to be, in vogue, in Sone Beel, causes concern. Lack of proper training of the fishermen in professional fishing regulations coupled with tremendous amount of proverty *vis-à-vis* not much rehabilitation measures being taken for them till date, seem to be some of the impelling causes for use of banned nets and devices by the poor fishermen. Further, in the absence of any large-scale welfare measures being undertaken for them till today, the poor fishermen are compelled to remain in the grip of unscrupulous 'middlemen'. Consequently, the per capita income (PCI) of the fishermen is low and ranges from Rs. 1.30-Rs.8.25 (Mean: Rs. 3.83 ± 2.14). Their poverty is further aggravated by their early marriage and big family size. Low level of life expectancy (*c* 55 years) and high rate of child mortality (*c* 10 per cent per year) also reflect their state of distress. Poor adult literacy (10-20 per cent) may be considered as the principal factor of their financial hardship. Juvenile literacy, though higher (*c* 60-70 per cent), is soon followed by increased percentage of dropouts. Consequently, responsible earning members are very meagre and the grip of poverty linger-on in the community.

The fishermen now realize the gross imbalance which exists between poor income from fish trade and the hard labour put in. In view of this, there is a burgeoning tendency among quite a large number (*c* 59 per cent) of the fishermen to motivate their children against taking-up fishing profession. As a result of this, it is found that, only a meager percentage (*c* 20 per cent) of the youth wish to keep faith with their parental avocation of fishing. Others aspire for varied types of profession like banking, agriculture, defence, medicine, engineering and teaching. As a remedy against such undesirable trend, a cross-section (*c* 12 per cent) of their Society advocates in favour of setting-up of 'Fish Farmers' Training Institute' in their locality which would help making fishing profession attractive towards the fisheryouths.

During the dry season, when Sone Beel shrinks to 409.09 ha, and fishing becomes less intensive; fishermen cultivate 'Boro' paddy (winter-spring paddy) within the Beel itself in order to tide over the adverse situation. The area of their cultivable land ranges from 0.07-4.2 ha (Mean 0.02 ha ± 0.105). The Kaibartas possess big quantum of land and Namasudras the least. *c* 7 per cent of the fishermen do not have any cultivable land. Nevertheless, during the off season, some of them associate themselves with livestock and poultry farming, pottery and carpentry, net and boat making, beverages, etc. Some of the affluent fishermen, who possess big areas of cultivable land, raise also the 'Sali' paddy (monsoon-autumn paddy) in addition to the 'Boro'. In this way, they involve themselves in paddy trade in a big way. In addition to paddy, they also grow vegetables and sungrass on a commercial scale.

Incidentally, religion forms a part and parcel of the daily life of the fishermen living around Sone Beel. Among the Hindus, there are both 'Vaishnavites' (*c* 33 per cent) and 'Shaktas' (*c* 30 per cent). The Muslims, however, belong to one sect only. Being illiterate, by and large, the fishermen have peculiar beliefs about super-natural things.

It is very amazing to note here that, there is an ancient Hermitage of a Muslim Hermit, called 'Sond Peer' or 'Hond Peer' (after whom the Beel had been named 'Sone Beel') at the village called Mokam tila (Anandapur) along the east shoreline of Sone Beel. There is also an equally ancient temple of Goddess Kali situated just adjacent to the Muslim Hermitage. Both Hindus and Muslims are said to pay equal homage and are said to be equally regardful to both the religious institutions.

Inference on the Socio-economics of the Fisherfolk

The fishermen of Sone Beel have been struggling against poverty since a couple of decades. In view of the nature of economic constraints faced by the fishermen, it is strongly felt that, elimination of unscrupulous middlemen by organizing credit facilities, marketing of product and purchase of domestic as well as production requirements through Co-operatives would go a long way in the emancipation of the fisherfolk. Further, encouragement of IMC culture in the confined 'bundhs' (Fig.6) at the DSL of the Beel could lead to a potential IMC fishery. Concomitantly, fast conveyance of the catch from the fishing centres to the fish landing stations and the urban markets would accelerate fish trade. Provision of cold storage facilities in the landing stations and the urban markets would be an added advantage to prevent loss of flesh due to decay. Moreover, popularization of the recent fish culturing and fishing techniques; and, proper rehabilitation of the fishermen could enable them to give up nomadism and paddy cultivation in the event of hardship and help gear up the upliftment of the anglers in the long run (Dey and Kar, 1989 c).

In Chatla Haor

This is a seasonal floodplain wetland (locally called 'Haor') which retains water generally from May to October. It does not have any DSL and is almost completely dry during the winter season. *c* 15,000 fishermen, belonging mainly to the Kaibarta community, live in 22 villages around the Haor. The fishermen belonging to other communities fishing in the Haor include some 'Meitei'

(Manipuri) and some 'Behari' who are mostly labourers of the surrounding Tea Gardens. The Kaibartas, by and large, belong to the Professional category of fishermen while the Meiteis and Beharis mostly belong to the Part-time and Occasional categories of fishermen.

Unlike in Sone Beel, there is not much diversity of fishing gears in Chatla Haor. The Kaibartas are generally seen to fish with Lar barshi, Dori, Paran, Patan jal, Maha jal, Ghuran jal, Jhaki jal and Dal jal. The Meiteis are seen to operate mainly the Patan jal and the Jhaki jal while the Beharis generally operate the Jhaki jal. Further, the Pelain (small trawl) is seen to be operated in Chatla Haor by all the three communities of fishermen. As in Sone Beel, the socio-economic condition of the fishermen in Chatla is also very poor and needs upliftment.

In the Anuas

The 'Anuas', though not very enormous in dimension (as compared to Sone Beel and Chatla Haor), are highly potential in fish production mainly because of their great depth (comparable to that of river Barak), having been part of the course of river Barak once upon a time. Fishermen belonging to different communities, notably, the Maimals, Patnis and Meiteis live in the villages around the Anuas. They operate mainly the Lar barshi, Patan jal and the Jhaki jal for fishing in the Anuas. Socio-economically, they are poor and most of them live hand to mouth. However, the Anuas are generally leased out by the government to fish traders who are rich and employ the poor fishermen as labourers in fishing without giving much benefit to them.

In the Lotic Systems

In River Barak

The fishermen living along the bank of river Barak in Assam belong to all the four principal communities, *viz.*, Kaibarta, Patni, Maimal and Namasudra. They fish with different types of fish catching implements and devices, some of which are specific to river Barak and its big tributaries, like the Surma and the Kushiara, *e.g.*, the 'Chairon jal' ('Clap net') which is used to catch only the Hilsa. The fishermen operating the 'Chairon jal' belong mostly to the Maimal community. They live in the boats for days together and operate the gear day and night almost during the whole rainy season.

The fishermen, in general, are poor and mostly live hand to mouth. The operators of almost all kinds of gears, like Chairon jal, Leua jal, Basta jal, Lar barshi, Patan jal, Pan jal, Dheki jal, etc., after the day's catch, generally, sell out their catches to the nearby fish markets and buy the essential commodities required by them. Sometimes, fish traders from the markets buy the catches from the fishermen on the river bank itself. However, the operators of 'Maha jal' are generally affluent Lessees who employ a number of fishermen to fish with his 'Maha jal ' under his employment without giving much benefit to them.

The fisherfolk living along the bank of river Barak in its upstream region within the province of Manipur belong to different tribal communities, notably, Meitei, Hmar, Kuki, Naga, etc.; and, they fish with different types of indigenous fishing devices.

In the Major Tributaries of River Barak

The fishermen living along the banks of the downstream stretches of the major tributaries of river Barak, in Assam, notably, Jiri, Chiri, Madhura, Jatinga, Sonai, Rukni, Dhaleswari, Surma, Kushiara, Baleswar, Longai and Singla, also belong to the four principal fishing communities (*viz.*, Kaibarta, Patni, Maimal and Namasudra), as found along the banks of river Barak. However, except Surma and Kushiara, the upstream regions of all the other tributaries, as mentioned above, are inhabited by tribals belonging to different communities. Principal tribals living in the upstream stretches of these rivers include, the Meiteis, Nagas and Dimasas in Jiri; the Dimasas in Chiri and Madhura, the Dimasas and Hmars in Jatinga; the Dimasas, Hmars and Mizos in Sonai, the Mizos in Rukni, the Mizos and Riangs in Dhaleswari; the Karbis and Jaintias in Baleswar; the Mizos in Singla and the Mizos, Riangs and Jamatias in Longai. They fish in these rivers with the help of diverse kinds of fish catching devices; although, cast net forms one of the common fishing implements. These fishermen are economically poor on the average.

In Mizoram

In the Lentic Systems

In the Dils

All the Dils are generally under the joint supervision of Fishery and Tourism departments of the Govt. of Mizoram. As such, the riparian fishermen can not generally fish in the Dils in the way they would like to. The fishermen have to lead a very difficult life in the event of hardship.

In the Lotic Systems

The situation in rivers is different. Practically, there is not much stringent fishing regulations in the rivers. In Mizoram, fishermen belonging to different tribal communities, notably, the Mizo, the Mara, the Chakma, the Hmar, the Kuki and the Thadou fish in different rivers. The principal tribal communities living along and around the major rivers of Mizoram are as follows:

The Mizos in the rivers Tuirial, Tlawng, Tuirini, and Mat; the Maras in the river Kolodyne; the Chakmas in the rivers Karnafuli and Tuichong; the Mizos, the Hmars, the Kukis and the Vaipheis in the river Tuivai; and, the Hmars and the Mizos in the river Serlui. Moreover, there are a lot of tribal fishermen belonging mainly to the Hmar community along the bank of River Barak around Tipaimukh which is at the border of Manipur and Mizoram with Assam situated little far away.

In addition to the above, there are quite a few non-tribals, mostly belonging to the present Jharkhand province of India, who have settled in Mizoram, mostly around River Mat and River Tuichong; and, are seen to fish in these rivers. In addition to cast nets and gill nets, the tribals fish in these rivers with the help of different types of, mostly indigenous, fish catching devices and fishing implements.

The fishermen living along and around all these rivers are, in general, poor and live mostly hand to mouth. They generally do not get a rich catch even after fishing for 8-10 hours. They all need immediate rehabilitation and socio-economic upliftment.

In Tripura

In Tripura, the fishermen living in the plains are mostly Bengalees. They fish with the help of common fishing gears like the Jhaki jal, Patan jal, Lar barshi, etc. The fishermen living in the hills belong to different tribal communities, notably, the Riang, the Hrankhawl, the Jamatia and the Deb Barma. Also, there are few Bengali fishermen in the hills as well.

In the Lentic Systems

In the Sagars

'Sagars' (the big tanks constructed by earstwhile Monarchs of Tripura) are generally leased-out to the Fishermens' Co-operative Societies. The Societies exploit the Sagars through the local fishermen. Gill nets and encircling gears are generally used for catching the fishes.The catch is generally sold in auction in the local fish markets. Being Co-operative Societies, the members share the benefit almost equally amongst themselves. As such, the fishermen, here, seem to be comparatively little better off economically than their counterparts in Assam and Mizoram. They are quite knowledgeable in the progress in development taking place in the fisheries sector.

In the Lotic Systems

As already stated, in Tripura, the fishermen living along the river banks in the downstream stretches in the plains are mostly Bengalees; while, those living along the river banks in the upstream stretches in the hills are mostly tribals belonging to different communities.

Some of the stretches, mostly in the downstream sector, of certain rivers are sometimes leased out by the government to the Lessees who fish in those river segments employing the local fishermen without giving much benefit to them. The catches are generally sold out in auction in the local markets.

Chapter 17

Fishing Centres, Fish Landing Stations and Fish Marketing

In a lentic water body, fishing generally goes on in the entire waterspread area. Similarly, in a lotic system, fish catching usually goes on in the entire stretch of the river. However, in both lentic and lotic systems, there are certain interesting spots, which are significantly different from other areas of the water bodies with regard to possibility of fetching a richer catch. Such spots are more apparent at the DSL of the water bodies and are distinguished from others by certain features, like increased depth, favourable physico-chemical characteristics of water and so on. Such 'distinguished spots', where intense fishing goes on at DSL, are regarded as 'Fishing Centres'. On the other hand, the fish markets on the shoreline of the Beels or on the banks of rivers, where the caught fishes land for the first time, are called the 'Fish landing stations.' From the Fish landing stations, the fishes are marketed to other different markets in the district and province.

An account of the Fishing Centres (FC), Fish Landing Stations (FLS) and Fish Marketing (FM) is briefly given below:

In Assam

In Lentic bodies

In Sone Beel

Fishing Centres

During the rainy season, the entire water spread area of the Beel is used for intensive fishing at both domestic and commercial scales. However, at DSL, when the water spread area of the Beel shrinks, and the Beel becomes shallow, intensive fishing activities are restricted mainly to eight distinct deeper 'Fishing Centres', which are locally called 'Bundhs'. Every year, at DSL of the Beel, these 'Bundhs' are leased out principally to the Kaibarta fishermen living around the Beel, who fish in the 'Bundhs' mostly during the period November to April. Such Bundhs, lying at deeper portions of the Beel at DSL, are generally kept encircled with bamboo mat (locally called 'Khati bundh'), of height varying from 2.0-8.0 m; and, fishes are captured and marketed regularly, as required, during the entire period of operation.

The names of such 'Bundhs', with their Fishing Areas (FA), Months of Operation (MO) and Water Level (WL) as recorded at the site during the period of operation (Figure 6), are listed below:

Bundh	*FA (ha)*	*MO*	*WL (m)*
Belala	1.12	Aug-Apr	0.10-4.00
Chirakhaora	48.50	Nov-Apr	0.10-2.00
Nunail	6.50	Sep-Nov	2.00 -4.00
Putighat	2.00	Nov-Apr	0.10-2.00
Kurerkata	15.05	Dec-Mar	0.25-1.50
Kanagajar	50.00	Oct-Dec	1.50-3.00
Baroitila	105.90	Nov-Apr	0.10-2.00
Vashankuri	8.46	Dec-Mar	0.25-1.50

The 'Bundhs' witness rich harvest, particularly, on 'Paush Sangkranti' ('Pongol') Eve, when fishing goes on almost for the whole night in festive mood.

Fish Landing Stations and Transportation of Fish

There are 3 principal Fish Landing Stations (FLS) along the

Figure 6: 'Bundhs' (Fishing Centres) and Fish Landing Stations of Sone Beel

shoreline of Sone Beel. These are Debodwar FLS situated on the North-Eastern shoreline, Kalibaribazar FLS situated on the South-Eastern shoreline, and Phakhoagram FLS situated on the South-Western shoreline of the Beel. The position of the FLSs change with the changes in the water level of the Beel (Figure 6). An account of each FLS along with the ways of catch disposals is summarized below:

Debodwar FLS

Location, 10.6 km from Hailakandi town (district Hailakandi) by road, generally accessible throughout year, sometimes, on foot, with much difficulty.

Fish Transportation

Fishes in bamboo baskets (dia 0.5-1.0 m) are generally transported by road on omnibus, taxi, tricycle (rickshaw) and to a big extent on foot. Transported to Hailakandi town (mostly on headload on foot), to Karimganj town (mostly by bus), to Silchar town (mostly by bus), to Badarpur town (usually by bus) and to Lala Bazar and Katlichhara villages (generally by bus)

The 3 positions of Debodwar FLS, as shifted due to changes in the water level of the Beel, are indicated below:

Debodwar FLS-A: May-Sep; area, 700 m^2; Leasing amount (LA), Rs. 1401.00 (1979-80) and Rs. 1501.00 (1980-81). (LA= Leasing Amount)

Debodwar FLS-A^a: called Ichaganj; Oct-Dec; Area, 2800 m^2; LA, Rs.5595.00 (1979-80), Rs.7750.00 (1980-81).

Debodwar FLS-A^b: also called Ichaganj; Jan-Apr; Area, 1900 m^2; LA, as in A^a

The Lessee of the FLS collects a toll from each fisherman generally @ Rs.0.10 P per rupee earned.

Kalibaribazar FLS

Location: 12 km from Hailakandi town by road, generally accessible throughout the year; sometimes, on foot, with difficulty.

Fish transportation: Fishes in bamboo baskets (dia 0.5-1.0 m) are generally transported by road on omnibus, taxi, tricycle (rickshaw) and to a big extent on foot. Transported to Hailakandi town (mostly by bus and sometimes by taxi and also on headload on foot), to Karimganj town (mostly by bus), to Silchar town (mostly by bus), to Badarpur town (usually by bus) and to Lala Bazar and Katlichhara villages (generally by bus), to Ram Krishna Nagar village (mostly by bus and sometimes on foot

The 3 positions of Kalibaribazar FLS, as shifted due to changes in water level of the Beel, are indicated below:

FLS-B: Kalibaribazar: May-Sep; Area, 400 m^2; LA, Rs. 14,700.00 (1979-80) and Rs. 17,175.00 (1980-81).

FLS-B[a]: Kalibaribazar: Oct to mid-Nov; Area, 468.16 m^2; LA, as in B

FLS-B[b]: Kalibaribazar: Mid-Nov to mid-Apr; Area, 1490.78 m^2; LA, as in B

The Lessee of the FLS collects a toll from each fisherman generally @ Rs.0.10 P per rupee earned.

Phakhoagram FLS

Location: 31 km from Karimganj town (district Karimganj) by rail; accessible throughout the year by rail; little road services are available now; transportation is snapped, if railway line is submerged in floodwater due to devastating floods.

Fish transportation: Fishes in bamboo baskets (Dia 0.5-1.0 m) are transported by rail, by road transport and sometimes on foot with headload to Karimganj town and also to the nearby villages, like Baroigram, Kayasthagram, Nilambazar, etc.

Area of the FLS site: 182.40 m^2

Unlike the Debodwar and Kalibaribazar FLSs, Phakhoagram FLS is generally found to be located at the same site without any shift of its location. Since this FLS site belongs to the Indian Railways, no leasing amount or toll is collected from the fishermen by the Railway Authorities.

Fish Marketing

The marketing of fishes, which are of prime quality or of established acceptability, does not pose much problems. On the other hand, the marketing of the species, which are caught in large quantities having wide variations in their annual landings; or, which are caught in appreciable quantities but do not have much acceptability in the markets, pose much problems in marketing. The latter are called the 'lesser fish' or 'trash fish'.

The inland fish production, unlike marine fisheries (which comes only from capture fisheries), is contributed both by capture and culture fisheries from the freshwater resources. The marketing problems, which arise out of considerable variations characterizing marine fish production, creating glut and lean periods, are not of any significant dimensions in the case of production from the inland fisheries. Besides, the inland fish production, in general, is characterized by certain prize food fish species, which have established acceptability in the market and are sold generally in fresh condition mostly by weight system. The marketing of catch from culture fishery resources, netted out periodically during the whole year from perennial water bodies; and, during 2-3 months prior to monsoon from seasonal tanks, may not cause much problem; because the netted catch could be held in live condition in net enclosures for making phased supplies to the nearby markets. Problems may, however, arise in the case of capture fisheries from riverine and reservoir sources because of scattered landings at several places. It may, therefore, be necessary to identify the potential centres where such landings could be concentrated; so that, marketing conditions could be improved by providing marketing sheds, by keeping cold storage facilities and by providing fast transportation and good road conditions.

In Sone Beel, the fish traders generally buy the captured fishes from the fishermen at the FLSs, generally by paying a lump sum amount mainly on the basis of estimated weight. It approximately amounts to Rs.35.00/kg (on the average) for the small trash fishes, like the *Amblypharyngodon mola, Puntius* spp, *Rasbora* spp, *Chanda nama, Parambassis ranga,* etc.; while, it approximately amounts to 125.00/kg(on the average) for the medium-sized prize food fishes, like *Eutropiichthys vacha, Ailia coila, Mystus cavasius*, etc. The IMCs caught from the Beel, generally, fetch Rs.150.00/kg at the FLS itself. Further, at the FLSs, fishes like *Channa* spp, *Glossogobius giuris, Mastacembelus armatus, Macrognathus* spp, *Botia dario* fetch approx. Rs.55.00/kg (on the average); while, *Gudusia chapra* and *Salmostoma bacaila* fetch approx. Rs. 85.00/kg (on the average). Nevertheless, the live fishes like *Anabas testudineus* and *Clarias batrachus* generally fetch Rs.150.00/kg (approx.) on the average at the FLS itself. Interestingly, the fry of *Chanda nama* (locally called 'Bhujiya') caught generally in enormous quantity by Maha jal are sold out at the FLSs, @ Rs.20.00/kg (approx.) on the average. The big fishes, when caught in little large number (which is rare though) are, sometimes, sold out in auction at the FLS itself.

The following is the approximate conversion rate (as on 03.01.2006):

1 US$ = Indian Rs.50.00; 1 British £ = Indian Rs. 70.00

There is no well-built market sheds in any of the major FLSs in Sone Beel, except, few thatched huts which serve as market sheds. There is also no cold storage facility at the FLSs. It is only the fish traders, who sometimes bring ice from the towns in order to prevent decay of fish flesh. Fast transportation is a far cry even today. As such, the fish traders, after buying the fishes from the fishermen, make arrangements for immediate transportation of the fishes from the FLSs to the urban markets by whatever means of transport available at that point of time, *e.g.*, bus, taxi, rickshaw or even on foot carrying headload.

It is very sad to note here that, the poor fishermen, who generally live hand to mouth, pay tolls as taxes to the Excheqquer. But, inspite of this; and, further, inspite of the fact that, there is a Sone Beel Fishermens' Co-operative Society, these poor fishermen are deprived of the bare minimum facilities which should have been given to them for fish trade.

Notwithstanding the above, it appears to be little pertinent here to discuss little bit about the Sone Beel Fishermens' Co-operative Society (SBFCS) and the leasing pattern of the Beel.

The Sone Beel Fishermens' Co-operative Society

The Sone Beel Fishermens' Co-operative Society (SBFCS) is a registered Society managed by the fishermen living around the Beel. The Society has a Chairman (elected directly by the fishermen), a Vice-Chairman and few members. The Secretary of the Society is deputed by the Government. The Beels in Assam are under the administrative control of the Deputy Commissioners, they being considered as 'Khas' lands. In another context, the Beels in Assam are also under the administrative control of the Assam Fisheries Development Corporation (AFDC) for their 'so-called' management. The AFDC floats tender notices; and, in response to the tender notice, the SBFCS as well as certain affluent individuals try to obtain the Beel on lease from the AFDC. Except on very rare instances, the Sone Beel Fishermens' Co-operative Society generally gets preference over the individuals in obtaining the Beel on lease from the AFDC. The lease amount varies from Rs. 1,50,000.00 approx. (during 1978) to Rs. 5,00,000.00 approx. (during 2002). But, inspite of this, due to some amount of mismanagement during the tenure of certain

Managing Committee of SBFCS, the Society remains a 'defaulter' in not being able to clear the revenue share which it is supposed to pay to the Govt. Nevertheless, during the tenure of some other Managing Committee of SBFCS, the Society made a big amount of profit and cleared the revenue share which it was supposed to pay to the Govt. The SBFCS raises the revenue by leasing-out the 'Bundhs' to different groups and individuals. It may be mentioned here that, fishing restriction is imposed only in the 'Bundhs' at DSL of the Beel after the lease becomes effective. At FSL of the Beel, any fisherman can fish anywhere in the Beel. Sometimes, there occurs Law and Order problem, when fishing in Sone Beel (particularly in the 'Bundhs' at DSL) becomes a bone of contention between rival groups.

In Chatla Haor

Fishing Centres

Chatla Haor is a much smaller seasonal floodplain wetland as compared to Sone Beel. It does not have any DSL and is almost completely dry during the winter. There are only a few insignificant fishing centres (locally called 'Kheo' or 'Jheng') on the NW side of the Haor where fishing is to be completed and all the fishes are to be harvested within Autumn.

Harvest includes different types of small fishes and also big fishes, notably, the IMCs.

Fish Landing Stations and Transportation of fish

There are only two major Fish Landing Stations (FLS), one on each shore, *viz.*, the Rajpur FLS on the west shore line (towards North-West) and the Silcoorie Machhghat FLS on the east shore (towards South-East). The Rajpur FLS witnesses fish landing only in the morning shift (beginning around 6 am) while the Silcoorie Machhghat FLS witnesses fish landing in both morning (beginning around 6 am) and afternoon (beginning around 2 pm) shifts. The Machhghat FLS shifts its position, sometimes quite frequently, upto around island village Bairagitila (situated *c* 3 km away from original 1st position which is situated adjacent to the Silchar-Hailakandi Highway) due to changes in the water level of the Haor, particularly in the month of October.

Surprisingly, Chatla Haor is said to be not within the administrative control of the AFDC; and, as such, it is not leased-out on auction by the government. Rather, a portion of the Haor, called the'Upor Beel', is leased-out by the Management of the Silcoorie TE every year to private entrepreneurs, on quite lucrative sum of money (the lease amount varies from Rs.50,000.00 approx.during 1980 to Rs.7,00,000.00 approx.during 2002); because, the 'Upor Beel' area is said to be within the jurisdiction of Silcoorie TE. The other portion, locally called 'Lamar Beel', is not said to be within the jurisdiction of Silcoorie TE; and, here, fishing is free for all during the entire period the Haor has water (from around May to October every year). It may be mentioned here that, an embankment-cum-road (which is a State Highway connecting Silchar city with Hailakandi town with the Assam Central University in between), said to have been constructed by the E and D Department of the Govt. of Assam, has partitioned the Chatla Haor into 'Upor Beel' and 'Lamar Beel'. Interestingly, although, Chatla Haor does not naturally have any DSL and is completely dry during the winter, water is retained in the 'Upor Beel', by the Lessee, by constructing embankments after the lease becomes effective. Concomitantly, the fishes therein, are prevented from escaping by erecting tall' Khati bundh' (bamboo mat fencing). The lease period usually ranges from 15 April to 31 March of the following year. Having no specific connection with the Govt in leasing-out of the Haor, no toll/tax is seen to be collected from the fishermen/fish traders in any of the FLSs in the Haor, unlike the situation in Sone Beel.

Chatla Haor is connected to Silchar city by vehicular traffic through two main routes, *viz.*, (a) the Hailakandi-Silchar State Highway connecting Silcoorie Machhghat FLS with Silchar city and (b) the Kathal Bagan-Silchar city road connecting the Rajpur FLS with Silchar city. The Silchar city is situated *c* 6-10 km away from the FLSs along both the roads. As such, being not very far, fishes are transported quite fast from either FLS. From Silcoorie Machhghat FLS, fishes are generally transported by trucks and autorickshaws hired by the fishermen. Also, fishes are transported from here by buses. A good consignment of fishes from here are carried as headload on foot by Muslim fish traders to the markets situated at Sonabarighat, Nutanbazar, etc., covering a distance of *c* 10-15 km. From Rajpur FLS, the fishes are transported mainly by jeeps and sometimes by buses and rickshaws; and, also on foot. Silchar city, being not very far, provision of cold storage is maintained by procuring ice from the city.

Fish Marketing

At the FLSs in Chatla Haor, rates, generally in vogue, for the different kinds of fishes are as follows:

Small trash fishes like *Amblypharyngodon mola, Chanda nama, Parambassis ranga, Glossogobius giuris, Puntius* spp, etc.: @ Rs.30.00-40.00 per kg.

Medium sized prize food fishes like, *Mystus cavasius, Ompok bimaculatus, Ailia coila, Eutropiichthys vacha,* etc.: @ Rs.125.00-150.00 per kg

IMCs: @ Rs. 125.00-150.00 per kg

The live fishes like the *Anabas testudineus* and *Clarias batrachus*: @ Rs. 150.00-175.00 per kg.

Other fishes like *Channa* spp, *Mastacembelus armatus, Macrognathus* spp, etc.: @ Rs. 50.00-60.00 per kg

Fishes like, *Salmostoma bacaila* and *Gudusia chapra:* @ Rs. 80.00-110.00 per kg.

Fishermens' Co-operative/Development Society

Unlike in Sone Beel, Chatla Haor does not have a Chatla Haor Fishermens' Co-operative Society. Nevertheless, there are a number of small registered or unregistered 'clubs', all of which are found to be not very active and concerned about the socio-economic upliftment of the poor fishermen. A 'Chatla Haor Development Society' has been initiated as a Registered Body; but, the members of the Society do not seem to be very active for the development of the locality.

The Anuas

The Anuas are said to be leased-out by the DC/AFDC to private entrepreneurs. The lease amount varies according to the prevailing rate and dimension of the water body. The Lessee fishes in the Anuas employing generally the local fishermen but do not seem to give much benefit to them. Most of the Anuas are highly eutrophic and choked due to prolific growth of the weeds, notably, the *Eichhornia crassipes*. The Lessee, sometimes, also, has to shoulder the onerous responsibility of clearing the Anuas of enormous amount of weeds in order to fish, in them, properly. The Anuas are, in general, deeper than the Beels and Haors. There is also not much fluctuation in the water level between rainy and dry seasons. As such, fishing is done throughout the whole Anua without much discrimination of areas by depth. There is usually a market near the shore of each Anua or little away. The fishes, if in small amount, are sold on retail basis on lump-sum amount in the nearby markets. If in large quantity, the fishes are sold, on whole sale basis, through auction. They are also sometimes transported by taxi, bus, etc., to the nearby urban markets. The rates of fishes vary from Rs. 40.00 to Rs.50.00 per kg for trash fishes; Rs.80.00 to Rs.150.00 per kg for the medium- sized prize food fishes; and, Rs. 150.00 to Rs. 200.00 per kg for the big fishes. Activities of Fishermens' Co-operative Society, if any, is not very much evident in any of the Anuas.

In the Lotic Systems

Fishing Centres

Fishing is done throughout the whole length of a river. However, a considerably big river, in its downstream region, generally, has deep pockets, called 'deep pools' (locally called 'Dhar'), where big-sized fishes, notably, the IMCs, *Bagarius bagarius, Aorichthys* spp, etc may take shelter, may be in order to feed, breed and evade the gears. These spots could be regarded as 'fishing centres' in the rivers. The river Barak is leased-out, on auction/tender by the Deputy Commissioner (DC), Cachar, segment-wise, to different Lessees for varying periods on different lease amounts. The Lessee fishes in the river Barak, within the leased portion of his jurisdiction, by employing fishermen (without giving much benefit to them), mainly by operating Maha jal. Intensive fishing operation goes on generally during the period November to April.

An almost similar situation is in vogue in all the tributaries of Barak, notably, the rivers, Jiri, Chiri, Madhura, Jatinga, Ghagra, Sonai, Rukni, Dhaleswari, Kushiara, Surma, Singla, Longai, Baleswar, etc. Notably, portions of River Dhaleswari is leased-out by the Forest Department of the Govt. of Asssam and not by the DC's establishment.

Fish Landing Stations

There are fish landing stations (FLS) on or near the banks of the river Barak and all its tributaries in different sectors of their course. For example, there are FLSs in different parts of Silchar city where the landings include a substantive part of catches from river Barak. Other prominent FLSs along the course of river are situated in the places like Lakhipur, Baskandi, Sonai, Sonabarighat, Masimpur,

Srikona, Salchapra, Katakhal, Panchgram, Badarpur, etc. The prominent FLSs along the course of the tributaries of river Barak are given below:

FLS in village Jirighat near the bank of river Jiri

FLS in the villages Harinagar, Rajabazar and Joypur near the banks of river Chiri

FLS in the villages Thaligram and Udharbond near the banks of river Madhura

FLS in the villages Harangajao, Damchhara, Balachhara, Borkhola near the banks of river Jatinga

FLS in the villages Kathal bagan and Srikona near the banks of river Ghagra

FLS in the villages Maniarkhal, Amraghat and Sonai near the banks of river Sonai

FLS in the villages Hawaithang, Bhagabazar, Dholai, Kabuganj and Palonghat near the banks of river Rukni

FLS in the villages Gharmura, Katlichhara, Lalabazar, Matijuri, Mohanpur, and Katakhal near the banks of river Dhaleswari

FLS in village Bhanga near the banks of river Kushiara

FLS in Chargola near the banks of river Kushiara and Kachua (the outlet of Sone Beel)

FLS in different parts of Karimganj city near the banks of river Kushiara

FLS in villages Rongpur, Nivia, Dullavchhara, Anipur near the banks of river Singla (the inlet of Sone Beel)

FLS. in villages Rangamati, Bajarichhara, Lowairpowa, Patharkandi, Baroigram, Nilambazar near the banks of river Longai

FLS in villages Gumra, Jalalpur, Malidahar near the banks of river Baleswar.

Fish Marketing and Transportation of Fish

At the FLSs on the banks of different rivers, the prices of the different species of fishes at different sites are, generally, as follows:

The trash fishes, like *Puntius* spp, *Rasbora* sp, etc: @ Rs.35.00-40.00 per kg

The medium-sized prize food fishes like the *Ompok bimaculatus, Eutropiichthys vacha, Ailia coila,* etc,: @ Rs. 100.00-150.00 per kg

Fishes like *Salmostoma bacaila, Gudusia chapra,* etc: Rs. 80.00-120.00 per kg

The IMCs: @ Rs 125.00-150.00

Hilsa: @ Rs.70.00-100.00 per kg

The fishes are transported generally by trucks, buses, taxis, trains, etc., mainly to the different urban markets.

In Mizoram

In the Lentic bodies

In the Dils

As indicated earlier, the Dils are under the joint supervision of both Fishery and Tourism Departments of the Govt. of Mizoram. They are mainly meant for generating revenue for the State Exchequer through tourism. Not being much exploited commercially, the aspects of fishing centres, fish landing stations and fish transportation and fish marketing are not very much relevant in the case of Dils.

In the Lotic bodies

In the rivers

The rivers are fished all along their length. Fishing is done at different spots. Fishing spots are chosen according to the microhabitats a particular species prefers to live in. Strategy of fishing is determined according to the microhabit and the fish species.

There is atleast one FLS on the bank of each river or a little away from it. FLSs are not too many along the banks of the rivers but are restricted to only those places which are accessible by man.

The location of some of the prominent FLS on the banks of different rivers in Mizoram are given below:

FLS at village Tuirial on the bank of river Tuirial

FLS at village Sairong and at village Bairobi on the bank of river Tlawng

FLS at village Seiling little away from the bank of river Tuirini

FLS at village Ngopa little away from the bank of river Tuivai

FLS at village Mat bridge on the bank of river Mat

FLS at village Kalchaw on the bank of river Kolodyne

FLS at Demagiri town on the bank of river Karnafuli

FLS at village Chemphai on the bank of river Serlui and at Belkhatri a little away from river Serlui.

The approx. price of different fish species (on the average) at the FLSs on the bank of different rivers is given below:

Neolissochilus hexagonolopis: @ Rs.150.00-175.00 per kg

Crossocheilus latius latius: @ Rs.100.00-150.00 per kg

Garra spp: @ Rs.100.00-125.00 per kg

Semiplotus semiplotus: @ Rs.90.00-120.00 per kg

IMCs (caught out of seeding done by the Fishery Department of the Govt.of Mizoram): @ Rs. 100.00-150.00 per kg

Puntius spp: @ Rs. 100.00-120.00 per kg

Danio spp: @ Rs.100.00-125.00 per kg

The fishes caught from the rivers, are transported by the fishermen, generally, on foot with headload to the FISs.

In Tripura

In Tripura, it is a happy blending of rheophilic and plainwater situations

In Lentic Systems

In the Sagars

Fishing operations are quite easy in the Sagars which are not affected by severe eutrophication, *e.g.*, the Amar Sagar, Fatik Sagar, etc. On the other hand, fishing seems to be quite difficult in the Sagars which are affected by severe eutrophication, *e.g.*, the Rudra Sagar. The Fishermens' Co-operative Societies fish in the Sagars by operating different types of gears.

There is a FLS on or near of shore of each Sagar. The fishes caught, are, generally, sold-out on auction. Some of the prominent FLSs on or near the shore of each Sagar are as follows:

FLS situated around village Melagarh near the shore of Rudra Sagar

FLS situated around Amarpur town near the shores of Amar Sagar and Fatik Sagar

In addition to the above, a huge quantity of fishes are caught from the Dombaru Reservoir created through damming of the river Gomati at villages Mandirghat/Tirthamukh. Although there is a lot of insurgency problem in those areas, the fishes, after being caught from the reservoir, are transported mainly on foot with headload (sometimes, by jeeps, if available) to the FLSs at Jatanbari and Nutanbazar.

Fishes are transported from the Sagars, generally, on foot with headload or in tricycles to the FLSs situated on or near the shores of the Sagars.

The approx. price (on the average) of some of the fish species at the FLSs near the shores of the Sagars are as follows:

Salmostoma bacaila: @ Rs. 60.00-80.00 per kg

Puntius spp: @ Rs. 40.00-60.00 per kg

Mystus spp: @ Rs. 55.00-70.00 per kg

Amblypharyngodon mola: @ Rs. 40.00-50.00 per kg

Rasbora daniconius: @ Rs. 40.00-50.00 per kg

Danio spp @ Rs. 55.00-65.00 per kg

In the Lotic Systems

In the Rivers

The rivers are generally fished all along their length. In the rheophilic upstream stretches of the rivers, where mostly the tribals (with some Bengalis) fish, the fishing spots are selected according to the preference of the fishes for a particular kind of microhabitat. Likewise,the *modus operandi* for fishisng is determined according to the nature of the microhabitat as well as the fish species type.

As in the case of the lentic systems, in the lotic systems also, there is, usually, a FLS (sometimes more than one) in each town, situated on the bank of each river. Some of the prominent FLSs in some of the rivers are given below:

FLS at village Chha Manu and at Manughat on the bank of river Manu

FLS at village Ganganagar, at village Chakmaghat and at Teliamura town on the bank of river Khowai

FLS at village Jatanbari, village Natunbazar, Amarpur town, Udaipur town and at village Melagarh on the bank of river Gomati.

FLS at Sabrum town on the bank of river Feni.

FLSs at different parts of Agartala city on the bank of river Howrah.

The approx. rates (on the average) of some of the fish species at different FLSs on the banks of some of the rivers are given below:

Puntius spp, *Danio* spp, *Rasbora daniconius, Mystus* spp, etc.: @ Rs. 50.00-70.00 per kg

Ompok bimaculatus, Ailia coila, Eutropiichthys vacha, etc.: @ Rs.100.00-150.00

IMCs: @ Rs.100.00-150.00 per kg

Aorichthys spp @ Rs.100.00-150.00 per kg

Chitala chitala @ Rs.125.00-150.00 per kg

Bagarius bagarius: @ Rs.125.00-150.00 per kg

The fishes are transported from the FLSs generally to the urban markets usually by tricycles, trucks, buses and sometimes on foot carrying headload.

Ornamental Fishes

Colourful and peculiar fishes are generally called the 'ornamental fishes.' These fishes, kept in aquarium, are used for decorating the house. Of late, they have occupied a significant place in the commercial trade, particularly in earning foreign exchange. A brief account of some of the informations on ornamental fishes is given below:

In India, aquarium trade mainly depends on the exotic fish in order to fulfil the demands of the hobbyists. *c* 288 species of exotic ornamental fishes have been introduced in India. Nevertheless, in our country, particularly in the Biodiversity Hot Spot North-Eastern region, a significant number of ichthyospecies are found in the natural water bodies, particularly in the rheophilic stretches, which have immense ornamental value due to their splendid colour, shape, manageable size, compatibility, longevity, etc. A huge bulk of indigenous ornamental fishes have high demand in the overseas markets, particularly, in Singapore, Thailand, Japan, The Netherlands, etc. This was revealed from a survey in Kolkata as well as through interactions with the exporters. In different countries, lucrative trade with ornamental fishes, like the *Colisa* spp., has been going-on through farming and captive breeding (Mahapatra *et al.*, 2003).

Further, attempts have been made to study *in vitro* sexual dimorphism of certain ornamental fish species in NE India. Extraneously discernible sexual dimorphism in certain ornamental Daniids, *viz.*, *Brachydanio rerio, Danio aequipinnatus, Danio dangila* have been studied. These fishes generally dwell in the fluviatile wetlands of NE India. Body shape, mouth profile, fin characteristics and colouration are the principal morphotraits identified and significant for *in vitro* sex differentiation in these species. Such revealations, hitherto remained non-descript, could help accentuate the ornamental fish trade amongst the aquaculturists. In addition, mass awareness is essential to check anthropogenic stress including extra-intensive fishing, habitat destruction and efforts for replenishment of fast depleting natural stock (Sarmah and Dey, 2003).

Chapter 18

Coldwater Fishery

Fisheries in streams and lakes situated in high altitude regions of the country comprise the 'coldwater fisheries'. Such a fishery comprise mainly of the Mahseers and generally the fishes belonging to the family Salmonidae.

The mountains in and around the Indian Sub-continent had been the abode of Saints and Naturalists since time immemorial. However, of late, these mountains have become a place of recreation, often leading to damages of the fragile ecosystem. It may be noted here that, *c* 10 per cent of the global population, belonging mostly to the developing countries, like Asia, Africa and South America, live in mountainous regions. Increasing man-made pressure resulted in fast rate of eutrophication of the water bodies. Many upland lakes in Kashmir, Uttaranchal, Western Ghats and the North-East have undergone dereliction beyond any repair.

Our knowledge of aquatic ecology and aquatic Biodiversity date back to the Vedic period. The first discernible account of modification in fishes due to rheophilic habitat could be traced back to the 'Manu Smriti' and the 'Sutras', the two integral components of the Vedas. Mentions of 'Mahasalka' (*Tor putitora*) and Chakratunda (*Garra gotyla*) are found in the Epics (Hora, 1951). Distribution of coldwater fishes in the Himalayas and other uplands find mentions in the Susruta Samhita (Hora, 1953). Susruta further described the body modifications undergone by the coldwater fishes in the form of flat abdomen as an adaptation to rheophilic condition in the fast mountain streams. Mention of Mahasalka finds a place in Manu Smriti (III, 268) which relates to fishes possessing large scales ('Mahseers'). The golden Mahseer (*Tor putitora*) had been regarded as a privileged fish which used to be relished by Saints and Brahmins, especially for appeasing the souls of deceased ancestors for an endless period. Somesvara discussed the details of use of rods, hooks, etc.for angling of Mahseers.

During the pre-independence period, studies on coldwater fisheries, generally, remained confined to the systematics, ecology and biology of the fishes in the coldwater regions along with the impact of the river valley projects on the fish resources. The impact of changing environmental conditions in the upland waters *vis-à-vis* the need for conservation of fish bioresources was realized during the 1940's. This trend continued till around 1960's. The establishment of a 'Coldwater Fisheries Research Centre' by the Indian Council of Agricultural Research (Ministry of Agriculture) was, perhaps, the first step towards beginning of scientific research in the field of coldwater fisheries. This had also encouraged the Universities in the Himalayan belt to take up research works in the field of coldwater fisheries.

Study of geological history of the Indian Sub-continent reveals that, Tethys Sea had occupied the region which is now occupied by the Himalayas and the Indo-gangetic plains. An enormous landmass, known as the Gondwana land, had existed towards the south of this sea, a portion of which represents the Deccan Plateau today. This gradually led to the formation of the mighty Himalayas. With the advancement of time and continuous upheaval, the Himalayan region attained further heights. The Himalayas, consisting of chain of mountains, can conveniently be classified into three zones, *viz.*, (a) the Greater Himalayas containing the highest peaks in India, (b) the Lesser Himalayas which includes the spurs and the highly dissected uplands, and (c) the Siwaliks which constitute the southernmost section containing rocks and broken chain of mountains. The Peninsular is, similarly, classified into (a) the Malwa Plateau in the North and (b) the Deccan Plateau in the south. From the coldwater fisheries point of view, the Western Ghats include the Sahyadri, the Nilgiris, the Annamalai and the Cardamom Hills.The principal watershed of the region is constituted by about

19 big rivers, of which, the Ganges, the Brahmaputra and the Indus forms the major component. On the other hand, the principal rivers in the south are the Cauvery, the Krishna and the Godavari.

The following are some of the major lentic systems from the coldwater fisheries point of view:

1. The Manas Sarovar: area *c* 350 km^2 situated in the Greater Himalayan ranges south of the Kailash mountain
2. The Rakas: area *c* 260 km^2 situated also in the Greater Himalayan ranges south of the Kailash mountain
3. The Kumaon Lakes in UP situated in the Lesser Himalayan and Siwalik ranges
4. The Dal and Wular lakes in Kashmir
5. Certain small lakes in Ooty and Kodaikanal
6. Lotic and lentic systems in the NE Hill region.

The Govindsagar and the Pong are the important Reservoirs in the Himalayan region from the point of view of coldwater fisheries.

Ichthyodiversity

Coldwater fishes have been mentioned in the ancient Hindu literature. Hamilton Buchanan (1822) is regarded as the first person to prepare a scientific account of the fishes of the Ganges and its environs. Day (1878, 1889) described about the fishes found in the mountain streams and lakes. Before that, Heckel (1844) dealt with the fishes of Kashmir and Ladakh. Subsequently, many workers have been working on different aspects of coldwater fishery (Hora, 1932, 1939, 1940; Hora and Silas, 1952; Menon, 1954; Misra, 1959). In addition to the above, attempts have been made to classify the coldwater fishes on the basis of their rheophilic adaptations (Menon, 1954) as well as on the basis of their distribution (Hutchinson, 1939; Jhingran and Sehgal, 1978). Works had also been directed towards the evolution of upland fishes (Hora, 1937; Hora and Menon, 1952). The outcome of these studies portrayed that, the evolution of Schizothoracids in the Himalayas had, probably, occurred during the first Inter-glacial period. The tremendous amount of proliferation of the Schizothoracinae had, probably, occurred during the second and subsequent inter-glacial periods when favourable environmental conditions were supposed to have prevailed with the formation of East-West drainage systems; thus, enabling migration of fishes westwards to Kashmir and even beyond. Conversely, with emphasis on coldwater fisheries, *c* 102 species of fishes have been listed from the Western Ghats (Day, 1865; Hora and Law, 1941; Sehgal, 1991). These studies led to the splitting of ichthyofauna of the Western Ghats into three divisions, *viz.*, the Deccan Trap (south of Tapti to Goa), the Central Division (Goa to Karnataka) and the Southern Division (Tamil Nadu to Kerala).

Notwithstanding the above, six species of Mahseers have been recognized (Sen and Jayaram, 1982). Further, efforts have been made to revise the valid species of Schizothoracinae.(Misra, 1976; Talwar, 1978). Studies have been conducted on the distributional pattern of the fish species from Garhwal Himalayas along with description of some new records (Badola and Dobriyal, 1987). Also, various expeditions had been made, after Heckel (1844), for the revision of the fishes of Kashmir (Silas, 1960; Talwar, 1978). In addition to the endemic coldwater fish species, there are some amount of information on the exotic coldwater fish species of the uplands of India (Sehgal, 1987).

There had not been much detailed systematic work to assess the ichthyoresources and the fish yield per unit area in the Indian uplands. Attempts had been made to investigate the ecology and ichthyoresources of the river Jhelum with special emphasis on Schizothoracids (Sehgal, 1967), followed by estimation of fish resources of Himachal Pradesh (Sehgal, *et al.*, 1971), Kerala, Tamil Nadu and Karnataka in Western Ghats (Sehgal, 1971).

Research Works Based on Ecomorphology

With the establishment of Zoological Survey of India (ZSI) during the early part of 20th century, there had been attempts to study the ichthyofauna of the hill streams (Day, 1878; Annandale, 1918; Hora, 1921, 1930, 1938; Hutchinson, 1939; Saxena and Chandy, 1966). Third Netherland Karakoram Expedition (Hora and Mukherjee, 1935) and Yale North India Expedition (Hora, 1936; Mukherjee, 1936) described the then ecological conditions prevailing in the lentic and lotic systems. These studies brought to light the ecomorphological adaptations developed in these fishes in response to their life in the rheophilic conditions. Hora *et al.* (1955) discussed certain unique characteristics of aquatic fauna of the Kashmir valley.

Further, the ichthyogeography of the torrential fishes of the Himalayan region had been studied by various workers, notably, Menon (1954), Sehgal (1974, 1988), Joshi *et al.* (1978).

Categories of Cold Water Fishery

Notwithstanding the above, on the basis of certain physical characteristics of the streams, *viz.*, origin, permanence, size, gradient, current velocity, substratum, thermal regime and biological diversities, cold water (snow melting) streams could be classified into two zones, *viz.*,

Rhithron

Head portion of the stream which generally has a monthly mean temperature of 17.3°C, DO of 10.0 mg/lit, current velocity of 0.9-1.8 m/sec, substratum consisting mainly of boulders and cobbles with pools here and there; and, inhabited mainly by the stenothermic trouts and snow trouts.

Potamon

Tail portion of the stream which generally has a monthly mean temperature of 22.1°C, Dissolved oxygen 8.1 mg/lit, current velocity 0.5-0.7 m/sec having mostly laminar flow, substratum consisting mostly of gravels and fines and less boulders, occurrence of patches of aquatic vegetations and inhabited mostly by eurythermic or warm-stenothermic minor carps, barils and mahseers. Nevertheless, there are, sometimes, certain amount of overlappings between the ichthyofauna of the tail end of rhithron and head end of the potamon zone. Sehgal (1988), therefore, made another attempt to classify the streams based on geographical location, abiotic and biotic characteristics of the particular section of the river and the predominant fish species. He had compartmentalized the stream into three sections which can be applied either in the same river system or in different rivers of any mountain, say the Himalayas. These sections are:

1. The Greater Himalayan section
2. The Middle Himalayan section
3. The Siwalik streams

Contrary to the above, the trophic levels of oligotrophic and near mesotrophic lakes in the high altitude, especially in the Kashmir

Himalayas, depict poor biological productivity and low nutrient status (Zutshi *et al.*, 1970). Also, the lakes in the Kumaon Himalayas generally portray a gradual degradation in ecological parameters; thus, rendering these water bodies uncolonisable by the fishes (Sehgal,1975; Sharma and Pant, 1984). On the other hand, investigations into the coldwater reservoirs, till 1970s, were restricted to few in the Western Ghats, notably, the Nilgiris and the Munnar High Range. Most of these reservoirs are either oligotrophic or mesotrophic (Sreenivasan, 1968; Jhingran and Sehgal, 1978).

Fish Biology

Studies on the fish biology of the coldwater species are mainly restricted to food and feeding habits, maturity and fecundity, developmental stages and age and growth. Studies revealed highest fecundity in *Tor putitora* and *Labeo dero* ranging from 52,000-85,000 and from 9,150-95,000 ova respectively (Sehgal, 1992). With regard to food and feeding habits, *Schizothorax richardsonii* have been found to subsist on benthic micro-algae growing over rocks and boulders (Sehgal, 1979). Studies on the age and growth of *Tor putitora* with the help of scales and bones had revealed that opercular bone annulus was very clearly marked.

Natural Fishery

With regard to coldwater fishes, studies on the natural fishery was confined mainly to streams, lakes and reservoirs in different regions of the uplands. In the hill streams, the natural fisheries are generally of the subsistence type; while, in the lentic systems, fishing could go on a commercial scale. The generally poor fish yield in the turbulent upstream sector could be attributed to the various physical and biological parameters. The physical factors which control the propagation and well being of the coldwater fishes in the upland streams, are generally related to the geographical conditions which prevail in the rheophilic stretches. 'U'-shaped narrow valleys, deep gorges, turbulence, non-availability of large number of pools, etc, make the stretches less accessible. Further, low temperature regime generally results in poor growth rate and yield. Moreover, the fish catch per unit effort (CPUE) in the upland waters is very poor.

Culture Fishery

In the hilly regions of our country, fish culture is generally carried out in ponds and lakes and sometimes in the running water raceways. Some of the natural rivers and streams, particularly in Mizoram and Tripura, are also seeded with the seeds of the IMCs. However, aquaculture programmes generally have a longer gestation period when compared with fishing in natural waters or other forms of food production. Nevertheless, the activities of culture fisheries in the hill regions have, by and large, remained confined to the individual States in their respective coldwater regimes. Earlier workers who had dealt with introduction and culture of coldwater fishes include Day (1878), Wilson (1909), Mackay (1944), Sehgal (1970), etc.

The aquaculture of Mahseer fishes (notably, *Tor khudree* and *T. putitora*) could be traced back to the 1970s. There have been some amount of success in the induced breeding of the Mahseer fishes in Lonavala and Bhimtal fish farms (Kulkarni, 1990; Sehgal, 1991). Recently, a moderate size hatchery with flow system had been devised at NRC on Coldwater Fisheries at Bhimtal in connection with induced breeding of *Tor putitora*. There have been an almost three fold increase in terms of hatching and survival in comparison to the conventional methods of hatching and seed rearing. Likewise, there had been some successful attempts towards evolving technology for aquaculture of certain species of schizothoracids (Raina *et al.*, 1986).

Sport Fisheries

The art of angling or sport fishing was well known to the ancient Indians (Hora, 1951). Today, as a recreational activity, sport fishing is gaining importance almost all over the world. Historical resume' of sport fishing includes the works of Thomas (1873), Dhu (1923), Hora (1943), Sehgal (1987), etc. The status of sport fishery of brown trout in the Himalayan region is not very encouraging. The account on trout fishing in Kashmir valley during 1955 fishing season revealed that, 66.0-100 per cent fishes hooked in Kashmir valley are < 250 mm in length and 400 g in weight.

The other fishes of importance in Sport fisheries are the Mahseers, notably, the *Tor tor*, *T. putitora* and the *T. mosal*. The status of Mahseer fisheries as a sport is understood from the catches recorded by the important angling associations in the country. The Transworld Fishing Expedition caught *T.putitora* in the Himalayan waters whose sizes ranged from 1.0-12.0 kg on the average (TWFE, 1984). The biggest catch of 20.0-32.0 kg was hauled by the Wildlife Association of South India. Angling activities, particularly angling of the Mahseers, in the NE Himalayan rivers in Assam, specifically in Northern part of Assam, is managed largely by the Assam (Bhorelli) Angling Association who had reported *Tor putitora* hauls ranging in size from 5.0-25.0 kg.

Brief Description of Trouts in Different Coldwaters

A brief description of the trouts in the Himalayan and in the Peninsular waters is given below:

Trouts in the Nilgiris

The pioneering attempt to transplant trouts in India was made in the Nilgiris located in the Western Ghats. The first two attempts to introduce trouts (*Salmo trutta*) from England was made by Sir Francis Day in 1863 and 1866; but, the attempts were not successful. Wilson succeeded not only in transplanting trout in the Nilgiris; but, also in establishing trout culture in that area on a scientific basis. Menon (1954) had reported the occurrence of rainbow trout (*Salmo gairdneri Shasta*), among others, as a principal species component among the Nilgiri trouts.

Trouts in Kashmir

Trouts had been introduced in the Kashmir waters independently of the Nilgiris. The first shipment of trout ova to Kashmir was said to have been done during 1900 AD which, however, did not meet with much success. Not much attempts had been made to introduce trouts in Kashmir waters till 1912. The trouts in Kashmir had been surveyed by Sehgal *et al.* (1971). They afford a sport fishery which attract anglers from almost all over the world. The brown trout seems to be more common though rainbow trouts are also available in certain streams.

Trouts in Himachal Pradesh

Brown trout was first introduced in the river Beas in Kulu valley in 1909 by Howell (1916). This stock, which got established there, is the source of trout fry introduced in the Beas and other streams. *c* 5,000 eyed eggs of the rainbow trouts were introduced in Kulu waters from Kashmir in 1919. Some of the important trout streams in

Himachal Pradesh include the river Baspa and its tributaries pertaining to the Sutlej system and the river Pabar of the Yamuna system.

Chapter 19

Diseases in Freshwater Fishes

The teleosts are said to inhabit an alien world and their utilization by man for food or sport has, until recently, been virtually dependent on the hunting of wild species. It is only in the last 100 years that intensive cultural methods for a few species have evolved. As such, the diseases which result from the intensive cultural techniques have been a major constraint on the economic development of aquaculture.

The aquatic environment encompasses a wide variety of parameters, virtually all of which influence the maintenance of homeostatis, essential for growth and reproduction of fishes. These, if altered beyond acceptable limits, may pre-dispose to, or actually cause, disease.

All surface waters may contain species of wild fish which could act as reservoirs of infectious disease. Further, toxin-producing algae are found in freshwaters in addition to marine and brackish waters. Under suitable environmental conditions, they grow to considerable cell densities, called 'blooms' or 'tides'; and, during or after blooms, toxins may be produced which are lethal to fish.

The anatomy and physiology of fish is modified principally towards the two major ecological factors which control their existence; *viz.*, the aquatic environment and the poikilotherm's inability to control its temperature. These factors are of overriding significance in dictating the chain of events following any pathological attack, such as, microbial infection, traumatic damage or nutritional deficiency.

In fish disease, more readily than in disease of any other farmed or wild species, it is possible to recognize the significance of 'stress' factors. The word 'stress' has different meanings for different groups of workers. As originally defined by Selye (1950), it was the sum of all the physiological responses by which an animal tries to maintain or re-establish a normal metabolism in the face of a physical or chemical force. Brett (1958) gave a definition which is correlated more readily with the fish disease situation when he suggested that, 'stress' is a stage produced by an environmental or other factors which extends the adaptive responses of an animal beyond the normal range; or, which disturbs the normal functioning to such an extent that the chances of survival are significantly reduced. The changes which occur in response to environmental stress are termed as the General Adaptation Syndrome (GAS). They are said to be non-specific physiological and biochemical changes, which take place in three phases:

1. The alarm reaction
2. The stage of resistance during which adaptation to achieve homeostasis under the changed circumstances is taking place.
3. The stage of exhaustion, when adaptation has ceased to be adequate and homeostatis is not achieved.

In addition to the above, specific immune system is one component of the protective system of all vertebrates which enables the individual to survive and maintain its homeostasis in an environment which is innately hostile. The property which distinguishes the immune defence system from other defence systems is its specificity and ability to 'remember' a particular infectious agent; and, thus, confers subsequent resistance to an individual who has recovered from an infection. Thus, a salmon recovered from furunculosis may be refractory to a subsequent infection by *Aeromonas salmonicida*; but, this immunity is irrelevant to subsequent infection by *Mycobacterium piscium*. In fish, generally, the thymic tissue develops embryologically from primordial associated with the epithelium of the pharyngeal pouches. Lele (1934) made a comparative study of the morphology of thymus in fishes and found the organ to have a variable origin and variable duration in different species of fish.

Diseases in FW fishes are generally studied under the following heads:

1. Parasitology of teleosts
2. Bacteriology of teleosts
3. Mycology of teleosts
4. Virology of teleosts

Parasitology of Teleosts

Many phyla of the Animal Kingdom have representatives which are parasitic on fish. The number of species of fish parasites reported till date number in thousands and many more remain to be discovered. Very few are seriously harmful to fish. Most individual fish in wild or cultivated populations are infested with parasites; but, in the great majority of cases, no significant harm appears to be caused to the host fish. Although, there are surprisingly few reports of parasites causing mortality or serious damage to wild fish populations, such effects often perhaps go unnoticed. Parsites in wild fish are frequently only remarked upon when they are so obvious as to lead to rejection of fish by fishermen or consumers.

No attempt could be made to present a detailed classification because of the great diversity of taxonomic groupings of the parasites found in fish and the dispute which surrounds their taxonomy.

Protozoan Parasites

Ichthyophthriasis or white spot disease is caused in IMC by the protozoan ciliate, *Ichthyophthirius multifilis* which infects different regions of the body externally and causes simple hyperplasis of the epidermal cells around the site of infection causing formation of pustules. Tripathi (1955) experimentally introduced *Ichthyophtoirius* in the fingerlings of *Labeo bata* and *Cirrhinus mrigala* in oder to study the effect.

The symptoms are the presence of small whitish cysts (*c* 1.0 mm in diameter) on the skin, gills and fins.

Affected fish to be dipped in 1 : 5000 formalin solution for an hour for 7-10 days; or, in 2 per cent NaCl solution for more than 7 days (Gopalakrishnan, 1963, 1964). Affected ponds be disinfected with salt or quicklime (Hora and Pillay, 1962). Examples of other ciliate infections observed in India are as follows:

1. *Trichodina indica* infect different species of fishes
2. *Scyphidia pyriformes* affect rohu, catla, mrigal, etc. (Tripathi, 1954)
3. *Cyclochaeta* spp infect pond fishes (Hora and Pillay, 1962)

   Both *Trichodina indica* and *Scyphidia pyriformes* are killed in 5-10 minutes by 2-3 per cent common salt solution
4. Costiasis caused by *Costia necatrix* is the most common mastigophoran infection observed in IMC. Symptoms include presence of a bluish coat on the skin of the host fish and presence of large amount of mucosa. The parasite causes irritation and disturbs respiration. Bath in 3 per cent NaCl solution for 10 minutes appears to be quite effective (Gopalakrishnan, 1964).

   Other flagellate parasites recorded are as follows:

   Bodomonas rebae in rohu, catla, mrigal (Tripathi, 1954); *Trypanosoma clariae* in *Clarias batrchus* (de Mello and Valles, 1936); and, *Trypanosomae punctati* in *Channa punctatus* (Hasan and Qasim, 1962)
5. Trichodinosis is caused by *Trichodina, Trichodinella, Chilodinella*, etc. These organisms affect carp skin and gills. Multiple injections are commonly recognizable as spherical organisms ovoid or long with rows of cilia. Chelated copper compounds, like 'Argant', 'Aquavet', etc., have recently been found to be effective against protozoan parasites.
6. Myxosporidians constitute typical fish parasites known to produce cysts on different regions of the body and viscera. Some of the common myxosporidian genera are: *Leptotheca, Myxobolus, Myxidium*, etc.

The general symptoms include weakness, emaciation, falling of scales, perforation of scales, raising of scales along the posterior margins, loss of chromatophores, etc. (Tripathi, 1952). Cysts on the scales show an inner and an outer fibrous layer of epidermal origin in fishes affected with *Myxobolus mrigalae* (Chakravarty, 1939).

However, myxosporidian infections have been commonly observed among the IMCs in waters having < 400 ppm of chloride content (Basu, 1950). Gill myxoboliasis among the IMCs (rohu, catla, mrigal) caused large scale mortality in Bangladesh.

Myxosporidian cysts are not easily eliminated by chemical treatment. Disinfection with Candy's fluid or a weak solution of NaCl, to some extent, controls *Myxobolus mrigalae* affecting mrigal fingerlings (Sarkar, 1946). Concomitantly, myxosporidian infections, encountered in over-stocked ponds, are controlled by thinning the population and using yeast pellets at 1g/kg of feed (Pal, 1975).

Helminth Parasites

Worm diseases are caused by trematodes (monogenetic and digenetic) cestodes, nematodes, acanthocephalans and hirudineans. A brief account of the helminth parasites in fish is given below:

Monogenetic Trematodes

Among the various monogenetic trematodes reported in fishes, *Gyrodactylus and Dactylogyrus* (Chauhan, 1953; Tripathi, 1957; Gupta, 1961; Gopalakrishnan, 1968) causes serious infections. The former infects skin and gills; while, the latter affects only the gills, which is further reported to feed on the blood of the thin capillaries of the branchial region (Jain, 1959).

Gyrodactylus attack causes fading of colours, falling of scales, accumulation of mucous on the caudal peduncle region and fins, peeling of skin, etc. When the gills are attacked, the fish dies due to the hypersecretion of mucus on the gill surface brought about by irritation. Infected fishes may be seen rubbing their body against hard surface in order to get rid of the parasites.

Bath, alternately, in 1: 2000 acetic acid and NaCl solutions have been reported to be effective in IMCs Dip treatment in 5 per cent NaCl for five minutes have also been found to be effective (Gopalakrishnan, 1964).

Digenetic Trematodes

Diplostomiasis or black spot disease is caused by *Diplostomum* spp. This disease is generally prevalent in some rearing and stocking ponds. Symptoms generally include presence of small black nodules almost all over the body of the fish having average diameter of *c* 1.3 mm.

Dip treatment of isolated infected specimens for about an hour in 3: 1,00,000 picric acid has been found to be effective in controlling the parasite (Gopalakrishnan, 1963).

Cestodes

Fish mortality due to cestodes had once been reported in a tank at Nagpur (Chauhan and Ramakrishna, 1958). Affected fish generally appear dull, sick and part of their gut becomes generally swollen or completely choked by cestode cysts. In some instances, the gall bladder is also affected.

Bothriocephalus infection affects intestinal epithelium. *Ligula* and *Schistocephalus* sometime take shelter in the peritoneal cavity of carps.

Nematodes

One of the widely-occurring nematodes, which affect the scale pockets of common carps in Europe, is the *Philometra lusiana.* However, some of the non-piscian nematode parasites include *Camallanus, Cucullanus, Heliconema, Spiroptera,* etc. Not much information is available on the details of nematode infection and their remedies.

Acanthocephalaus

Among them, *Acanthocephalus, Centrorhynhus,* etc., have been reported to affect the fishes (Tripathi, 1956).

Hirudinean Parasites

The IMCs are sometimes attacked by the leeches (Khan, 1944). The carps have been reported to be affected by *Hemiclepsis marginata* (Saha and Sen, 1955). The leeches feed on the blood of the host fish which causes irritation and abnormal movements of the host.

Dip treatment in 1 : 1,000 solution of glacial acetic acid and disinfection with 1:10,000 $KMnO_4$ has been reported to be quite satisfactory (Khan, 1944). There have also been suggestions of dip treatment in 2.5 per cent common salt solution for about 30 minutes.

Crustacean Parasites

Argulus (carp lice), *Lernaea* and *Ergasilus,* belonging to Class Crustacea, are considered as important ectoparasites on fishes. The parasitic copepods, *Argulus* and *Lernaea* attach themselves to the body of the fish with their body buried into the scale pockets and with a paired egg sacs protruding free. *Argulus*, particularly, attaches itself to the body of the fish by means of suckers and hooks; but, it can also swim freely in water. There have been many instances of fish mortality in Indian ponds due to argulosis. The affected fish become very weak and emaciated.

Prominent symptoms include stunted growth, peeling off of the scales and appearance of red spots at the sites of infection.

Treatment measures could include the following:

Ponds showing severe *Argulus* infection be drained and dried. Sometimes, vertical wooden bamboo poles are fixed inside the water, so that, the affected fish may rub their bodies against them to get rid of their ectoparasites. Lime be applied in the affected pond @0.1 to 0.2 g/lit after all the fish are removed and the pond bottom be exposed to sun atleast for 24 hours before the application of lime (Hora and Pillay, 1962).

Argulus sp reproduce rapidly at a temperature range of 20-28°C. Positive results in argulus treatment could be obtained by the application of Lindane (benzene hexachloride) @ 0.02 mg/lit

In addition to *Argulus*; *Lernaea*, also known as anchor worm, is a minute rod-like ectoparasite, which attaches to the host fish anywhere on the body by means of its anchor-like appendages present on the head of the parasite. Many carps are attacked by this parasite particularly during the summer.

When only a few fishes are affected, mechanical removal of the parasite could be possible by pulling them out of their anchorage in the host body with the help of a pair of forceps. This could be followed by a bath in weak permanganate solution for 2-3 minutes (Alikunhi, 1957).Dip treatment for a short period in 5 ppm solution of $KMnO_4$ is suggested in the event of large number of fishes being affected (Gopalakrishnan, 1963).

Tripathi and Chaturvedi (1974), recently, recorded an isopod parasite, *Ichthyoxenus jellinghausii* from *Labeo bata* and *L. gonius* of Pariat Lake in Jabalpur. This parasite does not generally cause much harm to the host. Both dip treatment and group treatment are generally recommended against this parasite.

Bacteriology of Teleosts

Bacterial diseases are responsible for heavy mortality in both wild and cultured fish. The actual role of these microbes may vary from that of a primary pathogen to that of an opportunist invader of a host. Water, especially, where organic loads are high, is an environment, in which, many genera of bacteria could thrive. It has been shown by many workers that the normal bacterial flora of fish is direct reflection of the normal bacterial flora of the water in which they swim. However, there seem to be few bacterial species which appear to be obligatory parasites and are unable to survive for long outside the fish host.

The 'Red spot of eels' is said to be the first bacterial disease to be described. According to Hofer (1904), it was first described by Bonaveri as early as 1718 from the Cammachio lagoons of the Adriatic coast of Italy.

During the last century, two factors have played significant role in the development of knowledge of bacterial diseases of fish. The first was the establishment of a Parliamentary committee of Enquiry into Furunculosis in Scotland in 1929, a disease, which had caused severe losses in Scottish salmon rivers. This Committee had produced three detailed reports in 1930, 1933,1935 which, even today, provide the basis for our knowledge of this disease; and, also established the criteria for subsequent critical studies.

In many fish diseases, bacteria have been found to be associated with the host, generally, as secondary invaders; and, sometimes as primary causative agents. Sometimes, disease diagnosis becomes complicated. For example, pathogenic *Aeromonas* spp have been associated with infectious dropsy in IMCs. Also, the disease could be reproduced by inoculation of pure bacterial cultures. Further, the possibility of viral involvement can not be easily ruled out. Diagnosis of ulcerative skin conditions and fin or gill rot is particularly difficult because a wide range of fungi, bacteria (*Aeromonas, Pseudomonas* and *Vibrio* spp) and protozoans may be present. Such ulcerative conditions and fin rots have recently been described in common carps and rohu in the Punjab. The conditions have been termed as haemorrhagic septicaemias and are generally complicated by the severe infections with the fungus, *Saprolegnia*. Further, the pathogen, *Aeromonas punctata* had been isolated from fish blood and dried ulcers and held to be the primary cause. Rohu were particularly severely affected with lesions in the head region. The condition had occurred during the coldest time of the year (about 17°C). It may be noted here that, most of the bacterial pathogens in fishes are Gram-negative bacteria.

A bath treatment for 30 minutes in 20 ppm proflavine hemisulphate is generally quite satisfactory for treatment of external lesions. However, if vibrios are suspected to be involved, a nifurpirinol (nitrofurazone-furnace) bath for one hour at 10 ppm (active ingredient) is recommended. Mycobacterial gill diseases are best treated with baths. QACs usually one hour in the range 1-4 ppm. If other compounds are not available, a dip in 500 ppm of copper sulphate for 30 seconds to 1 minute is worth trying. Further, kanamycin sulphate is said to be effective against certain Gram + ve and Gram–ve fish bacterial pathogens including *Aeromonas, Vibrio, Flexibacter* and mycobacterial infections.

A brief account of some of the bacterial diseases found to occur among the freshwater fishes of India is given below:

Fin and Tail Rot Disease

This diseasse affects both adults and youngs. In the early stage, the infection appears like a white line on the margin of the fin, spreading and imparting frayed appearance to the appendage which eventually putrifies and disintegrates. The disease may spread through contact and may cause heavy damage.

Treatment of the disease may include bath in 1:2000 solution of copper sulphate for 1-2 minutes till the fish shows signs of distress (Hora and Pillay, 1962). Some amount of cure could be there by painting the site of infection with concentrated solution of copper sulphate (Pal and Ghose, 1975).

Ulcer Disease

Haemorrhagic ulcers may be caused due to *Flexibacter columnaris* which shows red white plaques, often with reddish peripheral zone.

As a treatment, at first, the badly affected fishes be destroyed and the pond water be disinfected with 0.5 ppm of $KMnO_4$. In the case of fishes showing early stages of infection, dip treatment in 1:2000 copper sulphate solution for 3-4 days could be useful.

Carp Erythrodermatitis (CE)

It was, once upon a time, considered as a part of infectious carp dropsy syndrome. A severe skin ulceration in common carp stock was reported from West Java during 1980. Although not very much recommended, terramycin mixed with feeds @ 5-7 g/100 kg fish daily for 7-10 days could be useful to some extent. (Jhingran and Pullin, 1988).

Dropsy

Protrusion of scales, haemorrhagic ulcers on the skin and fins, inflammation of the intestine, exophthalmic condition and accumulation of fluid within the body are some of the symptoms of Dropsy.

As a treatment measure, disinfection with 1 ppm potassium permanganate solution or dip treatment in 5 ppm of the same solution for 2 minutes have been found to be useful (Gopalakrishnan, 1963). No food be given during the treatment process (Van Dujin, 1956)

Eye Disease

A variant of the bacterium *Aeromonas liquefaciens* causes an eye disease, which often assumes epidemic dimension, affecting medium and large-sized catla. The affected sites, generally, are the eyes, the optic nerves and the brain of the fish. (Gopalakrishnan and Gupta, 1960). Further, *Channa marulius* showing an eye disease, also in epidemic form, caused by *Staphylococcus aureus*, has been recently encountered, which causes cataract-like symptoms in fish. Initially, the cornea becomes vascularised and later opaque. Subsequently, the eye ball gets putrefied and results in death of the fish.

Treatment with $KMnO_4$ (dose: 0.1 ppm) for disinfecting the environment followed by application of lime (300 ppm) gives useful results (Kumaraiah *et al.*, 1977).

Mycology of Teleosts

A wide variety of Phycomycetes and fungi imperfecti have been known to be associated with diseases in fishes. However, in many cases, there is no direct evidence of a primary aetiological role. The integumentary mycoses associated with members of the Order Saprolegniales are, by far, the most significant fungal diseases in teleostean fishes.

Integumentary Mycoses

Saprolegniasis

This term is often used to describe fungal infections of skin and gills. Although, strictly speaking, the term be used only after definitive identification of the fungus as a member of the Order Saprolegniales, it could involve a variety of fungi. Clinically, similar lesions could occasionally be formed by the genus *Pythium*. Willoughby (1970) had also isolated *Leptomitus lacteus* from such lesions.

Considerable interest had developed, towards the end of 19th century, on an epizootic in Atlantic salmon caused due to fungal invasion (Huxley, 1882; Murray, 1885). *Saprolegnia ferax* was thought to be responsible for the condition at that time. Later, it was felt that, these workers were studying the final stages of a condition which is now known as 'Ulcerative Dermal Necrosis' (UDN). In 1932, Kanouse had described the sexual form of *Saprolegnia parasitica*, which she had obtained on a hemp seed substrate after original growth on fish eggs. Seymour (1970) had presented a comprehensive account of the genus *Saprolegnia*.

Saprolegniaceae are 'water moulds' of the Class Oomycetes, which possess profusely branched, non-septate mycelium, appearing as cotton wool-like tufts in water. The hyphae vary considerably in diameter between species but all contain cellulose.

A number of pre-disposing factors are involved in the development of fungal infection in fish: the factors which affect both the fish and the fungus: and, it is a combination of factors rather than any single condition which ultimately leads to infection. It has since long been believed that, the fungi responsible for saprolegniasis are secondary pathogens and lesions are commonly seen after handling and after any traumatic damage to the skin which could be caused due to overcrowded condition and in conjunction with pollution or bacterial or viral infection.

Primary saprolegniasis had been reported by Hoshima *et al.* (1960) in cultured eels without any visible prior injury to the fish. Tiffney (1939) found that macroscopic injury significantly increased the likelihood of fungal infection. Richards and Pickering (1978) had shown that, in outbreaks of saprolegniasis in spawning brown trout, a form of *Saprolegnia* with a low degree of homothallic fungus appears incapable of producing sexual structures despite prolonged incubation on a variety of media, except to a limited degree at low temperature. This supports the findings of Willoughby (1968), who had consistently isolated a similar sterile form of *Saprolegnia* from lesions of UDN.

Temperature plays a significant role in the development of *Saprolegnia* infections. Infection following trauma may occur at any temperature compatible with fish life, most of the epizootics occur when temperature is low. Hoshina *et al.* (1960) pointed out that a saprolegniasis of eels ceased when the when the water temperature rose above 18 C

Clinical features of saprolegniasis are as follows:

Saprolegnia lesions are focal grey-white patches on the skin of the fish which, when examined under water, have a cotton-wool-like appearance where the hyphal filaments extend out into the water. The early lesions are often almost circular and grow by radial extension around the periphery until lesions merge. At this later stage, the fungal patches are often dark grey or brown in colour as the mycelium traps muds or silt. Skin and gill lesions are by far the most frequently observed but there have been reports of infection of internal organs. Agersborg (1933) reported intestinal infection in fingelings of brook trout with *Saprolegnia ferax* and a similar infection with *Aphanomyces* spp was reported by Shanor and Saslow (1944).

Some of the histopathological features of saprolegniasis are given below:

The fungus usually establishes itself focally, invading the stratum spongiosum of the dermis and then extending laterally over the epidermis, eroding it as it spreads. Relatively superficial invasion of the dermis rapidly leads to fluid imbalance and peripheral circulatory failure due to inability to maintain circulating blood volume.

In Haematoxylin-Eosin stained sections of the skin infected with *Saprolegnia*, numerous hyphae are seen on the skin surface enmeshing cellular debris and material trapped from the water by the hyphal strands.

With regard to isolation of the fungus, a variety of methods may be used to obtain bacteria-free colonies, all of which involve the use of agar media.

A brief account of prophylaxis and treatment is given below:

Maintenance of the fishes in good husbandry conditions is a pre-requisite. Correct feeding, avoidance of crowded conditions and maintenance of good water quality are also essential. Should the fish develop Saprolegniasis, a variety of external disinfectants may be used. These include use of Malachite green, Copper sulphate, Potassium permanganate, Formalin, etc.

Other Fungal Diseases

Systemic Mycoses

Fijan (1969) had described a condition in channel catfish, in which, skin ulcers were seen having 2-15 mm in diameter and upto 5 mm deep, with no gross evidence of inflammation. Nodules upto 25 mm in diameter were found in association with adhesions and peritonitis, suggesting both haematogenous spread and local extension.The disease had been reproduced in the channel catfish. In experimental fish, abscesses containing hyphae and mixed purulent and caseous material were found.

Cerebral Mycetoma

Carmichael (1966) described a *Phialophora*-like fungus which had caused epizootics of so-called cerebral mycetoma in cut-throat trout. The organism was named *Exophiala salmonis*, the lesion being a chronic non-suppurative granuloma with the presence of numerous giant cells in the brain and cranial area.

Scolecobasidium humicola Infection

Ross and Yasutake (1973) had described a systemic mycotic infection in coho salmon held for experimental purposes. In affected fish, an enlarged abdomen was usually seen, often together with skin lesions. Ascitis, adhesions and grey areas in internal organs, especially in the kidney were also often seen. The disease could not be transmitted experimentally by incorporating the fungus into the normal diets; but, transmission was achieved when ground glass was also added to the diet. Doty and Slater (1946) had described a species of *Heterosporium* pathogenic to Chinook salmon.

Sphaeropsidales Infection

Phoma herbarum, a fungal plant saprophyte, had been isolated from three species of diseased salmonids in Washington State and

Oregon (Ross *et al.*, 1975). An earlier outbreak was briefly described by Wood (1968) in Chinook salmon. Detailed morphology of the fungus was described by Boerema (1970). It has branched septate hyphae and young cultures on Sabouraud dextrose agar are light buff in colour, changing through light pink, with ageing, to black, with the formation of pycnidia, which produce hyaline unicellular conidia.

In outbreaks, morbidity is rarely > 5 per cent and the disease usually affects fry and fingerlings. When the intensity of the disease is high, fishes are seen to swim abnormally and unable to maintain their equilibrium. They often develop swollen vents with haemorrhagic fin and skin lesions. Early internal lesions are confined to the swim bladder and are small (1-2 mm) white areas in the anterior end of the organs; the pneumatic duct area is, probably, first affected. In more advanced stages of infection, the lumen becomes filled with mycelium and the epithelium of the swim-bladder becomes hyperplastic. The wall is rapidly destroyed and adjacent internal organs are affected. There is an extensive acute inflammatory response or a chronic granulomatous reaction. PAS-and Giemsa-positive hyphae are usually evident and are approximately 50-100 mm in length and 2-3 mm in width.

Pure cultures of the organism could be obtained by aseptically removing material from the abdominal cavity and plating out on to Sabouraud dextrose agar. The disease was reproduced by Ross *et al.* (1975) but the organism appeared to be only weakly contagious and the condition was found to be related to diet.

Ichthyosporidium (*Ichthyophonus hoferi*)

This organism was assigned to the Haplosporidia and named *Ichthyosporidium gasterophilium* when it was described by Caullery and Mesnil (1905). Later, Laveran and Pettit (1910) had recognized a similar organism as a fungus and had named it as *Ichthyophonus hoferi* after Hofer, who had first described it as a haplosporidian affecting the flounder, sea trout, etc. Sprague and Vernick (1974) described the appearance of the organism, *Ichthyosporidium*, under electron microscope and classified it among the microsporidia.

Its clinical features, briefly, are as follows:

The disease is a systemic granulomatosis and is found in both freshwater and marine fish of many species. It has been the cause of death in large number of Atlantic herring in frequent epizootics along the East coast of America. Outbreaks in herrings usually occur in winter and spring and infection is evident as a roughened skin texture described as the 'sandpaper effect' occurring principally on the lateroventral tail region. This effect is caused by the loss of epithelium over proliferating dermal fungal granulomata. Further growth of the fungus causes local necrosis and results in the formation of either abscesses or ulcers. In certain other species of fishes, the internal organs are more frequently affected than the skin; and, such internal infection is evident as raised white nodules very similar to the granulomata or tuberculosis.

Its histopathology, briefly, is as follows:

Most frequently observed is the 'resting stage'. Germinating spore is frequently seen in section and consists of a cytoplasmic elongation bounded by the inner spore wall which pushes through the thicker outer wall. Heart, liver, muscle, kidney, spleen and even brain may be affected and signs will obviously vary according to the extent of the damage and the organ or organs implicated. Host response to the parasite is variable but a severe granulomatous response is the usual finding with large number of epitheloid cells and macrophages and occasional giant cells. In the early stages, cells of the inflammatory series are seen in large numbers. The granulomata usually have a well-developed capsule of connective tissue and occasionally spores are found surrounded only by a capsule of fibrous tissue.

The isolation of the fungus is briefly described below:

The organism may be cultured in Sabouraud dextrose agar slants with 1 per cent bovine serum added (Sindermann and Scattergood, 1954). Growth is abundant in 7-10 days at an optimum temperature of 10 C.

The prophylaxis and treatment, briefly, could be as follows:

A method of treatment of aquarium fishes using phenoxethol is quoted in Reichenbach-Klinke and Elkan (1965); but, this is only thought to be effective in the early stages of the disease.

Branchiomycosis

This disease, otherwise, known as gill rot, is characterized by areas of infarctive necrosis in the gill due to intravascular growth of

Branchiomyces spp of fungi. Two species of *Branchiomyces* spp are said to be involved in the disease.

The first record of branchiomycosis was reported by Plehn in 1912. Carps are the most frequently-affected fishes. *B. sanguinis* is usually localized in blood vessels of the gills and non-septate branched hyphae have been recorded.

Histologically, hyperplasia, fusion of gill lamellae and areas of massive necrosis, resulting from thrombosis of vessels by fungal hyphae. Affected fish may succumb as rapidly as two days after infection and morbidity of upto 50 per cent may occur.

Schaoerclaus (1954) suggested that, the disease was encouraged by waters rich in organic fertilizers, algal blooms and temperatures exceeding 20°C.

There is said to be no suggested treatment of the disease. Prevention could only be achieved by strict hygiene; removal of dead fish; and, avoidance of over-feeding especially at high water temperatures. Increase of water supply could prove beneficial during attacks. Eradication of the disease may be attempted by draining out the affected water followed by liming of the affected ponds.

Aphanomyces

The genus *Aphanomyces* belongs to family Saprolegniaceae and class Oomycetis. Today, Oomycetes are not always considered as true fungi; but, fungi-like protists. They are now often classed along side diatoms, brown algae and xanthophytes within the phylum Heterokonta as part of the third botanical kingdom, the Chromista. They are sometimes called Pseudofungi either as a general term or as a formal taxon (Cavalier-Smith, 1987). Nevertheless, some authors still regard them as 'fungi'.

The pathogenic *Aphanomyces piscicida* had been isolated from MG-affected fish in Japan. An *Aphanomyces* fungus was subsequently obtained from RSD outbreaks in Australia in 1989. *Aphanomyces* sp had also been independently isolated from EUS-affected fishes in SE Asia.

Virology of Teleosts

The study of virus infections in fish has been; and, in many instances, still, is, a poorly developed subject. It is only when viral infections cause a disease condition in fishes which hampers economy, health and nutrition in man, informations comparable to those available with viral infections in human beings and his domesticated animals, become available.

Nature of Viruses

Viruses are very small infectious agents which multiply only within the living cells of a host. The distinctive characteristics of viruses are their simple structure and their mechanism of replication. These are briefly summarized below:

Small Size

Viruses usually range from 20-300 nm; while, bacteria generally range from 200-2000 nm. The particle size of any one virus type is effectively constant.

Genetic Material (Genome)

This may be DNA or RNA but never both together.

Replication

Viruses can not multiply on simple media but only within and with the assistance of the components (*e.g.*, ribosomes) of living host cells. Inside the host cell, the virus genome directs the production of new virus. Virus particles do not grow in size, neither do they undergo binary fission.

Viral Morphology and Structure

There is a limited range of morphological types and these are used as a basis for classification. However, as the number of distinctive morphological features is small, viruses with very different biological properties may appear morphologically indistinguishable.

Structurally, a 'virus particle' or a 'virion' consists of a core containing the nucleic acid surrounded by a coat largely or wholly composed of protein. Survival under adverse environmental conditions depends on the toughness of this coat. The coat protein is antigenic and specific for each virus type. The components of a virion, briefly, are, as follows:

Capsid: Protein Coat

Capsomeres

The protein structural units of which the capsid is composed

Core

Inner structure containing nucleic acid.

Envelope

Some viruses are surrounded by a lipoprotein envelope containing viral antigens but also partially derived from the membrane of the host cells.

Virions show three basic types of shape or symmetry, *viz.*,

Cubic

Here, the virions resemble small crystals. More precisely, the shape of the virion is icosohedral. The capsid is constructed by assembling capsomeres according to a rigorous geometrical pattern, as in the adenoviruses.

Helical

Here, the virions are long rods, the capsomeres being arranged round a spiral of nucleic acid, as in the myxoviruses. In addition, a lipoprotein envelope of one or more layers may surround a nucleocapsid of either icosohedral or helical symmetry

Complex

Where the basic geometry is not immediately apparet, as in the pox viruses.

Classification

As yet, no single scheme of virus classification is entirely satisfactory. Almost all the following characters are generally used for virus classification:

1. Nature and number of strands of nucleic acid
2. Symmetry of capsid
3. Nature of capsid, naked or enveloped
4. Number of capsomeres and diameters of capsid and period of helix or helical symmetry.

   Further, for more diagnostic characterization, the following characters could be useful:
5. The molecular weight of the nucleic acid
6. Proportions of nucleotides in the nucleic acid

7. Number, molecular weight and antigenicity of viral proteins
8. Liability to chemical and physical agents, *e.g.*, acids, alkalis, solvents and heat.

However, till date, only a few viruses isolated from fish have been characterized in sufficient detail to allow classification according to the scheme for other animal viruses. Among the fish viruses, only the rhabdoviruses and channel cat fish virus are sufficiently characterized. Some of the viruses found in fishes are briefly tabulated below:

Classification of Viruses of Teleostean Origin

Virus Name	*Morphology*	*Size (nm)*	*Chemistry*	*Replcn. Site*	*Classification*
IPN	Icosohedral	55-65	RNA, protein	Cytoplasm	Unknown
VHS	Bullet-shaped	180 × 60-70	RNA, protein, lipid	Cytoplasm	Rhabdo-virus
IHN	–	160 × 90	—	—	—
CCV	Complex Icosohedral	175-200	DNA, protein, lipid	Nucleus	Herpes-virus
Herpes-virus salmonis	-Do-	175	DNA, protein	Nucleus	Herpes-virus
Lympho-cystis	-Do-	130-260	DNA, protein, lipid	Nucleus	Unknown

Virus Replication

The interactions between viruses and host cells are not very understood for many viral infections even today. The exact details vary, particularly, according to the nature of the nucleic acid of which the viral genome is composed.

The normal 'productive ' virus cycle, which forms the basis of acute infections. It is demonstrated by inoculating a suitably chosen tissue culture monolayer with sufficient virus to ensure that cells are infected; adequate time is allowed for the virus particles to attach to the cells; then, the excess inoculum is removed; nutrient medium is added; and, the cultures are incubated at an appropriate temperature.

Initially, there is a period (called the 'latent period'), during which, infectivity remains at a low level, representing 'residual inoculum'. Then, follows a period of exponential increase in infectivity (the 'rise period') and finally a point is reached, after which, there is no further increase in the yield of virus (the 'plateau'). The length of these phases, and the size of the final yield, depend on the virus, on the cells and on the conditions of incubation.

A unique feature of viral growth is that, during the early period (the 'eclipse'), the virus becomes undetectable, or, atleast, barely detectable, either microscopically or in terms of infectivity. The nature of this phenomenon becomes clear when one remembers that the essence of the virus is contained in its nucleic acid; the capsid serving merely as protection for this genetic material between generations. New virus is produced not by growth and division of parent organisms; but, by assembly from pools of sub-units, whose synthesis is directed by the infecting genome. This time between the end of the eclipse and the start of the rise period is known as the 'maturation phase' and indicates the time necessary for the assembly and release of new virus.

The initial attachment of virus to host cell is not dependent on temperature and is probably based on electrostatic interactions. The subsequent engulfing of virus is often called 'viropexis', a process similar to pinocytosis. Infecting virus may often be seen in the cytoplasmic vacuoles, though the entry of membrane-bound viruses (*e.g.*, myxoviruses) may involve fusion of outer layer of the virus with the cell membrane, only the viral core (nucleocapsid) being released into the cell.

From this point on, events are geared to the fulfillment of two major requirements:

1. The replication of the viral genome to give daughter copies exactly like the original
2. The provision of viral structural proteins, which will form capsids round these daughter genomes.

Viruses have, in common, an inherent ability to synthesise proteins; ribosomes and much of the other apparatus for this process must be supplied by the host, as must the relevant amino acid and energy resources. The degree of host dependence is, however, virus-specific, some agents having a greater capacity than others to provide their own enzymes and to direct autonomous metabolic activities. The latter appear to make good their deficiency by modifying host cell enzymes.

An account of the different diseases caused by viruses in fish are briefly discussed below:

RNA Viral Diseases

Infectious Pancreatic Necrosis (IPN)

M'gonigle (1940) described a disease of salmonid fingerlings in Canada which bears a striking resemblance to the disease which is now called IPN. There was a controversy about its cause (viral or nutritional). It was resolved when Wolf *et al.* (1960) had isolated a virus and demonstrated it to be the aetiological agent of the disease. IPN was first recognized in Europe by Bellet (1964) in French hatcheries.

Morphology and Chemical Composition

The IPN virion is icosahedral, non-enveloped and *c* 70 nm in diameter. The single protein shell encloses an RNA genome consisting of two segments of large size class, double- stranded RNA. The virion contains three size classes of polypeptide (large, medium and small). Classification of IPNV is controversial. It can not be included in the family of reoviruses because it contains only two RNA segments as against 10-15 in other animal viruses with double-stranded RNA genomes.

Antigenic Properties

Several serotypes of the virus exist as revealed from the neutralization tests. Of these, four have been identified in North America and two in Europe. Strong cross reaction between the twoAmerican strains and the European strain indicates considerable similarity. However, the sensitivity of the European strains to freezing and thawing distinguishes them from the American strains. Further, the virulence of the strains as well as their ability to produce CPE in tissue culture vary. The reference strain held in the American Type Culture Collection (ATCC) is called VR299

Viral Multiplication and Development

Multiplication of the virus occurs in the host cell cytoplasm where the virus can be observed under Electron Microscope (EM) as individual particles and sometimes in crystalline arrays.

Laboratory Cultivation and Diagnosis of the Virus

The virus grows in a variety of permanent cell lines (RTG-2, FHM and BF-2) and has also been reported to grow in several other poikilothermic primary cell cultures. At temperatures 15-20°C, CPE is evident from 1 to 20 days post-inoculation depending on the size of the inoculum and strain of the virus. The CPE starts as focal areas of necrosis, which is shown by the formation of 'blisters' of detached, 'stringy' foci of cells in the monolayer. Cells lose their fusiform shape and develop a granular cytoplasm and complete destruction of the monolayer follows. Identification of the virus is usually done by the neutralization technique using pooled antiserum but complement fixation and fluorescent antibody tests have also been described.

Infection and Disease

Infection can apparently occur at any age in most salmonid species of fish which has not been previously exposed to virus. In young farmed fish, this results in the majority suffering from a persistent chronic infection throughout the life. The virus is demonstrable in many organs, notably, the kidney. Clinical disease may not necessarily follow infection. However, where it does, it is largely restricted to first feeding fry upto 20 weeks after first feeding. Field observations suggest that, stress, such as, resulting from overcrowding, poor environment or bad husbandry, is an important disease precipitating factor.

Clinical Signs and Gross Pathology

Clinically acute or chronic disease condition occurs in very young cultured salmonids. Older fishes are affected only exceptionally. Variations are revealed in the clinical signs of the outbreak. Usually, the largest fry succumb first. Accumulation of abnormal number of such fry against exit grids is usually the first sign observed. The affected fish are generally darker in colour and display characteristic spiral type of swimming behaviour associated with nervous signs, exophthalmos and abdominal distension.

Histopathology

The only consistent histopathological (HP) feature of this disease is the focal or generalized destruction of exocrine pancreatic acinar tissue. The necrotic changes involve rupture of infected cells with release of zymogen granules. EM shows membrane-bound areas of assembly of virus. Kidney and liver have been reported as occasionally developing necrotic foci. Microscopic lesions have been consistently reported to be found in the intestine.

Mortality Pattern and Antibody

The mortality pattern depends partly on the husbandry, age and environmental factors; and, partly on the serotype of the virus strain involved. A significant proportion of the fish surviving an acute outbreak could slowly develop degeneration-fibrosis changes in the pancreas resulting in poor growth performance.

A neutralizing antibody is elicited in surviving fish but virus could still be detected indicating multiplication in sites protected from humoral antibody and presumably cell-mediated immune mechanism. IPN virus has been reported to induce interferon production in FHM cells but no reports of *in vitro* interferon production in fish are recorded.

Geographical Distribution and Host Range

The virus has been isolated from North America, Europe and Japan. It has been demonstrated from many species of salmonid fish, *e.g.*, brook trout, rainbow trout, brown trout, cut-throat trout, Arctic char, Atlantic salmon and coho salmon.

Transmission

Tranmisison of IPN could occur vertically, with the virus inside the egg; or, could occur horizontally via the water; or, via fish consuming other moribund fish. The resistance of the virus to the low pH of the stomach probably aids infection via roal route. Nevertheless, the most important reservoir of infection is the infected fish. Piscivorous birds may act as mechanical vectors although their significance in the epizootiology of the disease is doubtful.

Control

There are no means of treating the infected fish to eliminate virus or alleviate the severity of the disease. Avoidance is the only effective control measure. This requires that water supplies are free of infected fish and that all eggs and fish introduced to such waters are similarly free from IPN virus. Eradication of virus from infected farms has been successful when all infected or presumed infected fish are slaughtered and holding facilities adequately disinfected, provided a wild fish source was not present to cause re-infection.

Viral Haemorrhagic Septicaemia (VHS) (Egtved Disease)

This disease is said to be first detected in rainbow trout in 1949 in a village called Egtved in Denmark. However, conclusive evidence regarding its viral aetiology was first given by Jensen (1963) by isolating the virus for the first time. The name 'Viral Haemorrhagic Septicaemia'was recommended by the Office d'Epizootie (OIE) in 1966 in view of confusion of its nomenclature in various countries.

Morphology and Chemical Composition

This virus displays typical rhabdovirus morphology, the bullet-shaped particles measuring 180x60-70 nm with one end rounded, the other end flat. The virion has a central hollow axis which is surrounded by a spiral core. In turn, it is surrounded by a lipoprotein envelope. The viral genome is composed of a single-stranded RNA.

Antigenic Properties

Neutralisation tests distinguish only two serotypes of VHS virus. The reference strain is called F1. There is said to be no difference in pathogenicity between the two serotypes (Vestergaard-Jfrgensen, 1974).

Viral Multiplication and Development

Virus matures at the cell surface, budding off from the cell membrane. Cytoplasmic inclusion bodies can be observed, consisting of helices which may represent viral nucleoprotein. Viral antigens could only be detected in the cytoplasm, indicating this to be the site of virus replication.

Laboratory Culture and Diagnosis

The virus grows in both the RTG-2 and FHM cell lines. In the RTG-2 infected cultures, the cells swell and become spherical and granular, before detaching from the glass. Virus infection of tissue culture cells and subsequent CPE occurs only within a narrow pH range of 7.6-7.8. The virus develops at a temperature range of 6-18°C, the optimum being 14°C, at which, CPE may occur within 48-72 hours. The virus is identified usually by the neutralization technique. However, fluorescent antibody technique has also been described.

Infection and Disease

Infection of previously non-infected rainbow trout results in an acute disease within 7-15 days under experimental conditions. However, infection may occur upto one year later in the case of whole population exposed to virus. 'Temperature effect' is generally attributed as the cause of delay in the clinical appearance of the disease. The outbreaks usually occur below 15°C. It indicates that, in trouts, this disease is quite common during winter and ceases to occur as the temperature rises towards summer. The virus could be detected in trouts which are about to suffer; in those which reveal clinical symptoms; or, in those just recovering.

Clinical Signs and Gross Pathology

The clinical development of an outbreak in a previously disease-free population of rainbow trout usually follows a triphasic pattern (Ghittino, 1965). The first phase is characterized by high mortality, with affected fish becoming dark, lethargic and showing haemorrhages at the base of the fins and in the gills. The second phase is one of chronic disease, where the predominant feature is very black colouration of the fish body and anaemia. The gills become very pale with exophthalmos as a common feature. The third phase is associated with the cessation of mortality; but, the affected fish display nervous signs manifested by a looping swimming behaviour. There is also swelling and discolouration of the kidney. Virus is not generally isolated at this phase.

Histopathology

The microscopic features are those of an endotheliotropic virus disease. Haemorrhages may be found in any tissue but could obviously be more frequent in highly vascular tissues and are associated with haemopoietic necrosis. The anaemia of the chronic disease results in heavy deposition of haemosiderin in the melanomacrophage areas.

The VHS virus is apparently only weakly antigenic in trout and recovered fish often show no detectable humoral antibody response. In trouts, the virus is said to invoke an interferon response which protects passively immunized fish from VHS infection and also protects tissue cultures from infection with IPN virus.

Geographical Distribution and Host Range

The virus has been isolated from disease outbreaks in various countries like Denmark, Germany, France, Italy, Switzerland and Scandinavia; but, it had, perhaps, never been isolated from the British Isles. It is also said to be unknown in the American continent and the Far East.

Transmission

Carriers are said to be significant reservoir of infection as extensive circumstantial evidence suggests a persistent latent infection. Eggs from infected fish is said to initially carry the virus on the egg surface; but, it is lost after several days of incubation in running water. Water, thus, could act as a 'mechanical vector' because of significant maintenance of infectivity by virus in water.

Control

There is, as yet, said to be no means of treating the infected fish to remove the virus or alleviate the severity of the disease. The only effective control measure is, perhaps, to ensure that, the water supplies are free of virus; and, that, all fish and eggs introduced are free from VHS virus.Eradication of virus from infected trout farms could be successful, only when all the infected fishes are slaughtered; the water bodies are disinfected; and, other quality control measures adopted.

Infectious Haemopoietic Necrosis (IHN)

A disease causing serious losses in sockeye salmon was shown by Rucker *et al.* (1953) to have a viral aetiology. Similarly, a disease of Chinook salmon had recurred since hatcheries on the Sacramento river were, at first, established in 1941. This was shown by Ross *et al.* (1960) to also have a viral aetiology. Although clinically and pathologically similar, the cross-infectivity between the two viruses was low and the two were considered to be different until 1967. Then, another virus, similar in many respects; but, low in corss-species infectivity, was isolated from rainbow trout (Amend *et al.*, 1969). Subsequent investigations (Amend and Chambers, 1970) have shown that the three viruses are morphologically and biochemically similar. However, each has a specificity for the species from which it was isolated. It is now considered that, all three are strains of one virus type. However, in deference to its pathology, it is called Infectious Haematopoietic Necrosis (IHN) virus.

Morphology and Chemical Composition

The virus shows rhabdovirus morphology, appearing as a bullet-shaped particle measuring *c* 160 × 95 nm with one end rounded and the other end flat. The virion has an axial pore 20 nm in diameter, a core 60 nm in diameter and an outer lipoprotein coat 15 nm thick with fine spicules showing on the surface. The viral genome is composed of single-stranded RNA.

Antigenic Properties

No antigenic difference has been detected by serological methods between any of the viruses isolated from the three species of fish

Viral Multiplication and Development

The virus apparently matures at the cell surface budding from the plasma membrane. Assembly occurs in the cytoplasm.

Laboratory Culture and Diagnosis

The virus replicates and causes CPE in Chinook and sockeye salmon, steelhead trout and FHM cell cultures. Most but not all isolates grow and produce a characteristic CPE in RTG-2 cells, which round up and form clusters. Growth occurs over the range 4-18°C, the optimum temperature being 12-15°C. The IHN virus is usually identified in tissue culture by serum neutralization test.

Infection and Disease

Infection occurs in previously non-infected fish and results in a persistent infection when the virus is normally latent and undetectable but is isolable from sexually mature fish, particularly post-spawning. Disease could occur whenever water temperatures are suitable and the virus is present.

Clinical Signs and Gross Pathology

The clinical features include lethargy and sporadic hyperactivity, dark colouration, exophthalmos, anaemia, abdominal distension, and haemorrhages at the base of fins

Histopathology (HP)

The main feature of HP of this disease is focal haemorrhage and degeneration of all cell types in the haemopoietic tissue. The early cytological changes, characterized by swelling and nuclear pyknosis, are soon overtaken by necrosis and dissolution of wide areas of haemopoietic tissue. In addition, the Eosinophilic Granule Cells (EGC) of the sub-mucosa of the intestine are also destroyed.

Geographical Distribution and Host Range

The virus is endemic in wild salmonid fishes of the Pacific coast of North America. Disease in hatcheries is common and Williams and Amend (1976) showed that it could cause epizootics in wild sockeye salmon. The disease has been found to occur in Chinook and sockeye salmon and in rainbow trout. Coho salmon has been reported to be resistant to this disease. They are said to act as natural reservoirs of infection.

Transmission

This virus could be transmitted by direct contact via water from infected fish and by feeding infected carcasses.

Control

There is, perhaps, no method of eliminating virus from infected or diseased fish. Acute disease in infected stock could be prevented by elevating the water temperature above 15°C; but, the infection persists and may occur as soon as the water temperature drops. The only effective measure could be to keep the stocks free of virus as well as to ensure that water supplies are not contaminated.

Spring Viraemia of Carp (SVC)

The aetiology of the disease complex of carp known as infectious dropsy has been the subject of considerable dispute for many years. The first rationalization of the confused clinical picture was achieved when Fijan *et al.* (1971) had isolated a virus. Fijan, subsequently, separated the dropsical syndrome, as he(Fijan, 1972), recognized it into (a) SVC, a condition caused by a virus named *Rhabdovirus carpio;* and, (b) Carp Erythrodermatitis(CE), a condition said to be of putative bacterial origin.

Pike Fry Virus Disease (PFVD)

A new rhabdovirus was isolated by Kinkelin *et al.* (1973) from an outbreak of disease in juvenile Northern Pike in the Netherlands. Infection experiments (Bootsma *et al.*, 1975) have shown that this virus is the agent of the condition known as hydrocephalus or red disease (Bootsma, 1971).

DNA Viral Diseases

Channel Catfish Virus (CCV) Disease

This virus was isolated by Fijan (1968) and found to be a herpes virus (Wolf and Darlington, 1971).

Morphology, Chemical Composition and Development of the Virus

This virus is *c* 175-200 nm in diameter overall; but, inside an outer lipoprotein envelope, there is a symmetrical icosahedral structure of *c* 100 nm made up of 162 capsomeres. The genome is composed of DNA. The virus is produced in the cell nucleus and enveloped as it passes through the nuclear and cytoplasmic membranes.

Laboratory Culture and Diagnosis

This virus grows well in the Brown Bullhead (BB) cell line at 29°C, giving rise to syncytia. Plaques are formed when CCV-infected BB monolayers are overlaid with agar. The virus is characterized by serum neutralization.

Infection and Disease

The disease occurs in fry and fingerlings during the summer months when water temperatures rise to 25-30°C. Fishes who survive the disease are suspected of harbouring a persistent latent infection.

Clinical Signs and Gross Pathology

Clinical signs include loss of equilibrium, spiral swimming movements and hanging vertically in the water. The main gross pathological features are haemorrhages of gills, skin and internal organs and abdominal distension.

Histopathology

The lesions appear to begin in the posterior kidney with an increase in number of lymphoid cells and proximal renal tubular necrosis. Extensive necrosis of the renal haemopoietic tissue also occurs and EM study has shown that there are large amount of viruses within the renal cells (Wolf *et al.*, 1972). Subsequently, focal necrotic lesions develop in the liver, spleen and the digestive tract. The pancreatic acinar cells generally remain unaffected which is quite in contrast to almost all other generalized fish viral diseases (Plumb and Gaines, 1975).

Both artificially and naturally infected adult catfish produce neutralizing antibody, the former within 10-15 days of infection (Plumb, 1973). However, virus may not be detected subsequently.

Geographical Distribution and Affected Species

This virus has been reported only in the Southern USA in the cultivated channel cat fish. Many ictalurids may be reservoir of infection in view of ready growth of the virus in a continuous cell line (BB) of another ictalurid fish.

Transmission

The virus is shed by the diseased fish. Persistent infection, though suspected, is yet to be proven. Diseased fish, those surviving disease or those suspected to have got contact with the virus, are the major vectors of spread; but, the portal of entry is unknown.

Control

There is no, as yet, means of treating infected fish. As such, control by avoidance is, perhaps, the only possible control measure. Eggs be derived from brood fish free from virus as the disease principally occurs in the young fish. Further, virus-free water be used for growing beyond the fingerling stage.

Lymphocystis Disease (LD)

This viral disease results in chronic verrucose skin lesions and occurs in many species of wild and cultured marine and freshwater fishes. Lowe (1874) made the earliest record of the condition in flounders.

Wessenberg (1914), on the basis of transmission studies, was the first to postulate that the cause was a virus. However, it was Walker (1962), who had described the virus under EM followed by its isolation by Wolf in 1966 whence its viral aetiology was confirmed beyond any dispute.

Morphology and chemical structure

The virion is hexagonal in outline with no recognizable capsomeres. However, its form had been described as icosahedral. Depending on the host, the size of the virus differs, varying from 130 nm to 260 nm. The nucleoid core is a tangled mass of DNA threads and there is a recognizable capsid. It is likely that there could be a lipid component as lipid solvents rapidly destroy infectivity. No information is available on antigenic variation between LDV isolates.

Diseases of Putative Viral Aetiology

Papillomatosis (Cauliflower Disease of Eels)

This disease was first described in 1910 among the eels from the Baltic coast. Subsequently, it had been reported elsewhere in the Baltic, around Denmark and in the coastal waters of Holland and Belgium where the incidence had increased dramatically over the last few decades. A virus of polyhedral symmetry with 52-56 nm diameter had been isolated from the blood of affected eels by Pfitzner(1969). Later, Koops *et al.* (1970) had isolated it in RTG-2 and FHM cells. It had been reported to be thermolabile and is generally known to lose infectivity rapidly in water. It is believed to replicate in the nucleus; and, is, therefore, regarded as a DNA virus.

Carp Pox

It was described as a disease among the cultivated carps about 400 years ago; but, its probable viral aetiology was demonstrated around 1952 (Schubert, 1966). Further, the suspected virus had not been isolated. However, electron micrographs show areas of chromatin within the nucleus; but, differ from the nucleolus, in which, a viral-like genome is apparently assembled and bound in a single-layered protein-like coat. The virion becomes membrane-bound as it passes out of the nucleus into the cytoplasm. This process of replication and the general morphology and size of the particle suggests that carp pox virus belongs to the herpes virus group.

Piscine Erythrocytic Necrosis (PEN)

Erythrocytic inclusions in poikilothermic vertebrates observed under light microscope (LM) had originally been believed to be protozoan parasitic infections. Subsequent EM studies had displayed that some of these inclusions are viral (Johnston and Davies, 1973).

Cytoplasmic inclusions of infected cod erythrocytes are made up of icosahedral particles adjacent to a viroplasm (a DNA centre where viral components are synthesized and assembled). The viroplasm is a granular-fibrillar spheroid which may contain other structures. The virion (average diameter 330 nm) has a well-defined outer envelope and a wrinkled inner envelope enclosing an electron-dense core, 230 nm in diameter. At present, PEN is classified loosely as an icosahedral cytoplasmic deoxyribonucleic acid virus (ICDV).

There is no direct evidence that cod PEN virus causes disease. Heavily affected fish appear quite normal. Only mature erythrocytes are affected; and, as the infection progresses, the proportion of infected erythrocytes rises as high as 99 per cent. Herring PEN virus infects both immature and mature erythrocytes, causing degeneration of the cell, the largest inclusions being found in the youngest erythrocytes. There is little information available on the transmission of PEN virus.

Chapter 20

Epizootic Ulcerative Syndrome (EUS)

Introductory

'Disease, *per se*, is not an entity of an end in itself. Disease is the end result of an interaction between a noxious stimulus and biological system. In fact, it is necessary to understand all aspects of the biology of the species in order to understand a disease. The interdisciplinary approach, which has been currently characterizing so many branches of scientific research, is also in evidence in the field of diseases of fish.' (Mawdesley-Thomas, 1972).

Fishes are poikilothermic aquatic animals. Their existence and performance are regulated by the quality of the environment in which they live. As such, success of fish rearing in both culture and natural water bodies depend on their quality and management; as well as, on the understanding of the biology of fishes. The conditions in large water bodies, such as, seas and oceans, are usually said to be quite stable and uniform. Conversely, the conditions in smaller bodies of water, *e.g.*, the hatchery ponds and aquaria, in which, the fishes are generally raised artificially, are, generally, influenced by a variety of conditions. However, all species of fishes perform the best under certain optimal conditions. The amplitude of such conditions is usually quite narrow. Understanding the relationship between the fishes, their environment, pathogens and parasites is a pre-requisite in achieving success in pisciculture; and, is also the basis for rational management.

Severe outbreaks of disease may result from the introduction of parasites, pathogens, from malnutrition, from chemical and physical alterations of the water, from the genetic make-up of the fish; and, from the interrelationship of any or all these factors. This could be presented graphically as a set of circles representing the fishes, pathogens or parasites and the environment (Fig.7). If the circles intersect, the conditions are favourable for an outbreak of disease. This relationship could also be expressed in the form of an equation as follows:

$$H + P + S^2 = D$$

where,

'H' represents the 'Host'; 'P', the 'Pathogen'; 'S', the 'Stress' caused by the environment; and, 'D', the resulting 'Disease'.

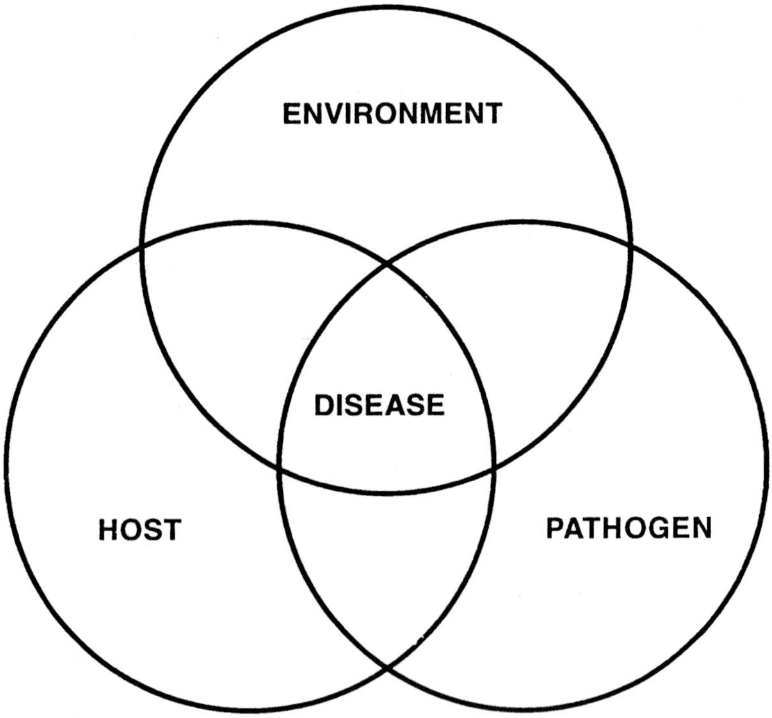

Figure 7: Interaction Between Disease and Environment

It may be noted here that, the stress caused by the changes in the environment may increase very rapidly after the acceptable range for a particular species of fish exceeds. As such, the effect of such a stress could be expressed as a quantity which increases exponentially; while, the concomitant changes in the environment could be represented arithmetically, *e.g.*, temperature by degrees; Dissolved Oxygen (DO) by mg/lit; and, so on. The stress created by such changes and the resulting mortalities are likely to increase geometrically or logarithmically.

Notwithstanding the above, intensive culture of aquatic organisms often encounters disease problems which pose a major threat to the Aquaculture industry.

Interest in fish farming continues to grow throughout South-East Asia, where extensive areas potentially suitable for aquaculture remain unexplored. On the other hand, intensive fish culture calls for high stocking densities, nutritious supplementary feeds, control of water quality and systematic management. Though there has been a growing interest in Polyculture (simultaneous culture of several non-competitive ichthyospecies), Monoculture is also a common practice. As such, high concentration of a single species encourage the spread of infection and infestation. Performance among nations for relatively few attractive species has stimulated International trade in fish seed with the concomitant hazard of spreading infectious diseases.

Fish forms a dependable source of animal protein for people of developing countries, like India, Bangladesh, Sri Lanka, and other countries of South and South-East Asia. Large-scale fish mortality often occurs due to environmental stress followed by parasitic afflictions. Fungi, bacteria and certain viruses are often found associated with such diseases which often become epizootic and epidemic resulting in cutaneous ulcers followed by haemorrhagic inflamed areas. Under such conditions, there occurs large-scale mortality among the fishes causing tremendous amount of loss to the nation.

Cutaneous ulcerative diseases, often involving a number of pathogens, are a common problem in wild and cultured fish. For the last two decades, a group of epizootic syndromes, all involving a severe ulcerative mycosis, have been reported from Australia (Rodgers and Burke, 1981); South-East United States (Noga and

Dykstra, 1986); and, Asia, stretching from the Philippines (Reantaso, 1990) in the East to India (Kar and Dey, 1988; Das *et al.*, 1990) in the West.

In addition to the above, with regard to more details, it could be said that, for over 25 years, outbreaks of an ulcerative disease, characterized histologically by mycotic granulomas, have affected freshwater and estuarine fishes over much of Asia and Australia. The disease has been given various names, but is most commonly known as 'Mycotic Granulomatosis (MG) in Japan, Red Spot Disease (RSD) in Australia and Epizootic Ulcerative Syndrome (EUS) in South-East and South Asia. MG, RSD and EUS have, in the past, been described separately as distinct conditions. However, recent studies of some workers have shown that, the same pathogenic *Aphanomyces* fungus is involved in each case; although, there are other workers who subscribe to its primary viral aetiology. However, it is now apparent that, an account of the history of EUS would remain incomplete without consideration of outbreaks in Japan and Australia.

An account of each of the three epidemic conditions are given below:

Mycotic Granulomatosis (MG)

The first report of an EUS-like condition came in Summer of 1971, in farmed ayu (*Plecoglossus altivelis*) in Oita Prefecture, Japan (Egusa and Masuda, 1971). The characteristic lesion, a granulomatous response to invasive hyphae, was described and the disease was named Mycotic Granulomatosis (Miyazaki and Egusa, 1972). It had rapidly spread to several other Prefectures and had affected various fish species, predominantly the cultured ayu and the goldfish (*Carassius carassius auratus*), the Formosan snakehead (*Channa maculate*) and the grey mullet (*Mugil cephalus*) (Miyazaki and Egusa, 1972). However, the common carp (*Cyprinus carpio*) was not affected. Hatai *et al.* (1977) had isolated the invasive Oomycete fungus from the affected fish and subsequently called it *Aphanomyces piscicida* (Hatai, 1980). *A. piscicida* is now known to be conspecific with the EUS pathogen, *Aphanomyces invadens* (Lilley *et al.*, 1997a). Serious MG epidemics had not been reported in Japan since 1973. However, outbreaks of various dimensions (not serious though) have been occurring time and again. Recently, Hatai *et al.*

(1994) have reported a similar disease in imports of ornamental dwarf gourami (*Colisa lalia*) from Singapore. Fungus belonging to genus *Aphanomyces* have been isolated also from such consignments (Lilley *et al.*, 1997 b).

Red Spot Disease (RSD)

In 1972, outbreaks of a cutaneous ulcerative condition, called Red Spot Disease (RSD) affected the estuarine fishes, particularly, the grey mullet (*Mugil cephalus*), in Queensland, Australia (McKenzie and Hall, 1976).The disease later progressed to affect the freshwater (FW) and estuarine fishes in coastal rivers in New South Wales (NSW) (Callinan, *et al.*, 1989), Northern Terrritory (Pearce, 1990) and Western Australia (Callinan, 1994 a).

A fungus of the genus *Aphanomyces* was isolated from diseased fish by Fraser *et al.* (1992) and was shown to reproduce the disease in fish using bath challenges; but, only when the skin of the experimental fish was artificially abraded (Callinan, 1994 b). Therefore, certain other factor(s) was considered to be involved in the disease outbreak process. Later, Virgona (1992) showed that RSD outbreaks in the estuarine fish in the Clarence river, NSW were associated with lower catchment rainfall. Subsequently, Callinan *et al.* (1995 a) related this to run off from acid sulphate soils. Ultra-structural examination of fish gills and skin showed that, the low pH and increased concentrations of monomeric aluminium (which are indications of estuarine acidfications), might induce significant lesion-formation in fishes (Sammut *et al.*, 1996). However, in aquarium trials, RSD was subsequently induced in fishes exposed to *Aphanomyces* spores sub-lethally to artificially acidified water at a pH value of both 3 and 5; as well as, at low concentrations of monomeric Aluminium (Callinan *et al.*, 1996; Callinan, 1997). As with *Aeromonas piscicida*, the pathogenic RSD-*Aphanomyces* has been shown to be same species as the EUS pathogen, *A. invadans* (Callinan *et al.*, 1995 a; Lilley, *et al.*, 1997 a, b).

Details

Origin of the Problem

The origin of EUS could be said to be a matter of controversy and speculation.

Following the report of outbreak of MG in Japan during the 1970s, EUS was reported in Australia and in South-East and South Asia during the subsequent decades.

Having considered the mycotic genus *Aphanomyces* as the root cause of initiation of EUS, the question arises, where did the fungus originate from ? Probably, in a tropical or sub-tropical region in South America. Africa could be the other main possibility. But, the fact that the African cat fish, which is widely farmed in Asia, is highly susceptible to EUS. This suggests that, EUS, most probably, did not originate from Africa.

Nevertheless, *Aphanomyces* fungus, probably, had entered Australia through import of ornamental aquarium fishes. There is bulk movement of ornamental and cultured fish around the Australian region in connection with ornamental fish trade.

Notwithstanding the above, the story of EUS has an interesting parallel in Europe. It seemed that, apparently, healthy American crayfish were introduced to some waterways in Europe during the second half of 19th century. A fungus, probably a different species of *Aphanomyces*, other than *A. invadens*, which was brought-in along with crayfish consignment, exerted devastating impact and progressively wiped out the freshwater crayfish from the Continental Europe. Then, it had spread to Great Britain and Ireland followed by Turkey. The destruction had continued (Lehane, 1996).

Chronology, Status and Major Outbreaks in the World

Following the report of Mycotic Granulomatosis (MG) in Japan, sometime around 1972, the earliest report of EUS outbreak goes back to Australia during 1972 and Papua New Guinea during 1974; from where, EUS has been sweeping almost in a chronological manner through most of the South-East and South Asian countries, like Indonesia (1980), Malayasia (1979-83), Thailand (1985), Kampuchea and Lao PDR (1984); Myanmar (1984-85), Sri Lanka (1987); Bangladesh (March, 1988); until, EUS had reached India through the Barak valley region of Assam during July, 1988; and, has been sweeping the region, even today, causing large-scale mortality among the freshwater fishes.

The chronology of EUS, outlined briefly above, is discussed below in a little detailed way:

Following the outbreaks of MG and RSD, there was a progressive spread westwards, across Asia, of a syndrome, associated with dermal ulceration, and involving large-scale mortalities in a number of FW and estuarine fish species. The syndrome was called by various names, like Ulcerative Disease Syndrome (UDS), Fish Disease Syndrome (FDS), etc. Finally, the name Epizootic Ulcerative Syndrome (in short, EUS), was adopted in 1986 at the meeting of the Experts Consultation Committee on Ulcerative Fish Diseases in Bangkok (FAO, 1986). Outbreaks of EUS have been reported in 18 countries of the Asia-Pacific region.

In 1975-76, an ulcerative disease outbreak, believed to be EUS, occurred in the rivers of Southern Papua New Guinea (Haines, 1983). During 1982-83, there were high mortalities in gudgeon (*Ophieleotris aporos* and *Oxyeleotris heterodon*) from inland areas and mullets from estuaries in Northern Papua New Guinea (Coates, *et al.*, 1989). Introduced tilapia (*Oreochromis mossambicus*) are common in these areas; but, they proved resistant. Preserved affected fish were later examined by Roberts *et al.* (1986) and confirmed as pathologically identical to EUS.

In 1980, outbreak of an epizootic haemorrhagic condition had occurred in Java, Indonesia affecting primarily cultured cyprinid and clariid fish; although, whether this was EUS is uncertain (Roberts *et al.*, 1986). Typically, ulcerated snakeheads and catfish have subsequently been reported in the Indonesian provinces of Sumatra, Sulawesi and Kalimantan (Widagdo, 1990). Invasive hyphae have been identified from sand gobies (*Oxyeleotris marmoratus*) from eastern Kalimantan (Rukyani, 1994) and Bastiawan (1993) is said to have isolated *A. invadans* from an EUS-affected sand goby in Java.

Further, Roberts *et al.* (1986) discussed unconfirmed accounts of ulcerated walking catfish (*Clarias batrachus*) in Singapore in 1977. They had also discussed the subsequent occurrences of EUS. However, there had not been any detailed report of extensive losses to ornamental fish industry due to EUS inspite of Singapore being one of the biggest trading centres of ornamental fishes.

Nevertheless, there have been reports of high mortality among the fishes in southern peninsular Malaysia in 1979 (Shariff and Law, 1980). However, the first report of large-scale fish mortality due to outbreak of typical EUS was during December, 1980 among the rice field fishes in Northern Malaysia (Jothy, 1981). Ever since, these outbreaks had occurred almost annually, although sometimes to a lesser extent (Shariff and Saidin, 1994). Major species affected were the snakeskin gourami (*Trichogaster pectoralis*), stripped snakehead (*Channa striata*), Climbing perch (*Anabas testudineus*) and the walking catfish (*Clarias batrachus*).

In addition to the above, significant well-documented epizootics have been occurring in Thailand since 1981 (Ulcerative Disease Committee, 1983; Chulalongkorn University, 1983, 1985, 1987). The second (during 1982-83) and the third (during 1983-84). EUS outbreaks were particularly devastating as it had affected the intensive fish culture systems in Central Thailand as well as the wild fishes in the waterways. Some of the most severe mortalities had occurred in the farmed snakeheads in the rice fields. The original outbreaks had started towards the end of the rainy season (September) and persisted throughout the winter till March of the following year. At the moment, EUS outbreaks tend to be restricted to the cooler months of December and January. Roberts *et al.* (1993) had described the isolation of the pathogenic fungus, *Aphanomyces invadans* from EUS-affected snakeheads collected from the Suphanburi province of Thailand.

Major outbreaks of EUS from Cambodia (Kampuchea), Lao PDR and Myanmar had been reported during 1983-84 (Lilley *et al.*, 1992). However, subsequent EUS outbreaks were less extensive, *e.g.*, EUS outbreak had affected 35 Myanmarese towns during 1984-85 and only 11 towns during 1989-90 (Soe, 1990). However, the impact of EUS continued to be significant because of the importance of the susceptible fish species to the rural economy.

Further, several reports of EUS-affected fish have come from Vietnam, China and Hong Kong. The first report of EUS outbreak among the snakeheads came from the Mekong Delta during 1983 (Xuan, 1990). Ulcerated *Labeo rohita* were first observed at the Pearl River Fishereis Institute in Guangzhou, South China during 1982 (Lian, 1990). Clariid catfish were affected in the same area during 1987-88 (Lian, 1990). Also, reports of EUS outbreak among *Carassius auratus* were reported from much of Eastern China during 1989 (Guizhen, 1990). Wilson and Lo (1992) had reported *c* 70 per cent mortality among the snakeheads (*Channa maculata*) in Hong Kong during summer of 1988.

Laguna de Bay in the Philippines, had experienced a serious outbreak of EUS during December, 1985. *c* 5-40 per cent of certain fish groups, notably, the snakeheads, gobies, gouramies, catfishes, etc., were found to be ulcerated; while, certain other fish groups, *e.g.*, the milk fish, the big head carps and the tilapia were found to be unaffected (Llobrera and Gacutan, 1987). The disease had continued to spread to at least 11 other provinces in the Philippines affecting wild fishes in the lakes, rice fields and swamps; as well as, the cultured fishes in the ponds (Bondad-Reantaso *et al.*, 1994). Mullet, goatfish (*Upneus bensai*), croaker (*Johnius* sp.), *Psettodes* sp., and *Scanthophagus argus* in a lagoon in Cagayan Province had suffered an outbreak in 1990 which was confirmed as EUS through histopathological examination. (Reantaso, 1991). The occurrence of EUS among these brackishwater and marine species might provide an explanation regarding the *modus operandi* of the spread of EUS between the islands. Several *A. invadans* isolates had been isolated from EUS-affected fishes in the Philippines (Paclibare *et al.*, 1994).

A major outbreak of EUS had occurred among the FW and estuarine fishes of Western Sri Lanka during December, 1987 (Costa and Wijeyaratne, 1989). It is suspected that, the disease was imported from SE Asia through shipments of imported fish, which might have had infected angel fish (*Pterophyllum scalare*), some of which were ulcerated and had suffered high mortalities (Balasuriya, 1994). Large necrotic ulcers in the affected fishes, *e.g.*, the snakeheads, were the most prominently visible signs of EUS. However, the tilapia was found to be refractory to the disease.

Since last about 15 years, EUS had been having serious impact on the fisheries occurring throughout the mainland of SE Asia.

In Bangladesh, EUS was first reported from Chandpur district during February, 1988. The disease had quite rapidly spread throughout the country being aided possibly by the flood of September, 1988 (Barua, 1994). Ulceration was observed in many wild species, notably, the *Channa* spp., the *Puntius* spp., the *Clarias* sp., the *Mystus* spp., the *Macrognathus* spp., and the *Mastacembelus* sp. The IMCs in the culture system were also very widely affected; although, large-scale mortalities due to EUS were probably restricted to their juveniles (Roberts *et al.*, 1989). There was a little decline in the severity of outbreak of EUS temporarily. However, intensity of EUS was again on the rise since 1993. *Aphanomyces invadans* had been isolated from farmed IMC (notably, *Labeo rohita*) in NW Bangladesh as well as from the wild fishes in the wetlands of NE Bangladesh during 1993.

In India, widespread initiation of outbreak of EUS started from Barak valley region of Assam since July, 1988 (Kar, 1999, 2005 d; Kar and Dey, S, 1988; Kar and Dey, SC, 1988 a, b; 1990 a, b, c; Kar *et al.*, 1990 a, b; 1993, 1994, 1995 a, b, c; 1996 a, b, c; 1997; 1998 a, b, c, d; 1999 a, b; 2000 a, b, c, d; 2001 a, b; 2002 a, b, c, d; 2003 a, b, c, d, e, f, g; 2004 a, b, c; Patil *et al.*, 2003).Outbreaks of EUS in India have been comprehensively reviewed at various Fora (The Zoological Society of Assam, 1988; Jhingran and Das, 1990; National Workshop on Ulcerative Disease Syndrome in Fish, 1990; Kumar *et al.*, 1991; ICSF, 1992; Das and Das, 1993; Mohan and Shankar, 1994, etc.).

From Barak valley region of Assam (July, 1988), after its initiation in India, EUS has been spreading and sweeping, almost unabated and more or less in a chronological manner through other regions of India, notably, West Bengal (1989), Bihar (1989), Orissa (1989), UP (1990), MP (1990), Maharashtra (1991-92), Karnataka (1993), Goa (1993), Tamil Nadu (1993), Andhra Pradesh (1992-93), Kerala (1994-95), till EUS had reached Pakistan during 1998-99 and is said to be progressing further.

However, in Assam, particularly, in the Barak valley region, EUS has become almost endemic recurring almost every year between November and March. The detailed works done on EUS, since its inception in India, by Dr Kar and his associates of Assam (Central) University at Silchar, would be presented little later in this chapter.

Bhutan and the Eastern Terai of Nepal were first affected in 1989; and, by 1993, EUS had covered almost the whole of this Himalayan region including Pokhara and Kathmandu where even the coldwater species, like the *Tor* spp. were affected by EUS (Phillips, 1989; Shresta, 1994). It had been estimated that 20-30 per cent of the Nepalese pond fish production (*c* 3000mt) used to be lost due to EUS every year.

Pakistan, of late, was, perhaps, the latest place to be invaded by EUS. EUS was confirmed among the snakeheads in Punjab province during April, 1996; and, among the mrigals in Sind province during January, 1998 (DFID, 1998). The blotched or the mud murrel was, perhaps, the most commonly affected species with *Puntius* spp., *Labeo rohita* and *Cirrhinus reba* also reportedly affected by EUS. *c* 20 per cent of the fish farms in Sialkot Division of Punjab province had been affected by EUS with the incidence being higher in ponds which were inundated by the floods during 1996.(AAHRI, ACIAR, IoA and NACA, 1997). However, loss of fishes were not very high in certain parts of the Punjab province, possibly due to the extensive use of tubewell waters (with little higher salinity) for fish farms. EUS, now, seems to be quite well established in some places situated along the bank of the River Indus; and, given its apparent rapid spread across the country, there are fears of serious future impacts to fisheries and aquaculture development.

Efforts in Unravelling the Enigmatic EUS

The first sustained report on EUS, appeared during the early 1980s. Since then, EUS has been continuing to spread unabated and extend its range through the different countries with devastating effects on farmed as well as wild fish populations. Nevertheless, more is now known about EUS since these early outbreaks. But, the underlying cause(s) are still obscure. As such, much works remain to be done in unravelling the mysteries of this enigmatic disease. A compiled report entitled, 'A Review of Epizootic Ulcerative Syndrome (EUS) in Asia' was published, during 1992, as a collaborative effort between the Aquatic Animal Health Research Institute (AAHRI) of the Royal Thai Government and the Network of Aquaculture Centres in Asia-Pacific (NACA). Investigations regarding the cause(s) of EUS and the possible measures to control and contain the disease are still being done.

The history of serious research with the aim to try to understand EUS could be traced back to the early 1980s, when the Indonesian, followed by Malaysian and Thai scientists, had made significant studies on EUS as it had been sweeping through these countries. In 1985, FAO, through its Technical Co-operation Programme (TCP), had launched a major collaborative effort between scientists of the Royal Thai Government and the University of Stirling (UK), in order to investigate the cause(s) and spread of the serious ulcerative diseases in the Asia-Pacific region. This Project team had reported the results of their study on EUS to the Expert Consultative Committee on Ulcerative Fish Diseases in the Asia-Pacific region held at Bangkok during 5-9 August, 1986. Among others, the Expert Consultation Committee had recommended long-term studies of the environment in relation to outbreak of EUS. As a response to these recommendations, NACA had initiated a Regional study, during 1987, on the relationships between the environment and the outbreak of EUS, being supported by the United Nations Development Programme (UNDP) and the FAO through the NACA Project (RAS/86/047). This study, which had started with 18 researchers located in 11 countries in the region, had subsequently grown to form a network of 21 scientists in 12 countries. These scientists emphatically examined, in a quantitative manner, the interactions between EUS and the environmental factors. The results obtained from these studies, formed a significant basis and component of future studies and reports. Further, an important outcome of this Regional Research Programme has been the demonstrable benefits which could be obtained through collaboration of regional scientists working to resolve common regional problems. Unfortunately, it is said that, much of the previous efforts in fish disease researches in the Region had been futile mainly due to lack of co-ordinated efforts.

The 'shock' of EUS has also brought positive benefits. The concern generated among many Asian Governments as a result of EUS has led to an increased awareness of the importance of Fish Health Management in the protection of fishes in Aquacuture as well as in the wild. The Workshop on Fish Disease and Fish Health Management conducted by Asian Development Bank (ADB) in association with NACA (ADB/NACA, 1991), being collaborated by Overseas Development Administration (ODA), had demonstrated the importance of tackling the fish health management problems in a co-ordinated and concerted manner.

Notwithstanding the above, there have been many developments in the field of EUS research since the publication of the earlier AAHRI-NACA review of EUS in 1992. EUS, today, is a semi-global problem which has been studied both independently and collaboratively by many different workers. The early occurrences of MG in Japan and RSD in Australia, about 35 years ago from today, are now considered to have been outbreaks of the disease which has been subsequently named as EUS. Reference to the earlier works on MG and RSD has, therefore, been considered to be important and relevant in understanding the later outbreaks of EUS in SE and South Asia.

As indicated earlier, a number of different agencies have been supporting the works on EUS. As a summary, the initial outbreaks in SE Asia were investigated by a team funded by FAO and the name 'Epizootic Ulcerative Syndrome' (EUS) was later proposed at an FAO-convened Consultation of Experts in Bangkok in 1986. NACA has also been integrally involved in the studies on EUS. Much of the data on environmental parameters associated with EUS outbreaks had been generated by NACA's Regional Research Programmes on Ulcerative Syndrome in Fish and the Environment. The Department for International Development (DFID) of the United Kingdom (formerly, the Overseas Development Administration or ODA) and the Australian Centre for International Agricultural Research (ACIAR) had subsequently been funding major research projects on EUS. Both these organizations have also been funding various types of publications through the Fisheries Programme of ACIAR and the DFID SE- Asia Aquatic Animal Disease Control Project. The unabated spread of EUS may be due partly to the large-scale movement of fishes within the Asia-Pacific region. As such, the probability of further spread is high. Consequently, the probability of initiation of EUS in the countries, which are, as yet, unaffected, is high. This has been a concern for many countries. The need for development of effective strategies to reduce risks associated with the spread of significant aquatic animal pathogens is now widely recognized. And, in Asia, it is being addressed through a Co-operative FAO/ NACA/OIE Regional Programme for the development of technical guidelines on Quarantine and Health Certification and establishment of Information Systems for the Responsible movement of the Live Aquatic Animals in Asia. As such, the importance of co-operation among countries in Asia and its neighbours is important in controlling aquatic animal diseases and in promoting sustainable Aquacultural development.

Review of Current status on Research and Development on EUS

International Status

The earliest report of EUS outbreak, which has been causing panic among the people, because of its suspected health hazard, goes back to Australia (Mackanzie and Hall, 1976), from where it has been sweeping, in more or less a chronological manner, through almost the entire SE Asia including Java, Borneo, Sumatra, Indonesia, Malyasia (Shariff and Law, 1980), Kampuchea and Lao PDR (Boonyartapalin, 1985), Vietnam, Myanmar, Thailand (Tonguthai, 1985), Bangladesh (Baura, 1990) until EUS had reached India through the Barak valley region of Assam during July, 1988 (Kar 2004) and has been sweeping even today.

Various postulates have been put forward for this disease including physical factors, such as water pollution causing changes in physico-chemical characteristics and micro-nutrients. But these may just be the triggering factors. The origin of the disease is unknown and remains a matter of speculation even today. Attempts have been made in different laboratories to investigate the cause of EUS and to correlate the outbreak of the epizootic with environmental parameters. Rodgers and Burke (1981) stated the rapid seasonal depression of salinity and temperature as important environmental factors pre-disposing the fish to attack by EUS due to the stress condition created. No particular characteristic of soil fertility appear to be related to the occurrence of the EUS (Mackintosh and Phillips, 1986). *Aeromonas hydrophila* is one of the most commonly isolated bacterium involved in haemorrhagic fish diseases in the Indo-Pacific region (Roberts, 1989). A primary viral aetiology has been considered a likely possibility for the rapid and uncontrollable spread of EUS (Frerichs *et al.*, 1986; Ahne *et al.*, 1988).

Further, Roberts *et al.* (1993) reported the involvement of one or more invasive fungi and a range of opportunistic bacteria and certain viruses. Chinabut *et al.* (1995) reported the isolation of the invasive fungus, *Aphanomyces*, from EUS-affected fishes. Willoughby *et al.* (1995) stated that the primary cause of EUS is not yet known although there appears to be a close correlation between outbreak of EUS and temperature involving a range of microbes. They, further, stated that, previous works had failed to establish a strong correlation between the isolated mycotic flora and the disease outbreak. Leung *et al.* (1994) stated that, *Aeromonas hydrophila*, an opportunistic pathogen associated with haemorrhagic septicaemia, is an ubiquitous and heterogeneous organism which produces disease in fish under stress conditions or in contact with infection by other pathogens. They stated that, the pathogenesis of *A. hydrophila* is multifactorial and a variety of virulence factors work in concert to contribute to the overall virulence of this bacterium. Extra-cellular products, including toxins, haemolysins, proteases and acetyl cholinesterase appear to contribute to the establishment of *A.hydrophila* infection in fish. Cartwright *et al.* (1994) reported that, a virulent strain of *A. hydrophila*, associated with EUS, was used to produce monoclonal antibody which identified virulent strains of *A.hydrophila*. Lilley and Frerichs (1994) compared seven Rhabdoviruses, isolated from fish suffering from EUS, in terms of their morphology, cytopathogenicity, antigenic relatedness and structured polypeptide composition. Robert *et al.* (1994 a) reported about a number of Birna virus strains, a Reo virus and atleast eight Rhabdovirus isolates.

According to OIE (1997), EUS is a seasonal epizootic condition of wild and farmed fresh and brackish water fish of complex infectious aetiology. According to them, the most primary causative agent of EUS is the fungus *Aphanomyces invadans* often associated with bacterial septicaemia involving generally *Aeromonas hydrophila*. A variety of parasites have also been reported from the diseased fish; but, their presence is inconsistent. Associated viral infection also occur frequently. It is believed that, the fungus, by itself, can not normally invade fish; and, it is postulated that, certain co-factor, such as, epidermal damage (which could be initiated by an array of agents), severe environmental stress, or viral infection, is required to initiate this complex and exceedingly dangerous condition. They have reported that EUS is now endemic in South-East and South Asia.

Status of EUS in India

The records of EUS in India can be dated back to the year 1988. It is said to have entered India through the North-Eastern corridoor from the adjoining Bangladesh. The ICAR also started the investigation with the report of outbreak of EUS in India

The fishes of Barak valley region of Assam received the first attack of the disease being very adjacent to Bangladesh. Dr. Devashish Kar and his co-workers are among the few researchers who initiated this pioneering work in India.

In India, widespread initiation of outbreak of EUS started from Barak valley region of Assam since July, 1988 (Kar, 1999, 2005 d; Kar and Dey, S, 1988; Kar and Dey, SC, 1988 a, b; 1990 a, b, c; Kar and

Das, 2004 a, b; Kar *et al.*, 1990 a, b; 1993, 1994, 1995 a, b, c; 1996 a, b, c; 1997; 1998 a, b, c, d; 1999 a, b; 2000 a, b, c,d; 2001 a, b; 2002 d, e, f; 2003 a, b, c, d, e, f, g; 2004 a, b, c; Patil *et al.*, 2003).Outbreaks of EUS in India have been comprehensively reviewed at various Fora (The Zoological Society of Assam, 1988; Jhingran and Das, 1990; National Workshop on Ulcerative Disease Syndrome in Fish, 1990; Kumar *et al.*, 1991; ICSF, 1992; Das and Das, 1993; Mohan and Shankar, 1994, etc.).

From Barak valley region of Assam (July, 1988), after its initiation in India, EUS has been spreading and sweeping, almost unabated and more or less in a chronological manner through other regions of India, notably, West Bengal (1989), Bihar (1989), Orissa (1989), UP (1990), MP (1990), Maharashtra (1991-92), Karnataka (1993), Goa (1993), Tamil Nadu (1993), Andhra Pradesh (1992-93), Kerala (1994-95), till EUS had reached Pakistan during 1998-99 and is said to be progressing further.

However, in Assam, particularly, in the Barak valley region, EUS has become almost endemic recurring almost every year between November and March. The detailed works done on EUS, since its inception in India, by Dr Kar and his associates of Assam (Central) University at Silchar, would be presented in this chapter.

Among some of the other Indian workers who had worked on EUS problem, are, Bhaumik, Chatterjee and Pandit (1990) who had worked on the impact of Epizootic Ulcerative Syndrome on the Society. Chakraborty and Dastidar (1991) had repeatedly isolated chemoautotrophic nocardioform bacteria from fish affected by EUS. Chattopadhyay, Pal, Das, Das and Pal (1990) had done microbiological investigations into EUS in fishes. Das, Pal, Ghosh, Das, Joshi, Mukhopadhyaya and Hajra (1990) had prepared a comprehensive account of EUS. Goswami, Datta, Sharma, Sarma (1988) had done studies on certain aspects of prevention and treatment of fish suffering from EUS. Jhingran (1990) had worked on strategy for containing EUS. Prasad and Sinha (1990) had prepared a status paper on the occurrence of EUS in fishes of Bihar. Purkait (1990) had done some case studies on the EUS of fishes in Chanditala of Hooghly district in West Bengal. Rahman, Chaudhury and Barman (1988) had studied the role of EUS-affected fish on the health of ducks in Assam. CIFA (Central Institute of Freshwater Aquaculture) (ICAR) developed the CIFAX which has been claimed to control

EUS. Mohan and Shankar (1994), Vihswanath *et al.* (1997, a, b; 1998; Mohan *et al.*, 1999) did works revealing different aspects of EUS.

Major Species Affected and Pattern of Spread of EUS

In India, systematic study, conducted by us, since 1988, revealed wide scale attack among four species of fishes. These are:

Channa punctatus

Macrognathus aral

Mystus vittatus

Puntius conchonius

Study revealed a periodicity in the magnitude of the disease among the four most widely affected species

(a) From Around Mid-July to Mid-August, 1988

Mystus vittatus > *Channa punctatus* >
Macrognathus aral >*Puntius conchonius*

(b) From Around Mid-August to First Week of September, 1988

Mystus vittatus > *Macrognathus aral* >
Puntius conchonius > *Channa punctatus*

(c) From Around First Week of September to Around Last Week of September, 1988

Mystus vittatus > *Puntius conchonius* >
Channa punctatus > *Macrognathus aral*

(d) From Around Last Week of September to Around Third Week of October, 1988

Puntius conchonius > *Channa punctatus* >
Macrognathus aral > *Mystus vittatus*

Other species affected by EUS during the same period, but not very widely, included *Aorichthys aor, Amblypharyngodon mola, Catla catla, Cirrhinus mrigala, Heteropneustes fossilis, Labeo rohita, Lepidocephalus guntea, Notopterus notopterus* and *Salmostoma bacaila.*

Our continued study revealed that, during the period 1992-94, the following species had been found to be very severely affected by EUS: *Parambassis ranga, Chanda nama, Nandus nandus, Glossogobius giuris* (Kar *et al.*, 1994). Some of the other species affected by EUS to a lesser extent during this period, included *Mastacembelus armatus*, *Macrognathus pancalus*, *Xenentodon cancila*, *Colisa fasciatus*, etc.

Our recent studies (Kar *et al.*, 1995, a,b,c; 1997; 1998 a,b,c; 1999 a; 2000 a; 2001 a; 2002 a; 2003 a; 2004 a) indicated that the following species of fishes have been severely-affected by EUS, since 1995, particularly, during the period Nov-Feb, causing large-scale mortality among them:

Channa striata, *C. punctatus*, *Anabas testudineus*, *Clarias batrachus*

Thus, there has been a differential pattern of spread of EUS among different fish species during different seasons.

With regard to the Global scenario, more than 100 fish species have been reported to be affected by EUS (Lilley *et al.*, 1992); but, only relatively few reports have been confirmed by demonstrating the presence of MG in histological section or by isolation of the pathogenic fungus, *Aphanomyces* sp., from tissues underlying ulcers.

Similarly, some commercially important species are considered to be particularly resistant to EUS. But, not much studies have been done to confirm these observations and investigate the mechanism of resistance. Species reported to be unaffected by EUS outbreaks include the Chinese carps, tilapias and milkfish (*Chanos chanos*). Hatai (1994) experimentally injected catfish (*Parasilurus asotus*), loach (*Misgurnus anguillicausatus*) and eel (*Anguilla japonica*) with hyphae of *A. invadans* and found them to be refractory to infection. Wada *et al.* (1996) and Shariffpour (1997) experimentally injected common carp (*Cyprinus carpio*) with zoospores of *Aphanomyces* from MG and EUS outbreaks respectively; and, demonstrated that fungal growth was suppressed by an intense inflammatory response.

Some authors have commented that the most severely affected species in natural outbreaks are generally bottom dwellers (Llobrera and Gacutan, 1987; Chondar and Rao, 1996) or the fishes which possess air-breathing organs (Roberts *et al.*, 1994 b).

In the case of snakeheads, no particular size group appears to be more susceptible, with affected fish ranging from 40 g to 900 g (Cruz-Lacierda and Shariff, 1995). However, there is a possibility that, size or age may be significant in other species. For example, the IMCs suffer high mortalities as fingerlings (Roberts *et al.*, 1989); but, larger fish, although appear ulcerated, are not reported as dying in large numbers (AAHRI, ACIAR, IoA, NACA, 1997).

Notwithstanding the above, some of the EUS-susceptible species seem to have a wide range of geographical distribution, which could sometimes be beyond the current limits of EUS outbreaks. For example, several snakehead and clariid catfish species occur in Africa and Central Asia. This suggests that, there is potential for further spread of the disease to these areas. However, it could be noted that, the optimal temperatures for vegetative growth *in vitro* for *Aeromonas invadans* are in the range of 20-30°C (Fraser *et al.*, 1992; Lilley and Roberts, 1997); and, probably, for this reason, natural outbreaks of EUS, to date, have been found to be restricted to the latitudes between 35 N and 35 S. Experimental injection challenges of native European and American fish species have shown that, the pathogenic fungus, *A. invadans,* is capable of causing lesions in rainbow trout at 18 C; but is seems to be less infective in Stickleback (*Gasterosteus aculeatus*) and roach (*Rutilus rutilus*) at 11-16°C (Khan *et al.*, 1998).

Epidemiological Study

Epidemiology is the study of the distribution and determinants (*i.e.*, causes) of disease in populations. Epidemiologists typically take a wide view of causal factors, defining them as 'any event, condition or characteristic which play an essential role in producing an occurrence of the disease'. By contrast, many pathologists and microbiologists may consider, for example, a particular infectious agent to be the cause of a disease and may relegate all other contributions to 'contributing' or 'predisposing' factors.

For most diseases, including EUS, there is a strong evidence that, outbreaks occur only when a number of causal factors combine together. Many of the causal factors which have been identified or suggested, on the basis of reasonable evidence for EUS, may be represented in a causal web. It may be noted here that, there are several levels within the web; and, that, a number of factors may act at the same level, but not necessarily at the same time or intensity. It may, further, be noted here that, for EUS to occur, combinations of causal factors must ultimately lead to exposure of dermis, attachment to it by *Aphanomyces invadans;* and, subsequent invasion of the dermis and muscle by the fungus. The resulting mycotic granulomatous dermatitis and myositis may be considered as expression due to EUS (Lilley *et al.*, 1998).

The multifactorial nature of the cause(s) of EUS could also be represented by using the concepts of 'necessary cause, component cause and sufficient cause'. Each combination of various causal factors ('component causes') which together cause a disease, could collectively be regarded as a 'sufficient cause' for that disease to be initiated. It may be important to note here that, different combinations of 'component causes' may result in sufficient amount of causes for the initiation of a disease; which could, however, be different under different circumstances. Moreover, all sufficient causes for a particular disease have in common at least one component cause, known as a 'necessary cause', which must always be present, as expected, for that disease to prevail.

For EUS, studies have suggested that, there are a number of sufficient causes, each could make-up its component causes. It may be noted here that, each of these sufficient causes includes, amongst its component causes, one of the recognized necessary causes, *viz.*, propagules of *A. invadans* (Lilley *et al.*, 1998).

Notwithstanding the above, a unique feature of an epidemiologist is to test the aetiological hypothesis and identify the underlying causes (or risk factors) for the diseases (Park, 1997). A number of factors have been hypothesized as either factors or determinants of EUS outbreak in India. So, in order to find out the exact causative organism, or the infectious aetiology, it is the need of the hour to study the epidemiological aspects of the disease.

A standard proforma has been designed by following Park (1997). This proforma has been used to collect different information related to epidemiology of EUS in the Barak valley region of Assam, where EUS had made its initiation in India. It has been covered from the angles of socio-economic impact of it, epidemiology of the disease and its impact on auqacultural farm.

Since 1988, EUS has been considered to be one of the most serious diseases affecting fresh water fishes in India. Here, according to our survey, the highest mortality rate has been found among *Cirrhinus mrigala, Channa stirata, Puntius ticto, Labeo gonius, Lepidocephalus guntea*, etc. When analysis is planned to throw light on the aetiology, it is essential to use Specific Death Rates (SDR) (Park, 1997). Therefore, we had tried to determine the SDR due to EUS.

$$\text{SDR due to EUS} = \frac{\text{Number of deaths during a year}}{\text{Mid year population (Total number of species)}} \times 100$$

To represented the killing power of the disease, we have studied the Case Fatality Ratio (CFR), which is as follows:

$$\frac{\text{Total number of deaths due to EUS} \times 100}{\text{Total number of cases due to EUS}}$$

Further, a model related to the study of factors affecting the outbreak of EUS in farmed fish in Bangladesh, carried out by Ahmed and Rab (1995), in collaboration with ICLARM and the Department of Economics, Islamic University of Bangladesh at Kusthia, was replicated by us in the field in Assam University at Silchar, in which, three experimental ponds (*viz.*, ponds A, B, C) were prepared of sizes 15′ × 15′ × 3′ each. Lime was applied during the post-stocking, generally, in the proportion of 0.027 kg, 0.032 kg and 0.043 kg respectively. Pond A had been seeded with five species of fishes, *viz.*, *Cirrhinus mrigala*, *Labeo rohita*, *Ctenopharyngodon idella*, *Oreochromis mossambicus* and *Channa punctatus*. Pond B was seeded with only two species of fishes, *viz.*, *Cirrhinus mrigala* and *Channa punctatus*. Pond C was seeded with only a single species of fish, *viz.*, the *Cirrhinus mrigala*. No EUS-outbreak had been found to occur in any of the experimental ponds (*viz.*, A, B, C) during the period of Nov 2003 to Mar 2004.

Our study conducted during 2002-2004 (Kar *et al.*, 2004 a), the highest mortality rate has been found among *Cirrhinus mrigala*, *Channa* spp., *Puntius* spp., *Labeo* spp., and *Clarias batrachus*. Specific Death Rate (SDR) due to EUS = No. of deaths during a year/Mid-year population (total number of species) × 100 = (22/38) × 100 = 57.89 per cent.

In the same study, during the said period, the killing power of EUS has been represented by the Case Fatality Ratio (CFR)= Total No.of deaths due to EUS/Total No.of cases due to EUS × 100. CFR has been calculated specieswise and has been found to be high among *Cirrhinus mrigala* (83.72 per cent), followed by *Puntius ticto*

(75 per cent), *Channa marulius* (70 per cent), *Mastacembelus armatus* (69.23 per cent), *Anabas testudineus* (53.85 per cent), and, so on.

The Inter-continental and Intra-continental spread of EUS had already been described above.

Symptoms and Gross Pathology

EUS is characterized by the occurrence of large haemorrhagic or necrotic ulcerative lesions on the bases of the fins and other parts of the body which becomes larger inflamed areas with acute degeneration of epidermal tissues (Kar and Dey, 1988 a).

With regard to little details, Pin-head sized red spots develop on the body surface, head and fins, caudal peduncle, dorsum or operculum with no noticeable haemorrhages or ulcers. In the early stages, these may simply be areas of acute dermatitis forming rosacea.

The intermediate stage lesions are represented by small (2-4 cm) dermal ulcers, with associated loss of scales, haemorrhage and oedema. Roberts *et al.* (1989) noted that in *Puntius* spp., gouramies and other mid-water fish, ulcers are particularly dark and usually circular; often only one large superficial lesion occurs on the flank or dorsum. Most species, other than snakeheads and mullets, will die at this stage.

The advanced stage lesions appear on other parts of the fish's body and expand into large necrotic open ulcers; resulting eventually in death. Some affected species, *e.g.*, stripped snakeheads, can survive with much more severe chronic lesions which may have completely destroyed the caudal peduncle or eroded deep into the cranium or abdominal cavity; sometimes, even expose the swim bladder. Head tissue erosion is a particularly common feature of diseased stripped snakeheads; and, specimens have been found with exposed optic nerves or loosened articular bones, such as, maxillae and mandibles. Diseased stripped snakeheads, with moderately advanced lesions, placed in improved water-quality conditions often recover. Similarly, lesions on estuarine fish, such as, mullet appear to heal quickly when fish move into brackish or marine environments. Healing ulcers are characterized by a conspicuous dark colour caused by increased numbers of melanophores (Lilley *et al.*, 1992, 1998).

Progressive Diagnostic Symptoms of EUS

Behaviour

1. Reduced appetite
2. Swim with head out of the water
3. Floating lethargically
4. Death

External Signs

1. Appearance of small amber-coloured, red or grey erosions/ lesions
2. Expansion of lesions into large ulcers, loss of scales, haemorrhaging, formation of oedema.
3. Snakeheads may show severe erosion of the head or body cavity.

Clinical Signs

Studies on the pathology of EUS in Asia have tended to focus on the striped snakeheads (*Channa striata*) as this is the species most commonly and most severely affected. However, significant differences with other species have been noted. In general, lesions on EUS-affected fish could be distinguished into 3 groups on the basis of gross appearance (Viswanath *et al.*, 1997).

Clinical signs in the early stage of the disease are similar. Appetite is reduced or absent and fish becomes lethargic, either floating just beneath the surface or swimming with the head out of water.

Clinical Pathology

The initial features of an EUS outbreak in a new location are dramatic, whether in wild fishes in rivers, estuaries, lakes or in rice fields (where heavy losses often occur) or in farmed fish (Chinabut and Roberts, 1999). A significant proportion of the fish stocks are found floating on the surface; most of them already dead, with large red or grey shallow ulcers, often with a brown necrotic, fungus-covered centre. Generally, the ulcers are on the side, though they can occur on any part of the head and body. There is often remarkable symmetry of lesions, with the ulcer located in the same position on each fish of a particular species. This effect is very much noticeable among major and minor carps and barbs, such as, *Puntius* spp, all of which are highly susceptible to the disease. It is seen in both wild and farmed fishes. Often such lesions appear as brand marks or burn-marks.

Some fish species, such as, the snakeheads (Channidae), although highly susceptible, nevertheless take much longer time to succumb. Their lesions are, therefore, more extensive and can lead to complete erosion of the posterior portion of the body; or, to necrosis of both soft and hard tissues of the cranium, so that, the brain is exposed in the living animal.

When an outbreak occurs, although dead and dying fish will appear to have similar, extensive lesions, by examining fish from neighbouring ponds or rice fields, it could be possible to build up a clinical picture of the stages in the pathogenesis of the disease.

The first signs of affected fish are that, they become darker and do not feed. They may float just below the surface; or, in some species, with the head just breaking the surface. Sometimes, they may be hyper-active with a very jerky pattern of movement.

However, the initial lesions vary. In the IMCs and in *Puntius* spp., which, in some regions, appear particularly susceptible, a small area of darkened skin may be seen before they become moribund and die. In other fishes, there may be one or more reddish or amber-coloured areas of acute dermatitis, which gradually ulcerate to produce small (2-3 mm) ulcers which often displace a scale. Subsequently, these enlarge and develop a grey necrotic centre, with a white rim of exposed dermal collagen. Around this is usually a wide zone of darkly pigmented skin. In the centre of such larger lesions, there will generally be necrotic adherent mat, consisting of fungal hyphae of secondary species, such as, *Achlya*, haemorrhagic tissue and saprophytic protozoa, such as, *Epistylis* sp.; as well as, certain opportunistic bacteria.

Notwithstanding the above, our studies (Kar, 2003 b, 2004) revealed the following with regard to initiation of EUS:

EUS, in carps, begins with a small tubercle-like swelling, in which, bacteria and fungi could not be detected and isolated. Gradually, this tubercle gorws into a tuber-like size from which bacteria and fungi could be isolated. This faintly indicates that bacteria and fungi are not the primary pathogens for the initiation of EUS.

Further informations on progressive diagnostic symptoms of EUS (In India) (Viswanath *et al.*, 1997).

Type I: Early Lesions

Pin-head sized reddish or amber-coloured spots on the body surface. No noticeable haemorrhage or ulceration. Skin around the spots is normal with no discolouration. Sections show focal inflammatory changes. There are several nodular structures associated with fungal hyphae. Dermis and skeletal muscle are normal, without evidence of fungal invasion

Type II: Moderately Advanced Lesions

Approximately 2-4 cm in diameter, raised, circular, discoloured areas on the body surface. These areas are soft with relatively intact skin and scales. In sections, mycotic granulomas, seen in epidermis, dermis and skeletal musculature, associated with many non-septate fungal hyphae. Significant necrotizing dermatitis and myositis due to fungal invasion are also seen. In most of these lesions, scales and epidermis are not completely lost.

Type III: Advanced Lesions

Circular or oval, open dermal ulcers extending to the skeletal musculature. Characterised by large haemorrhagic and necrotic open ulcers on the body surface, devoid of epidermis and scales, with loss of dermis at the site of the ulcer. In most cases, the underlying musculature is exposed and largely replaced by fungal granulomas and host inflammatory tissue. Fungal hyphae extend in all directions from the focus of the dermal ulcer. Necrotic muscle fibres and fungal hyphae often found within granulomas.

Notwithstanding the above, EUS has been described in more than 30 species of FW fishes. Also, some species of brackish water fishes have been affected by EUS. As might be expected, there is great variation in the clinical picture between the different species; and, thus, also, in the underlying pathology. Some workers (Chinabut and Roberts, 1999) conveniently attempted to classify the clinical pattern, on the basis of their histopathology, into 5 types, as follows:

Type I

Peracute infection with minimal host response or secondary infection and rapid death.

Type II

Acute infection with some amount of host response and often secondary infection before death.

Type III

Chronic slowly progressive infection with considerable amount of host response; the outcome is generally fatal but invariably with secondary bacterial and/or fungal involvement.

Type IV

Chronic infection with host response sufficient to allow recovery unless there is secondary infection

Type V

Superficial response to adventitious infection in such fish species which are generally resistant to infection. Unless there is significant amount of secondary infection, such fish would almost invariably recover.

The above-mentioned classifications of clinical picture could only be considered as general in their application. Specific circumstances may affect the outcome; but, in general, the response of a particular species and age group could be consistent. This may allow a clear clinical picture for an outbreak to be defined and the basic diagnosis to be made based on the specific histopathology which underlies it.

Pattern of Investigation of EUS

Snieszko (1974) had stated that, an overt infectious disease could occur when a susceptible host is exposed to a virulent pathogen under stress. The influence of each sub-set could be variable; and, disease outbreak (*e.g.*, in this case, EUS) may occur, only if there is sufficient relationship among them. Although a filterable biological infectious agent is thought to be the primary cause of EUS outbreak, it is generally accepted that, certain abiotic factors which result in sub-lethal stress of the fish, are also important in initiation of disease outbreaks. Snieszko cited temperature, eutrophication, sewage, metabolic products of fishes, industrial pollution and pesticides as potential sources of stressful environmental conditions.

Life and disease processes of fishes are similar in many ways to those of other vertebrates, in that, most animals have muscles, skeletons, skins and internal organs which function in approximately similar ways. However, there is one major difference: 'fish lives in water'. As such, all of their physiological structures and functions are influenced by this fact. Thus, to understand what water means to fish is to lay a foundation for a more complete understanding of what fish needs for health and how disease processes are related to physiological and environmental requirements.

In order to understand the influence of environmental factors on fish disease, one has to realize that, in aquatic environment, there is much greater chemical and physical variability than exists in the terrestrial environment. In very large bodies of water, such as oceans, conditions may be relatively stable; but, in the coastal and estuarine areas, there is a greater magnitude of environmental change. In small bodies of water, this variability is even greater; while, in fish hatchery operations, many man-made stresses are added. Therefore, in the aquatic environment, life goes on under dynamic and unstable circumstances and fishes must continually acclimate to changes in population density, pressure, temperature, dissolved gases,light, pH, etc., the effects of which could be far more severe than those ever faced by terrestrial animals. Such environmental changes could impose stresses of considerable magnitude on the somewhat limited homeostatic mechanisms of fishes.

In view of hitherto unknown nature of EUS, Dr. Kar and his associates began the pioneering works on EUS fish disease in India (being situated in Barak valley region of Assam where EUS had made its initiation in India), with the attempt to ascertain the cause(s) of the disease through a number of different approaches. These are:

1. Whether outbreak of EUS is due to any severe organic pollution of water and soil
2. Whether it is due to any radioactivity
3. Whether it is due to any heavy metallic contamination
4. Whether bacterial and fungal flora are the causative factors
5. Whether there is any primary viral aetiology.

These are analysed in the following paragraphs:

Dr Kar (Kar and Dey, 1988 a) began his study with the ichthyological survey of the lentic water bodies in Barak valley region of Assam (24°10′ N–93°15′ E) with the identification of the four most widely-affected species of fishes due to EUS during July, 1988. These are:

Macrognathus aral, which portrays lesions on the skin generally in the abdomen and occasionally in the caudal region;

Channa punctatus, which shows lesions on the skin generally at the base of the caudal peduncle and sometimes show reddish swollen eyes (exophthalmos);

Mystus vittatus, which depicts amber-coloured lesions at the base of the rayed dorsal fin and sometimes at the base of the caudal fin;

Puntius conchonius, which exhibits amber-coloured lesions at the base of the dorsal and caudal fins and sometimes in other parts of the body.

Other species affected (Kar and Dey, 1988 a) by the present epidemic, during July, 1988, but not very widely, include the following:

Mastacembelus armatus, Macrognathus pancalus, Salmostoma bacaila, Gudusia chapra, Badis badis, Glossogobius giuris, Ailia coila, Lepidocephalus guntea, Clarias batrachus, Cirrhinus mrigala, C. reba, Parambassis ranga, etc. *Wallago attu* and *Hilsa(Tenualosa) ilisha* have not been found to be widely affected by the prevailing epidemics. Among the IMC, during July, 1988, barring a few *Catla catla* and *Cirrhinus mrigala*, none of the other IMC were found to be affected by EUS. Nevertheless, *Amblypharyngodon mola, Ctenopharyngodon idellus, Rasbora daniconius, Puntius ticto* and *Channa marulius* have been found to be affected only in certain areas of the district.

Our continued study revealed that, during the period 1992-94, the following species had been found to be very severely affected by EUS: *Parambassis ranga, Chanda nama, Nandus nandus, Glossogobius giuris* (Kar *et al.*, 1994). Some of the other species affected by EUS, but to a lesser extent during this period, included *Mastacembelus armatus, M. pancalus, Xenentodon cancila, Colisa fasciatus*, etc.

Our recent studies (Kar *et al.*, 1995, a,b,c; 1997; 1998 a,b,c; 1999 a; 2000 a; 2001 a; 2002 d, e, f; 2003 a; 2004 a) indicated that the following species of fishes have been severely-affected by EUS, since 1995, particularly, during the period Nov-Feb, causing large-scale mortality among them:

Channa striata, C. punctatus, Anabas testudineus, Clarias batrachus

Thus, there has been a differential pattern of spread of EUS among different fish species during different seasons, as has been indicated earlier.

A little details of the analysis under different aspects of study, as indicated above, are given below:

(a) Whether the Outbreak of EUS is due to any Severe Organic Pollution of Water and Soil

Field estimation of various water and soil quality parameters as well as other limnological parameters, conducted in various lentic bodies in Barak valley region of Assam, during the period July, 1988 to October, 1988 (*i.e.*, the initial peak period of the EUS outbreak) revealed the average values of the following parameters (Kar and Dey, 1990 a):

Water temperature (C°): 33.4

Turbidity (TU): 61.9

pH: 6.9

Dissolved oxygen (DO): 5.95 mg/lit

Free Carbon di-oxide (FCO2): 1.25 mg/lit

Total Alkalinity (TA): 7.5 mg/lit

Aquatic macrophytic biomass (AMB): 0.406 kg/m^2
(on the average)

Total Limnoplanktonic count: 144 u/lit
(On the average)

Similar field estimations conducted in the same water bodies in the same region during the pre-disease outbreak period, since January, 1979 revealed the range values of the above-mentioned parameters, as follows (Dey and Kar, 1987):

Water temperature °C: 32-34

Turbidity: 50-70 TU

pH: 6.0-7.2

DO: 3.0-6.0 mg/lit

FCO2:0.8-1.9 mg/lit

TA: 35-65 mg/lit

AMB (on the average): 0.308-0.706 kg/m^2

Total Limnoplanktonic count (on the average): 87-173 u/lit

The above data indicate similarity in the values of the parameters between the disease outbreak and pre-disease outbreak periods. These results indicate that, the lentic water bodies of Barak valley region of Assam have been largely maintaining their organically unpolluted nature and may not, therefore, be assigned as a reason for the outbreak of EUS in fishes.However, total alkalinity (TA) depicted alarmingly lower values during the EUS outbreak period as compared to pre-EUS outbreak period; thereby, could be serving as a pre-disposing 'stress factor' for EUS to set-in (Kar and Dey, 1990 a,b, c).

(b) Whether EUS is due to any Radioactivity

Studies conducted with Gamma-ray spectrometer indicated that the level of radioactive contamination recorded in the diseased fishes was very negligible to cause any harm in fish or in human beings who consume fish (radiation level: background, 770, 791; with healthy fish, 741, 716; and, background, 400, 460; with EUS-affected fish, 316, 306). As such, radioactivity may not be regarded as a possible cause of outbreak of EUS (Kar and Dey, 1990 a, b, c; Kar *et al.*, 1990 a).

(c) Whether EUS is due to any Heavy Metallic Contamination

Quantitative chemical estimation of trace elements (heavy metals) in water, soil and in the EUS-affected fish tissues by atomic absorption spectrophotometry (Vogel, 1978) did not reveal appreciable quantities of Mecury, Lead, Arsenic and Cadmium which could be of lethal proportion (average quantity, in mg/kg: Hg, Pb, As, Cd respectively: (a) in EUS-affected fish tissues: 0.99, 0.60, 0.0001, 0.0003; (b) in apparently healthy fish tissues: 0.63, 0.54, 0.00005,

0.00028) (Kar and Dey, 1990 a; Kar *et al.*, 1990 a; 1993; 2003 a). Therefore, effect of trace elements, being negligible, they may not be regarded as the possible cause of outbreak of EUS.

(d) Whether Bacterial, Fungal Flora are the Causative Factors

Bacterial

Bacterial culture (Cruickshank, 1972) from the surface lesions and gills of each of the four most widely-affected fish species revealed isolation of haemolytic *Escherichia coli, Pseudomonas aeruginosa* and *Klebsiella* sp preliminarily (Kar and Dey, 1990 a, b. c). All these bacteria have been found to be sensitive to Gentamycine, Nalidaxic acid, Furadentine, Ampicillin, Chloramphenicol, Tetracycline, Septran and Sulphamethazole (Kar et al,, 1993, 1994, 1995 a,b, c).

Further works have been done on the culture and isolation of bacterial flora from EUS-affected fishes belonging to nine species, *viz., Puntius conchonius, Notopterus notopterus, Cirrhinus mrigala, Glossogobius giuris, Channa punctatus, Anabas testudineus, Labeo gonius, Labeo calbasu* and *Amblypharyngodon mola,* which have been collected from water bodies in Barak valley region of Assam. The fishes portraying ulcerative lesions have been selected for the isolation of bacteria. The fishes were brought to the laboratory, cleaned properly and the parts showing lesions were sheared for taking swabs. Swabs from the lesions and from the different organs, such as gills, liver, viscera, kidney were streaked on to blood agar (5 per cent sheep blood) and McConkey's lactose agar (MLA). The plates have been incubated at 35°C aerobically for 24 hours. Distinct and isolated colonies were picked-up, purified and identified on the basis of their morphological, cultural and biochemical characteristics (Ellner, 1978; Roberts, 1978). All the isolates have been tested *in vitro* against nine anti-microbial agents by disc diffusion method (Ellner, 1978) with commercially available discs (HiMedia, India). Results showed that, nine species of fishes showing typical lesions of EUS, yielded different types of bacteria. In most cases, lesions and the internal organs yielded similar types of bacterial isolates. Mixed infection was common with *Aeromonas hydrophila, Pseudomonas aeruginosa, Klebsiella* sp. The most common isolate was *A. hydrophila* followed by *P. aeruginosa,Klebsiella* sp and *Staphylococcus epidermitis.* Strains of *A. hydrophila* were highly susceptible to Norfloxacin, Tetracycline and Chloramphenicol; while, *Pseudomonas* sp.were resistant to most of the anti-microbial agents (Kar *et al.*, 1999).

Still further works have been done on the characterization of bacterial flora in EUS-affected fishes. Bacterial cultures prepared form body lesions of live EUS-affected *Anabas testudineus* have been confirmed to be as *Aeromonas* spp by biochemical tests, which includes Indole test, oxidase test, TSI test, citrate test, MLA, Gas production, DNase test, slide agglutination with standard antisera AH_4 (2), etc. The bacteria were further confirmed to be as *Aeeromonas hydrophila* and *A. sobria* by PCR technique and PCR-RAPD analysis. The bacterial slants, prepared at the same time, were simultaneously tested for the presence of *A. hydrophila;* and, it has been found that the slants which were prepared from EUS-affected fishes, contain *A. hydrophila;* where as, the slants prepared from the corresponding healthy fishes do not contain *A. hydrophila.*

Fungal

EUS-affected fishes, *viz.*, *Cirrhinus mrigala* and *Puntius conchonius,* having pale raised lesions which have not yet completely ulcerated, have been chosen for fungal isolation works. The fish specimen has been killed by decapitation and pinned to a dissecting board with the lesion uppermost. The scales around the periphery of the lesion were removed and the underlying skin was seared with a red hot spatula so as to sterilize the surface. The fish with the board was then removed to a laminar flow cabinet with filtered air free from fungal elements. Then, the stratum compactum underlying the seared area was cut through using a sterile scalpel blade and a sterile, fine-pointed pair of rat-tooth forceps. This was followed by exposing the underlying muscle by cutting horizontally and reflecting superficial tissues. Four pieces of affected muscles, *c* 2 mm^3 in dimension, have been carefully excised up using aseptic technique and they were placed in Petri dish containing Czapek Dox medium with standard supplements. This experiment led to isolation of *Aphanomyces* sp with concomitant occurrence of the same fungal genus in histological sections of the same EUS-affected fishes (Kar *et al.*, 2000c).

It may be mentioned here that, the isolated bacterial and fungal flora have been living with the fishes in water since time immemorial without generally causing any harm to them. As such, bacteria and fungi may not be considered as the Primary aetiology for outbreak of

EUS in fishes. Further, intra-dermal inoculation (Dey and Dey, 1980) of these microbes into Guineapig (*Cavia* sp) revealed no lesion-formation in the latter's body for about two years.

(e) Viral

Absence of haemorrhagic septicaemia, (which is so characteristic of *Aeromonas* infection) in all but most of the ulcerated fish (Roberts, 1989); non-occurrence of these pathogens in the early part of the disease (Lilley *et al.*, 1992); as well as, close association of these bacteria with the fishes in water (as stated earlier) since time immemorial (Kar and Dey, 1990 a, b,c); suggest that, *Aeromonas hydrophhila* is unlikely to have any primary infective role in the pathogenesis of EUS. And, a primarily viral (?) aetiology (filterable biological agent) is always suspected to be associated with the initiation of EUS. (FAO, 1986; Kar *et al.*, 1990 a,b).

As such, in order to confirm the presence of virus, EUS-affected fish tissues have been subjected to virological examination. Ulcerated tissues from EUS-affected *Clarias batrachus* have been homogenized in PBS, clarified in high speed centrifugation (10, 000 rpm) for 10 minutes and decontaminated through membrane filtration (0.22 mm). Monolayer cultures of BF2 fish cell line have been grown in 96-well micro-titre plate at 28°C in Leibovitz L-15 medium supplemented with 10 per cent FCS and vitamin mixture. 50 ml aliquot of each dilution (upto 10^{-7}) of the filtered supertanant have been added to 80 per cent confluent monolayer of BF2 Fish cell line grown in 96-well micro-titre plate. After two hours of adsorption at room temperature, the cultures were overlaid with 150 ml of maintenance medium (L-15 supplemented with 2 per cent FCS) and incubated at 28°C. Progressive Cytopathic Effect (CPE) had appeared between 48-72 hours Post-Infection Day (PID) in 10^{-1} to 10^{-4} dilution. The infection was passable in the subsequent cultures indicating the isolation of virus (Kar *et al.*, 1993, 1994, 1999, 2003 a, 2004 a). Efforts are being made to characterize the virus and correlate the same with viruses isolated elsewhere (Frerichs *et al.*, 1989; Ahne *et al.*, 1988; Wattanavijarn *et al.*, 1985). Electron microscopic studies of the TCF showed occasional round particles unlike those of Roberts' group in Scotland (Frerichs *et al.*, 1986).

The above study confirms that EUS outbreak in FW fishes is initiated by primary viral infection followed by secondary bacterial and fungal contaminations. Viral presence is said to be only for a very brief period and the havoc is caused mainly by bacteria and fungi.

Electron Microscopic Studies

Preliminary Electron Microscopic (EM) works done with EUS-affected tissues of *Channa punctatus* and *Anabas testudineus* revealed the occurrence of virus-like particles (Kar et al; 1993, 1995 a). Detailed Electron Microscopic studies done with ultra-thin sections of EUS-affected tissues of *Channa striata* and *Anabas testudineus* revealed membrane bound and granular abnormal structures; hepatocytes with vacuolations; and, virus-like inclusion bodies in the nucleus (Kar,1999). Further, recent EM works done with EUS-affected tissues of *Wallago attu* revealed the presence of virus-like particles (Kar, 2004). Further works, in this direction, are in progress.

Histopathology

Detailed Histopathological (HP) studies done with tissues of EUS-affected fishes revealed focal areas of increased fibrosis and chronic inflammatory cell-infiltration in muscles; focal areas of fatty degeneration of the hepatocytes surrounding the portal triads in liver; hypertrophy of the muscle coat with mild chronic inflammatory cell infiltration in the gut; dense infiltration with chronic inflammatory cell in the interstitium of the kidney. The liver section further showed occasional infiltration by mononuclear cells in the sub-capsular region; while, skin of diseased fish showed epidermal necrosis and desquamation; stomach portrayed atrophy of the mucosa showing necrotic villi; and, the section of gill reflected infiltration by large number of mononuclear cells. Section through heart of EUS-affected fish showed coagulative necrosis of muscle fibres (Kar and Upadhyaya, 1998; Kar *et al.*, 2003 a).

Certain Observations on Tissue Cultural Works

Some of our observations on tissue culture works done during 2003-04 are as follows:

The BF2 fish cell lines have shown best results at room temperature (RT) and at 28°C. When BF2 cells were infected with filtered suspension (10-20 per cent) tissue extract, the cells started showing signs of rounding and detachment from second day post-infection in 24-well plates. This phenomenon persisted on serial passaging in 25 cm^2 flasks and CPE was observed from 2nd day onwards. The same was observed when the cells were grown in 75 cm^2 flasks. The cells showed bulging and rounding, which persisted till 7-8 days.

When another fish cell line, *viz.*, SSN1, were infected with the undiluted TCF collected from BF2 cells, at passage level 4, after 2 blind passages, the infected SSN1 cells started showing arrested growth with a few distorted cells. This effect persisted on further passage.

The infected TCF collected after 3-4 passages in BF2 cells was titrated. The CPE was observed upto the dilution of 10^{-6} to 10^{-10} in freshly- collected TCF as well as in stored TCF at -20°C with no loss of titre.

Test on various cell lines showed that BF2 and SSN1 cells are susceptible to the viral isolates. The CPE consisted of granulation, rounding and clumping followed by detachment of cells in the case of BF2. In SSN1, the cell growth was arrested while, another fish cell line, *viz.*, RTG2, was found to be resistant to EUS infection.

After inoculation of suspected virus/filterable biological agent, best infection result was obtained at 21± 1°C

In all the experiments with tissue culture, display of steady CPE was observed upto the 5th passage level on the average.

There has been a gradual decrease in CPE with increase in dilution.

Experiment was conducted to see the effect of UV radiation on susceptibility of cells *vis-à-vis* enhancement of sensitivity of the cells to viruses. Confluent monolayers of BF2 cells were exposed to UV radiation for various time period, *viz.*, 15, 35, 45, 60 seconds; and, then, the irradiated cells were infected with viral suspension. Non-irradiated cells were used as control. It was found that, UV-irradiated BF2 cells showed CPE in the 2nd passage instead of normally at 4th or 5 th passage. It could be due to the fact that, the virus was inhibited by certain cellular mechanism which is unlocked due to UV radiation.

Experiment was conducted to see the effect of heat on the activities of the virus (filterable biological agent). Aliquots of viral isolate were heated in a water bath upto 56°C. After 1, 2, 5, 10, 15, 30, 45 and 60 minutes of incubation, treated virus was inoculated into confluent monolayer of BF2 cell line. Unheated viral isolate was used as control. It was found that the heat-treated virus could not show CPE beyond 10^{-2}; while, the viral isolates not treated with heat showed usual CPE generally upto 10^{-6}

Further, effect of temperature on viral virulence was ascertained through experiments. In temperature sensitivity experiment, cells infected with virus and incubated at 20-25°C were found to show prominent CPE. In contrast, cells incubated at 28-30°C, after viral inoculation, does not show such prominent CPE or show late CPE. This indicates that, temperature plays a very important role in the pathogenesis of EUS. This lower temperature range corresponds with that of the ambient water temperature during Nov-Feb when natural outbreaks of EUS generally occur.

Haematological, Histochemical and Enzymological

Preliminary haematological studies revealed higher DLC and ESR but low Hb content in the blood of EUS-affected fishes as compared to the corresponding healthy fishes of the same species (Kar *et al.*, 1994).

Preliminary histochemical studies (Kar and Das, 2004 a) revealed interruption of glycogen synthesis in the liver and blockade of respiratory pathways in the gills. Further studies in this aspect are in progress.

Preliminary enzymological studies (Kar and Das, 2004 b) in the EUS-affected fishes revealed high values of Alkaline phosphatase, SGOT, SGPT and LDH. Further works in this direction are in progress.

Details of Clinical Symptoms and Histopathology

The initial signs of the disease are similar to many other fish disorders. Appetite, if any, is reduced and the fish become lethargic, either floating just under the surface or swimming with the head out of the water.

Small red or grey erosions then begin to develop on the head, fins (Menasveta, 1985), caudal peduncle, dorsum or operculum (Prasad and Sinha, 1990) In the early stages, these may be simply areas of acute dermatitis forming rosacea. Some authors have described the appearance of tiny red spots on the skin (Tonguthai, 1985; Shrestha, 1990).

Subsequently, lesions appear on other parts of the fish body and expand into large ulcers; with associated loss of scales, haemorrhaging and oedema; resulting eventually in death. Roberts *et al.* (1989) noted that, in *Puntius*, gouramies and other mid-water fish, ulcers are particularly dark and usually circular; often only one large superficial lesion occurs on the flank or dorsum. Snakeheads, however, are found surviving with much more severe, chronic lesions which may completely destroy the peduncle or erode deep into the cranium or abdominal cavity sometimes exposing the swim bladder. Head erosion is a particularly common feature of diseased snakeheads and specimens have been found with exposed optic nerves or loosened articular bones, such as, maxillae and mandibles.

Diseased snakeheads, with not too advanced lesions, placed in improved water quality conditions often recover. Healing ulcers are characterized by a conspicuous dark colour caused by increased number of melanophores.

With regard to details of Histopathology, despite difficulties in detecting early skin lesions, some samples have been observed and found to be principally areas of epithelial necrosis with surrounding oedema, haemorrhage of the underlying dermis and some inflammatory cell infiltration. EUS-lesions are often complicated by the presence of various common pathogens. It has not been possible to identify a positive fungal involvement in most of these early samples but a few have certainly harboured a small number of hyphae. Roberts *et al.* (1989) were able to study early lesions during an EUS outbreak in a captive population of IMC. They observed an acute necrotizing myopathy, more severe than is usually seen in wild fishes, spread over a wide area below the active skin lesion. The epidermis around the ulcer itself was degenerate and thickened, and only contained a very small number of fungal hyphae enclosed within an epitheloid capsule. The blood vessels of the dermis were very hyperaemic and some had a collar of lymphoid or myeloid cells which might be associated with virus infection although no viral inclusion bodies were detected.

Subsequent pathological developments in all infected fish species involve significant degenerative changes in skin-muscle tissue with the involvement of saprolegniaceous fungi and minimal disruption of internal organs. Tesprateep *et al.* (1983) noted that gill inflammation with talangiectasis of secondary lamellae was also a common feature.

Advanced lesions show large bacterial ulcerations with massive necrotizing granulomatous mycosis of the underlying muscle fibres, involving the distinctive branching aseptate oomycete. Fungal cells become progressively more covered with host epitheloid cells, and some areas may show evidence of myophagia and healing. Major carp with the most advanced lesions showed fungal hyphae invading the abdominal viscera and penetrating renal tubules and glomeruli, which would almost certainly be the ultimate cause of death. A large number of mycotic granulomata have been demonstrated in the kidney, liver and digestive tract of spiny eels (Chinabut, 1990).

The internal organs of diseased fish, other than those infiltrated with fungal hyphae, show only mild histopathological changes, which, Roberts *et al.* (1986) pointed out, may sometimes be the result of background pathology. Palisoc (1990) observed minimal tissue disruption in the kidney of *Channa striata* in terms of an increased number of melano-macrophage centres, haemosiderin pigments and few mitotic figures. Spleen sections showed a marked increase of whilte pulp production and the heart, liver and gills underwent mild histopathological changes, but none were observed in the stomach and intestine. Other kidney samples have shown tubular, vacuolar degeneration with granular occlusion and haemapoetic tissue degeneration or focal proliferation. Chinabut (1990) demonstrated that these features were consistently more severe in spiny eels. Pancreas samples occasionally show acinar necrosis with eosinophil and inflammatory cell infiltration. In the liver, mild focal hepatic cellular degeneration may also occur in advance of bacterial/fungal involvement. The only consistent haematological change in diseased fish is a significantly lower level of haemoglobin as a result of extra-and intra-vascular destruction of red blood cells (Tangtrongpiros, *et al.*, 1985).

Our View

Although the water quality/limnological parameters, by and large, was almost the same between EUS outbreak period and pre-EUS outbreak period, there has always been a sudden alarming fall

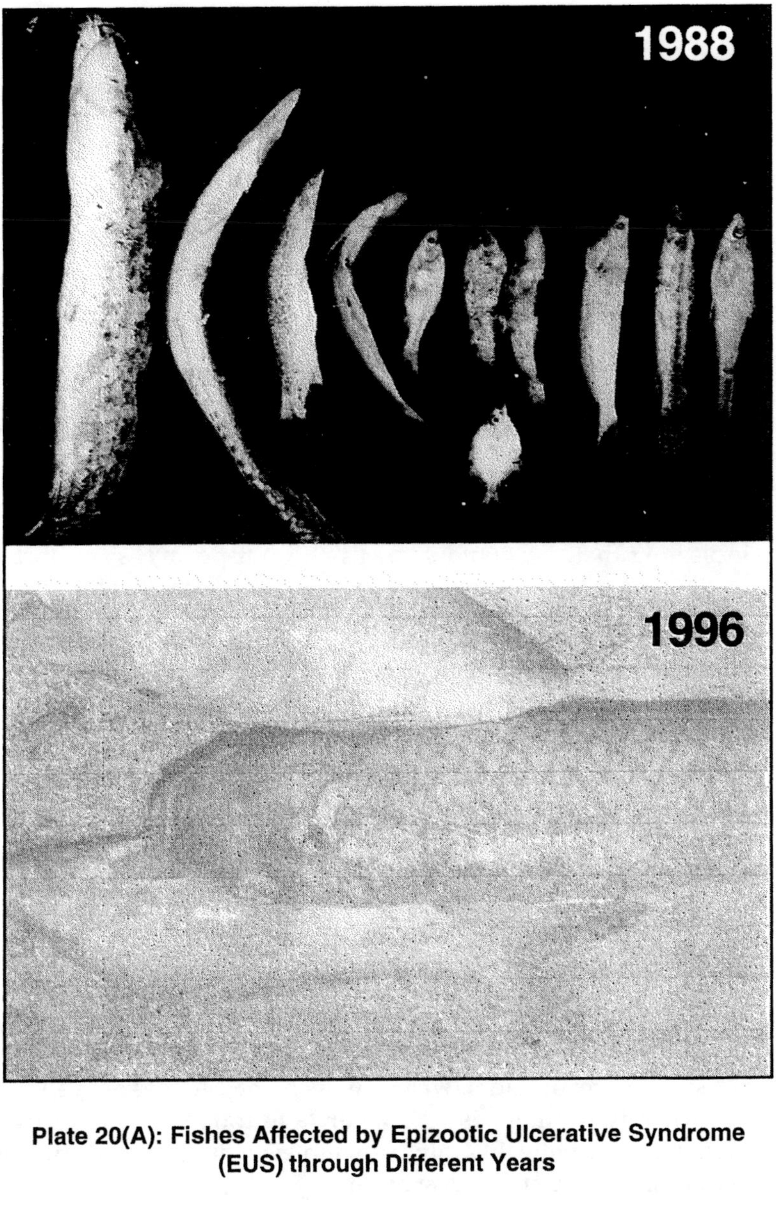

Plate 20(A): Fishes Affected by Epizootic Ulcerative Syndrome (EUS) through Different Years

contd...

Plate 20(A)–contd...

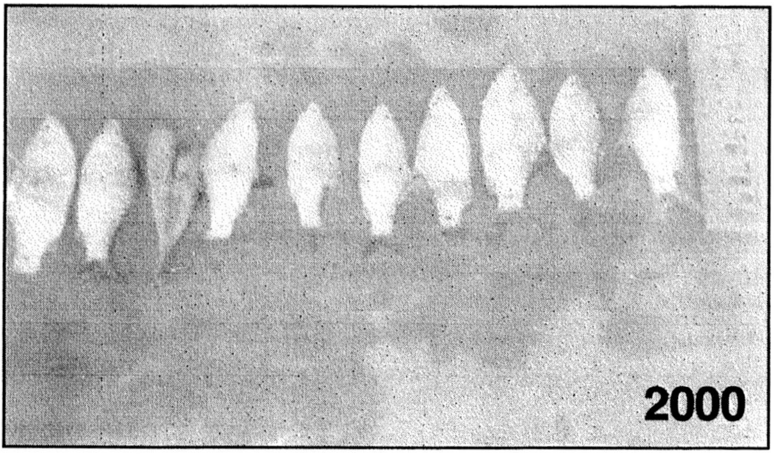

in Total Alkalinity (TA) values just before the EUS-outbreak starts; indicating that, TA could serve as a pre-disposing 'stress factor' leading to fall in immunity level in fishes; thus, enabling the viruses to attack the fishes 'primarily' for a brief period causing a trauma which makes way for the bacteria and fungi to infect the fishes secondarily and cause havoc in them.

The bacteria which have been isolated, are mostly acid-secreting and psychrophilic. Around Nov-Feb every year, when there is sudden fall in air and water temperature, these psychrophilic bacteria grow prolifically producing large amount of acids in water. High acidity (caused due to large bacterial population) lowers TA and makes the aquatic environment favourable for the 'viruses'to be activated. These viruses cause the trauma and disintegrate soon. The bacteria and fungi, then, attack secondarily and cause havoc in the affected fishes.

With the gradual rise in air and water temperature, with effect from March, the psychrophilic bacteria show decline. Hence, there is less secretion of acid in water. As such, the TA goes high and the aquatic environment is no longer favourable for the viruses. Therefore, currently, EUS almost completely disappears during the period March to October every year.

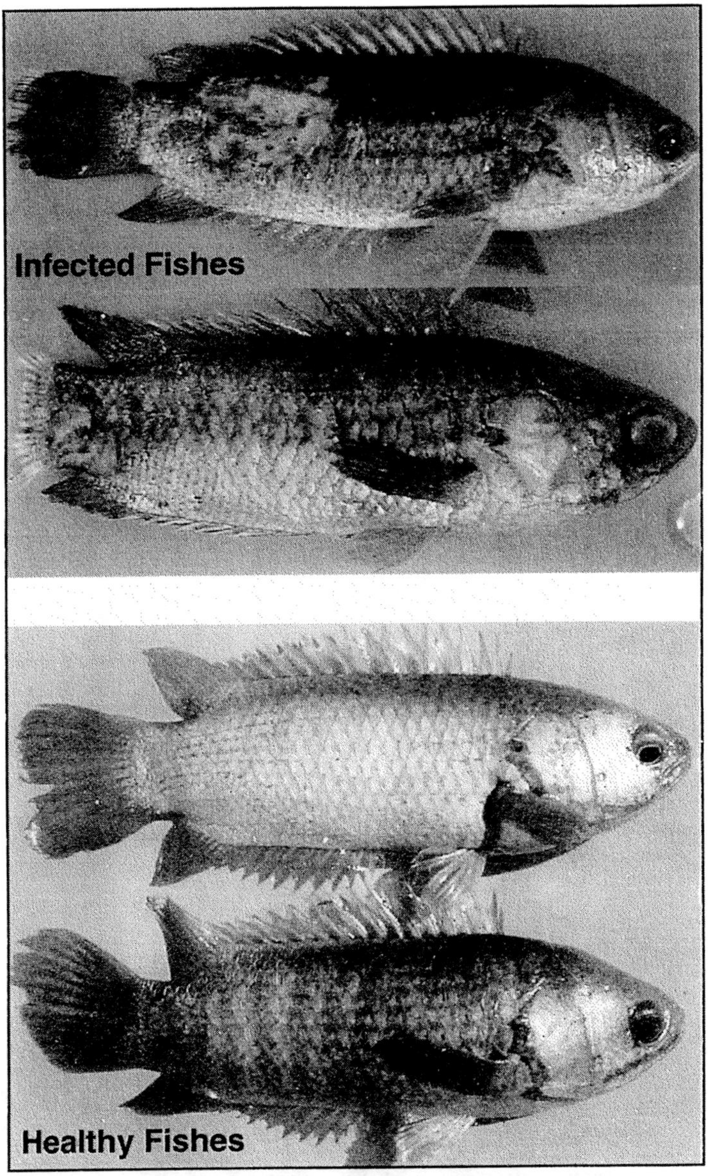

Plate 20(B): Fishes Severely Affected by Epizootic Ulcerative Syndrome (EUS)

contd...

Plate 20(B)–contd...

Infected Fishes

Infected Fishes

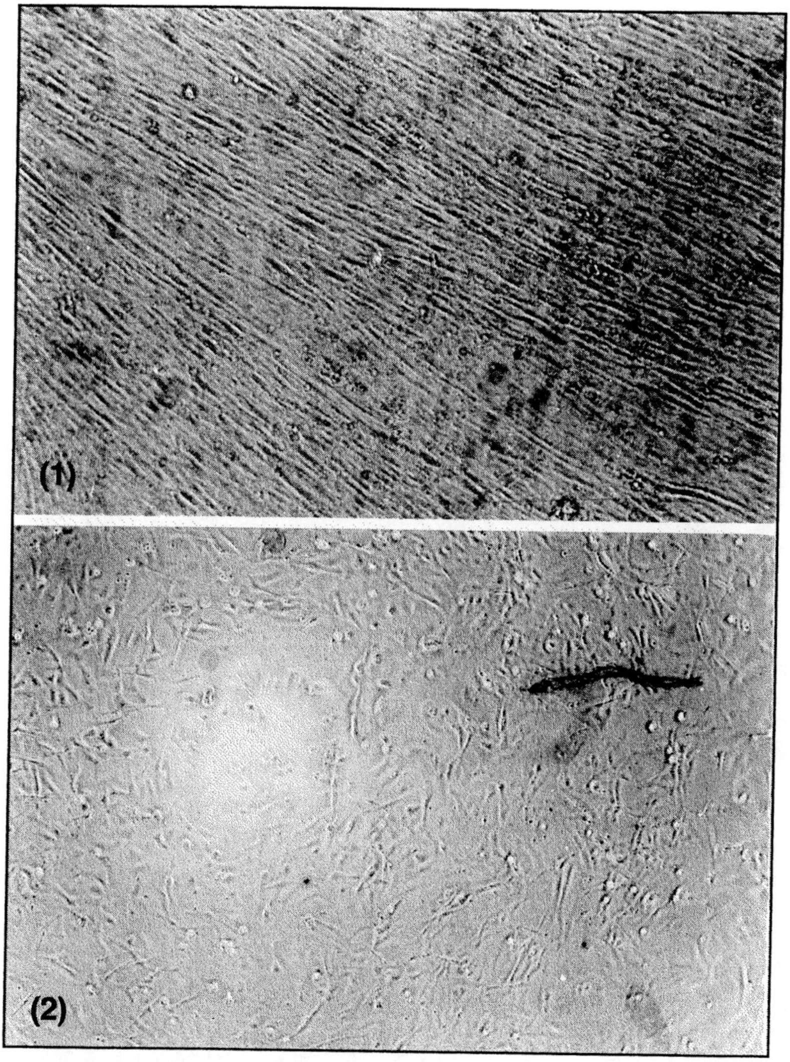

Plate 20(C): Molecular Level Works with EUS Fish Disease Pathogens

(1) Unaffected BF2 Fish Cell Line
(2) 1+ CPE after Inoculation of the Suspected Virus
(3) 4+ CPE with the Suspected Virus
(4) Bacterial Isolation from EUS Affected Fishes

contd...

Plate 20(C)–contd...

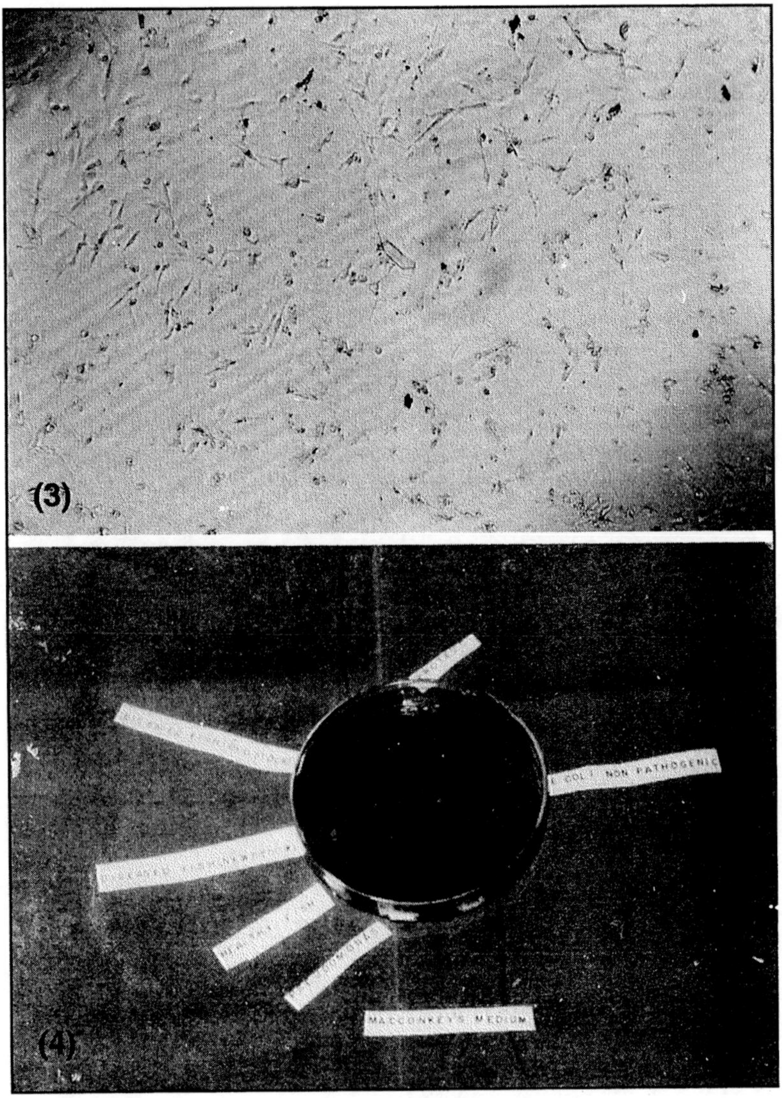

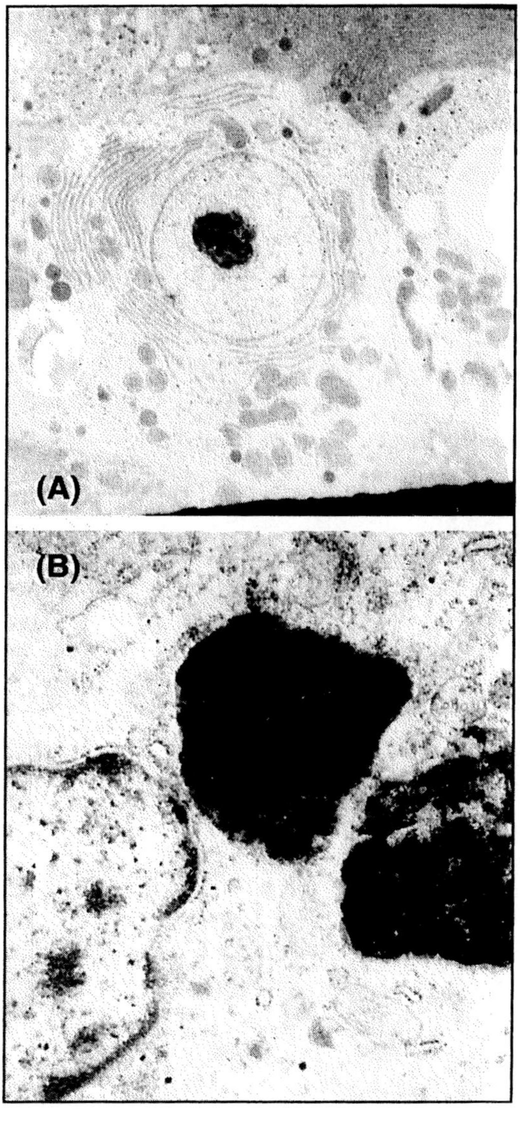

Plate 20(D): Electron Microscopic Studies of Different Organs of EUS Affected Fishes

(A) Cell Showing Normal Features
(B) Inclusions at Higher Magnification

Certain Informations Pertaining to EUS Obtained from Different Workers Round the World

Frerichs, Miller and Roberts (1986), in their paper, 'Ulcerative Rhabdovirus in fish in South-East Asia' stated that, electron microscopic examination of phosphotungstinic acid stained pellets obtained following ultracentrifugation of cell culture fluids from passaged samples, revealed bullet-shaped virus particles.

Ahne, Jorgensen, Olesen and Wattanavijarn (1988), in their paper, 'Serological examination of Rhabdovirus isolated from snakehead (*Channa striatus*) in Thailand with Ulcerative syndrome' had isolated a rhabdovirus from snakehead (*Channa striata*) in Thailand showing an ulcerative syndrome. The virus did not exhibit any serological relationship to VHSV, IHNV, RVC, PFRV or EVX

Tonguthai (1985) while preparing her report on 'Preliminary account of Ulcerative Fish Diseases in the Indo-Pacific Region, had stated the following:

1. While only limited pathological and clinical information is available, when the data from the various outbreaks of ulcerative, haemorrhagic disease which have been reported, of late, from the Indo-Pacific region, are assembled on a time scale, many typical features of a possible International epizootic may possibly be considered to become apparent.
2. Proper identification of the aetiology of each outbreak is essential, to ensure that, it is, in fact, a single epizootic; and, to allow development of National preventive and curative measures. This will certainly require international co-operation.

Frerichs, Miller and Alexander (1989), in their paper 'Rhabdovirus Infection of Ulcerated Fish in South-East Asia' published in 'Viruses of Lower Vertebrates (Ed: Ahne and Kurstak), stated that, the pathogenesis of the epizootic ulcerative condition has not yet been established by controlled laboratory studies. Initially, it was considered that, pollution of the waterways and ponds with herbicides, particularly paraquat, or organo-chlorine pesticides used in rice cultivation were responsible for the condition in SE Asia. Industrial effluents have also been incriminated in some instances. These chemical agents have now been effectively discounted, as outbreaks of EUS have occurred in areas where herbicides and pesticides have not been used and where there is no associated industrial development. Furthermore, areas where these compounds have been applied to crops for some years, or, where industrial effluents have been discharged into the waterways, have not always been affected by the disease.

The possibility of a primary bacterial or fungal infection has also been extensively investigated because *Aeromonas* or *Vibrio* species of bacteria are frequently associated with the lesions in FW and estuarine species respectively. Also, invasive *Saprolegnia* sp or *Achlaya* sp of fungi are commonly found in advanced cases of disease. However, the absence of septicaemic condition, which is characteristic of epizootic *Aeromonas* and *Vibrio* infections, and freedom from fungal involvement in the early stages of ulcer development, indicate that, these environmental micro-organisms are essentially opportunistic secondary pathogens. Nevertheless, the pattern of spread of the disease, particularly over the past few years, firmly indicates the role of an infectious agent (virus ?)as the possible underlying cause of a primary infection. And, the presence of virus-like particles of varying morphology were observed in the tissues of EUS-affected snakehead fish in Thailand in 1982 (Wattanavijarn *et al.*, 1983). Subsequently, a Birna virus (SGV) was isolated from EUS-affected sand gobies (*Oxyeleotris marmoratus*) in Thailand in 1984 (Hedrick *et al.*, 1986). Also, an unclassified virus, similar to IPNV in some respects, has been identified in ulcerated snakeheads by Saitanu *et al.* (1986). Although some agents can not be dismissed as potential primary pathogens, it is considered that the single, distinct species of Rhabdovirus, which has now been isolated from similarly EUS-affected fish in widely separated geographical areas at different times, is most likely to be the primary causal agent of the EUS. Further, the low rate of recovery of UDRV from clinically-affected specimens was a disappointing feature of these studies. This could have been due to a number of factors including the elimination of the virus from the affected fish at some stage in the disease process; the presence of anti-viral substances or defective interfering (DI) particles in the tissue extract; inappropriate selection of storageof tissues for examination; or, the use of sub-optimal isolation procedures. Substantive proof of the pathogenic role of any of the viruses, so far isolated from EUS-affected fish, must await the outcome of the experimental infection studies. However, as EUS appears to be mediated by naturally variable environmental conditions which can not be replicated in the Laboratory, this is likely to prove to be particularly difficult to fulfil. However, some workers have been trying to work in that direction (Kar, *et al.*, 2005).

Roberts, Frerichs and Millar (1992), in their paper, 'Epizootic Ulcerative Syndrome: the current position' opined the following: The EUS of SE Asian FW fishes has now extended as far west as Nepal. Studies on the aetiology and pathology of EUS disease have led to the conclusion that bacterial invasion, although frequent, is secondary to a highly invasive aquatic fungus of the Saprolegniaceae which is responsible for most of the characteristic pathology. The factor, which induces the fungal infection and allows invasion of the fish with very little host response, is not known. It was suspected that the rhabdovirus, isolated initially by Scottish team and subsequently by Thai and German workers from a range of outbreaks, might be the precipitating agent; but, evidence to confirm this has not been forthcoming; and, interest is now focussing on another, more recently-isolated; and, as yet uncharacterized, cytopahic agent, which may be suppressing host defence to cause secondary infections.

Roberts, Willoughby and Chinabut (1993), in their paper, 'Mycotic aspects of Epizootic Ulcerative Syndrome (EUS) of Asian fishes' stated that a survey of fish affected with EUS taken from outbreaks in countries throughout South and SE Asia showed that a morphologically typical fungus was consistently present within lesions. Although the majority of the fungal mycelium was dead in most lesions, it proved possible to isolate a very delicate and culturally demanding *Aphanomyces* from such lesions in a few cases. It also proved relatively easy to isolate other members of the Saprolegniaceae including *Aphanomyces* from the surface of lesions; but, these were considered saprophytes derived from background spore burdens in the water. Sporangium morphology of the putatively pathogenic isolates of *Aphanomyces* was different from that of saprophytic *Aphanomyces* strains and they also had a lower thermal tolerance. When a mycelium from these strains was placed below the dermis of healthy fish, it caused an inflammatory response and proceeded to migrate down into the tissues of the fish, inducing severe myonecrosis with chronic epithelial reaction. The saprophytic isolates induced a local host response followed by healing of the induced lesion, and destruction or expulsion of the mycelium. It is considered that the specific slow-growing, thermolabile *Aphanomyces* is the pathogenic fungus which causes so much tissue damage in this disease, although it may not be a primary pathogen in its own right.

Subasinghe (1993), published a paper entitled 'Effects of controlled infections of *Trichodina* sp. on transmission of Epizootic Ulcerative Syndrome (EUS) in naïve snakehead, *Ophiocephalus striatus* Bloch'.

Cartwright, Chen, Hanna and Gudkovs (1994), in their paper, 'Immunodiagnosis of virulent strains of *Aeromonas hydrophila* associated with Epizootic Ulcerative Syndrome (EUS) using monoclonal antibody', reported that a virulent strain of *Aeromonas hydrophila* associated with EUS was used to produce monoclonal antibodies (Mab) which identified various strains of *A. hydrophila*. Antibodies from a clone designated as F26P5C8 were found to identify the *Aeromonas hydrophila* serotype I isolates associated with EUS fish, and which were found to be virulent after subsequent inoculation studies. Immunodiagnosis of a large number of *A. hydrophila* from Australia and Japan showed some additional isolates to be identified by F26P5C8; but, the status of their virulence is not very well known.

Leung, Yeap, Lam and Sin (1994), while working on the 'Serum resistance as a good indicator of virulence in *Aeromonas hydrophila* strains isolated from diseased fish in South-East Asia', reported that, *A. hydrophila* strains isolated from diseased fish in SE Asia were studied for their virulence in naïve tilapia, *Oreochromis aureus* (Steindachner), and the blue gourami, *Trichogaster trichopterus* (Pallas). Five of the most virulent strains used in this study, shared a common resistance to the killing effect of naïve fish serum. Other factors, such as, autoagglutination in 0.2 per cent acriflavine, instability after boiling, production of an S-layer and proteases and haemlysins did not correlate well with virulence. In addition, serotyping could not group all the virulent strains. Therefore, serum resistance is considered as a good indicator for screening of *A. hydrophila* strains isolated from diseased fish in SE Asia.

Willoughby, Roberts and Chinabut (1995) worked on *Aphanomyces invaderis* sp.nov., the fungal pathogen of FW tropical fish affected by EUS.

Chinabut, Roberts, Willoughby and Pearson (1995), in their paper, 'Histopathology of snakehead, *Channa striatus* (Bloch), experimentally infected with specific *Aphanomyces* fungus associated with EUS at different temperatures', stated that, snakeheads, *Channa striata* (Bloch), were inoculated with a spore suspension of the specific pathogenic *Aphanomyces*, isolated from fish affected by EUS in SE Asia. Fishes were held in three different temperatures:19, 26 and 31°C. Histological changes induced by the infection have been described by them. In the early stages of the disease, degenerative changes were observed in all samples; but, inflammatory infiltrate was much more marked in fishes kept at 26 and 31°C. By 8 days post-injection, extensive mycotic granulomatosis was observed in the samples kept at 26 C and 31°C. The fish kept at 19°C developed a severe invasive myonecrosis with limited macrophage response. From 14 to 28 days post-injection, healing became well-established at 26 and 31°C; and, surviving fish kept at these temperatures, recovered completely by 28 days. The lesion was still progressing at 21 days post-injection in fish kept at 19°C; and, all such fish succumbed by this time. Thus, mortalities in the fish kept at 19°C were considerably higher than in the groups of fish kept at 26 and 31°C. These results could help to explain why mortalities from EUS occur when water temperatures are low.

Ahmed and Rab (1995), in their paper, 'Factors affecting outbreaks of EUS in farmed fish in Bangladesh', stated that, among 257 fish ponds investigated in Bangladesh, 46 per cent were affected by EUS, either fully or partially. Fish culture ponds containing Thai silver barb, *Puntius gonionotus* (Bleeker) were the worst affected by EUS; while, all Nile tilapia, *Oreochromis niloticus*(L.) monoculture ponds remained generally unaffected.

Anthropogenic Factors

The development of industry and intensive farming in some countries of the EUS-affected regions led workers to suspect the discharge of industrial effluents into waterways and the increased use of agrochemicals in rice-fields and nearby agricultural areas as the primary cause of EUS (FAO, 1986).

Several studies have been conducted to ascertain the links between use of pesticides and herbicides and the outbreak of EUS. Boonyaratpalin *et al.* (1983) detected residues of organochlorine pesticides, *e.g.*, dieldrin; and, herbicides, *e.g.*, paraquat in 87.5 per cent of the fish and in 98.3 per cent of the water samples taken from sites across Thailand. Later studies in Thailand and the Philippines have shown that, although occasionally high concentrations of organochlorine and organophosphorus insecticides can be found in water, no correlation can be made with the presence of residues and disease outbreaks. Work on tissue residue concentrations, in both infected and uninfected snakeheads from the Philippines, has shown no relationship between tissue concentration and EUS infection (Phillips and Kedie, 1990). One additional factor is that, some affected areas, such as, parts of the Lao PDR and Myanmar, have never been exposed to herbicides or pesticides of any kind and have no associated industry; but, yet have suffered severe epizootics of ulcerative diseases (FAO, 1986; Phillips and Keddie, 1990).

However, although agrochemicals are not the primary cause of EUS, the possibility exists that in some parts of the region, they cause sub-lethal stress and pre-dispose fish to disease. The application of urea fertilizer in Malayasia and its accumulation in paddy fields is thought to be a factor that has led to the country's few isolated cases of ulcerative fish disease in recent years.

Minor Elements

Earlier FAO (1986) and later (Phillips and Keddie, 1990) studies have shown that the concentration of metals and other potentially toxic minor elements in water samples have been generally low and within recommended guidelines, with occasional peaks above toxic thresholds. However, such peaks could not be correlated with disease episodes. Copper, zinc and mercury have been found at high residual levels in comparison with 'safe' values in some areas, particularly, India (Phillips and Keddie, 1990).

Macintosh and Phillips (1986) recorded consistently high levels of aluminium and iron in water samples from infected sites. High levels of aluminium may be released from soils after heavy rains (particularly if these rains follow a dry period) causing gill and epidermal damage in fish. Aluminium is particularly toxic to fish at a high pH range of 5-6 and is now recognized as an important toxin in even mildly acidic freshwater (Exley and Phillips, 1988). Its role in outbreaks of EUS, if any, remains to be elucidated.

Soil Conditions

No particular characteristics of soil fertility or organic content appear to be related to the occurrence of the ulcerative disease condition (Macintosh and Phillips, 1986). However, many infected areas have been found to have sediments of a slightly acidic nature and low calcium content. Such soils would account for the poorly buffered acidic water and high levels of aluminium and iron in water samples from infected sites. Outbreaks of red spot in Australia tend to be associated with the acid sulphate soils (Callinan, 1998). Acidity, together with high levels of aluminium and iron, is commonly flushed out of acidic soils when rainfall follows a period of drying, conditions that are sometimes reported to precede EUS outbreaks.

Basic Water Quality Parameters

Rodgers and Burke (1986) cited the rapid seasonal depression of salinity and temperature as important environmental factors predisposing fish to attack by Red Spot Disease (RSD). Callinan (1998) had postulated that these factors occur following periods of heavy rain when acidic compounds are washed out of the surrouonding soil, and it is the resulting low pH that triggers outbreaks of RSD.

Seasonally variable parameters have also been implicated in outbreaks of EUS. Several reports indicate that epizootics occur during periods of declining and/or unstable temperatures. Monitoring at affected sites in Bangladesh, China, India and Lao PDR during 1988 and 1989 found that outbreaks occurred during months when the mean daily temperature was below the annual mean daily temperature (Phillips and Keddie, 1990). However, outbreaks have also been recorded during the warmer months (Philippines and Thailand: Phillips and Keddie, 1990) suggesting no clear relationship between EUS and temperature. Diurnal temperature fluctuations of 10°C were recorded during later outbreaks in both Bangladesh and the Philippines (Phillips and Keddie, 1990). Such fluctuations, which would be of greater magnitude in shallow rice-field environments, are undoubtedly, sufficient to stress resident fish populations. Further detailed experimentation is required to understand the significance of temperature in EUS outbreaks.

Low alkalinity, hardness and chloride concentration and fluctuating pH also show a link with EUS occurrences. This is significant as waters with low alkalinity are known to be vulnerable to fluctuations in water quality, respond poorly to fertilization and result in increased toxicity to fish of metals and some organic pollutants. Low chloride concentration also make fish less tolerant to environmental toxins. Acidic water itself may damage the gill structure, and epidermis of fish and result in lowered resistance to infection. The importance of alkalinity (and chloride) has been further emphasized by reports from several countries of successful treatments of EUS by the application of lime (and salt).

Some sites have revealed that, high levels of carbon di-oxide and low levels of dissolved oxygen could be potentially harmful to non-air breathing fishes which would certainly have resulted in stress (Phillips and Keddie, 1990). However, sub-optimal levels of these parameters tended to persist for much of the year, and no correlation was found with outbreaks of disease.

Ammonia and nitrate are rarely recorded at excessive levels prior to an outbreak, indicating that toxicity of these compounds is unlikely to be a significant factor (FAO, 1986). Non-occurrence of EUS in water bodies having high concentrations of nitrogen- and also phosphorus-compounds suggest that the disease is not particularly associated with heavily fertilized or organically loaded waters.

Climate and Seasonality

The water quality parameters that best correlate with epizootics (low temperature, alkalinity and hardness) generally follow the period of greatest rainfall at the start of the dry season. In the latest survey, 72 per cent of the outbreaks occurred during the dry season and normally stopped at the onset of the next rainy season (Phillips and Keddie, 1990). However, the rainfall pattern in SE Asia is variable; whereas most of the Thailand experiences a distinct dry season from September to March, the equatorial areas of SW Malaysia and Indonesia have rain for most of the year.

The FAO (1986) report indicated that some outbreaks of EUS in estuarine fish of Singapore and Thailand were preceded by phytoplankton blooms. Such blooms are also very much related to weather, usually occurring in still, bright weather conditions following a period of rainfall. The potential significance of rainfall patterns requires a more detailed analysis.

General Husbandry Aspects

Menasveta (1985) reported that 50 per cent of diseased snakeheads recovered when moved to ponds with improved water quality; and Das *et al.* (1990) observed that ill-fed, over-stocked fish pond with poor sanitary conditions are more susceptible to infection by EUS. These reports illustrate the importance of good husbandry in controlling EUS. It is also advisable for farmers to try as far as possible to exclude the entry of pathogens from infected areas. Although there is no direct evidence indicating the route of infection, particular care be taken to isolate ponds from potentially 'contaminated' water, fish and equipment.

Disinfection

Destroying the stock, sterilizing the rearing facilities and starting again with disease-free fish, might be a means of eliminating EUS, particularly, in small, well-controlled environments like hatcheries. Care be taken to dispose of dead diseased stock by incineration or burial with disinfectants such as slaked lime. To prevent infection, or re-introduction of the disease after sterilization, nets and other equipments be dipped in, or sprayed with, disinfectant before being moved between ponds. Hygienic procedures could include the requirement of workers and visitors to wash hands and feet with chemical disinfectants, such as, Densol and Lizol. The wheels of vehicles can also also be disinfected if there is a risk of infection from nearby ponds.

The Water Supply

Farms receiving their water supply from irrigation canals or paddy fields have a high risk of infection; so, it is advisable, if at all possible, for ponds to be filled using tube-well water. If this is not feasible, ponds be filled with water collected from sources other than irrigation canals and their embankments be raised to prevent the entry of flood water. Wild fish, if any in the cultured pond, be preferably removed.

The Fish Stock

Farmers should ensure that fish from disease-free sources are stocked and on no account, should wild fish be introduced or be allowed to enter the culture pond. Hatcheries could maintain movement records of fry (including species, number, size and destination), so that, if disease does break out, sites at risk could be easily traced. Seed suppliers be encouraged to provide verification document(s) of the disease-free status of the fingerlings, and farmers may encourage this practice by demanding health certificates with each delivery. If feasible, a prophylactic treatment (of salt or malachite green, for example) be administered to new stock; and, if possible, they be isolated from the rest of the farm stock for 2-3 weeks.

High stocking densities are often the major cause of stress to cultured fishes, but optimal densities vary with species. If too high, water quality will deteriorate, stressing the fish further. If mortalities occur, it is important that dead fish are removed as soon as possible.

The reduction in severity of EUS outbreaks in subsequent years suggests that, resistance to the disease is gradually built-up over successive generations. Stocking of resistant fish stock could offer some protection, but is only advisable in areas where the disease has occurred as they may harbour latent carrier states that could infect fish in 'naïve' areas. The stocking of resistant species, such as, tilapia during the EUS season has been encouraged by some authorities.

Nutrition and Feeding

Malnutrition results in profound changes in the immune response of fish (Landolt, 1989); and, sub-clinical deficiencies of protein, vitamins, energy and certain minerals have been shown to reduce the ability of fish to resist disease. Menasveta (1985) gave anecdotal evidence that cultured snakeheads fortified with vitamin C stood less chance of being infected by EUS; and, recommended this procedure before the onset of the cool season.

As an alternative to trash fish, good quality pelleted feed, stored in a cool dry area and fed at an appropriate regime can provide all the nutritional requirements of intensively farmed fish. Although more costly than trash fish, Hardy (1990) argued that the improved nutritional profile and water quality conditions presented to fish on a fixed pelleted diet would result in a higher return to the farmer in more intensive farm conditions.

A drastic reduction in feed levels is recommended at the first sign of infection. No other organic inputs, such as, application of fertilizer, should be made during an outbreak, as these may result in additional stress to the farm stock.

Goswami *et al.* (1988) also noted that grass fed to cultured *Ctenopharyngodon idellus* is usually collected from submerged areas nearby and may, therefore, be a potential source of infection.

Ricefield and Pond Management Practices

In general, it was observed that, the occurrence of EUS in paddy field systems normally coincided with the end or towards the end of the paddy cultivation period. At this time, the water level is low, accumulation of certain types of fertilizer may take place and decomposition of organic matter, especially grass and aquatic weeds, is common. This pattern can be observed in the Philippines, Thailand, the Lao PDR, Vietnam and also Malaysia (Phillips and Keddie, 1990).

In a recent survey of the region (summarized by Phillips and Keddie, 1990), the most common fertilizers that were used in each of the above-mentioned countries, were, urea, TSP, NPK (mixed fertilizer), unspeicified inorganic fertilizers and amophos. The use of certain types of pesticides were also common. In Myanmar, dieldrin and DDT; in Thailand, malathion, endrin, and dieldrin; in India, dipterex; in Malaysia, paramine, 2-4-D, gramoxone, Ker Tex 200 and 2-4-D azodrine, diazinon and Machate are generally used in the Philippines. In general, pesticides and fertilizers were commonly applied in the irrigated paddy field systems. Careful examination of cropping patterns and application timing reveals no clear correlation between application of agro-chemicals and EUS outbreaks. In most cases, the application of pesticides was found to have no relation with the outbreak of EUS. Herbicides accounted for 75 per cent of the chemicals used; and, they are usually applied during the field preparation period or field clearing period. Certain sub-acute effects resulting from accumulation of these chemicals might occur but no direct effects seem to relate to EUS.

In pond systems, it has also been found that major incidences of EUS occur at the end of fish culture period. This was clearly indicated by data from Indonesia, Myanmar, Vietnam and the Lao PDR (Phillips and Keddie, 1990). Some degree of organic matter decomposition may occur from fertilizers (chicken manure, cow dung, ducks manure, pig dung, night soil) and also supplementary feeding. Several studies have also shown that, good management of pond water quality (correct fertilization, liming, correct use of feed, etc.) can reduce the incidence of EUS, suggesting that poor water quality does play a role in outbreaks of EUS, because it adds to the already existing stress conditions.

Public Health

Many countries have not experienced a fish disease epizootic on such a large scale as that of EUS before; and, therefore, not surprisingly, there has been a great deal of local apprehension as to the effect of consuming EUS-affected diseased fish or using affected waters for domestic or agricultural purposes. The so-called death of ducks, cattle, other animals and even man had, once upon a time, been attributed to the large-scale prevalence of EUS, which had later been found to be baseless; as there had been no scientific evidence in favour of animal or human death due to EUS (Kar and Dey, 1988 b). Rahman *et al.* (1988) were unable to induce any disease symptoms in ducks fed EUS-infected fish or even injected with *Aeromonas hydrophila* cultures. Therefore, it is important to take public educational measures and allay the natural fears of farmers, fishermen and consumers about any wider effects of EUS.

There is some evidence that the cases of patients suffering from diarrhoec symptoms caused by *Aeromonas hydrophila* increased during the 1982-83 epizootic in Thailand (Chumkasien, 1983), although Reinpraryoon *et al.* (1983) reported a fairly high carrier rate of the bacterium anyway in some rural villages. In any case, it must be stressed that good hygienic practices be adhered to. In particular, dead fish be not collected for sale or consumption; not because of ulcerative disease as such, but because bacterial or toxins present in any dead fish may cause illness in man.

The uncontrolled use of chemotherapeutants to treat EUS or other diseases in intensive culture systems is also a matter of public health concern. Chloramphenicol, for instance, is used in treating typhoid in man; and, there is risk that the built-up bacterial resistance in treated fish (Poonsuk *et al.*, 1983) may be transferred to man. Of greater concern to farmers is the possibility of severe allergic reactions affecting farm workers in contact with the drug. There is also the danger that consumers may be exposed to drug residues in marketed fish that had been hurriedly harvested before the recommended withdrawl period had been completed. Although these are issues that affect aquaculture, in general; the occurrences of EUS has underlined the need to develop appropriate guidelines and legislation to protect farmers and consumers against the indiscriminate use of chemotherapeutants.

Socio-economics

The potential social and economic impacts of a highly infectious disease affecting rice field fish in Asia are immense. It is estimated that 250 million families in the region depend on rice as a main crop and much of the incidental fish harvests from these paddies are an important part of the families' diet (Macintosh, 1986). Farmers involved with fish culture in rice fields and pond systems, stand to lose significant financial investments as well. It could be noted here that, the months for harvesting rice paddy-fish are generally from September to February (in different parts of Asia), the period when most ulcerative disease episodes occur. Therefore, fish producers face the problem of trying to market the fishes when the disease is likely to be most prevalent.

The outbreak, to date, affected, in a very devastating way, the intensive snakehead and catfish culture activities in different parts of Asia. It is estimated that the 1982-83 outbreak alone caused direct economic losses of around 200 million Baht in Thailand (approx. US$ 8.7 million) (Tonguthai, 1985). A widespread but unfounded fear of disease transmission to consumers led to a drastic decrease in market demand for food fish, including marine species that were unaffected by EUS. Confidence in FW fish farming, particularly, among potential investors and financial agencies, was badly affected.

Priyadi (1990) reported losses of common carp (*Cyprinus carpio*) broodstock in Indonesia in 1980 amounting to US$ 2.0 million; but, this was, probably, the result of an unrelated bacterial septicaemic condition.

In the Philippines, the average daily income of fishermen (roughly US$ 4.0) declined to US$ 1.50 during EUS outbreaks in Laguna de Bay due to the rejection of affected fish. There is also concern that repeated annual outbreaks of EUS, particularly, in Laguna de Bay, could lead to a significant decline in the natural fish population; and, eventually, result in depleted stocks of the affected species.

Bangladesh suffered severe losses from EUS in 1988 and 1989. Reduced consumption following the outbreak of disease resulted in initial price reductions of FW fish upto 75 per cent.

EUS did also have quite a severe impact on the newly developed fish farm industry in Nepal; and, on the nutritional welfare of the population; given that, there are no marine fish reserves. It has been reported that Nepal has already lost 15-20 per cent of total fish production or NRs 30.0 million (US$ 9,37,000) from just 328 ha of affected ponds in the Eastern Development Region.

The recent occurrence of EUS in epizootic dimension in ponds culturing IMCs, could have extremely significant economic implications for the whole of the Indian Sub-continent; as well as, a potentially crippling effect on the expansion and even viability of carp culture in some areas. In the latest reports from India, Bhaumik *et al.* (1990) found that, more than 73 per cent of the cultured ponds in West Bengal have been affected and most of these were reported to have lost between 31-40 per cent of their stock. Traders in that region have also been affected by the decline in fish consumption. Prasad and Sinha (1990) reported that from only 6 of the 39 districts which make up the earlier Bihar Province, fish worth more than Rs. 48.0 Lakh (US$ 2,50,000.0) have perished.

Callinan (1998) calculated the annual loss of FW and brackish water fish due to ulcerative disease on the East coast of Australia at approximately AUS $ 1.0 million (US $ 13, 30,000.0). This does not take into account the potentially irreversible damage to fish stocks which could have a significant negative influence on the sizeable commercial and recreational fishing industry in the future.

Reports from eastern China have indicated significant socio-economic damage from an ulcerative disease showing characteristics of the Australian condition. The relationship between the Australian and Chinese accounts of EUS remains to be established.

The potential of fish disease to cause major financial losses has vividly been drawn to the attention of National Authorities (ADB/ NACA, 1991). Therefore, one possible positive economic impact of EUS has been the increased funding allocated to fish disease research and diagnostic facilities in Asia by local governments and International organizations.

Control of EUS

Effective control and treatment of EUS is a major problem today because the primary pathogen has not yet been identified and characterized.

Control of EUS in large natural water bodies is, perhaps, impossible. However, in view of the fact that, TA of water acting as a predisposing stress factor, application of lime in low alkaline water bodies, helps in controlling the spread of EUS (Kar, 2004).

Treatment of EUS

The effective treatment and control of EUS is hindered by several major problems; in particular, the disease's widespread occurrence among wild fish populations and its still as yet unknown aetiology. It is virtually impossible to control pathogens in large bodies of water containing wild FW fish. However, fish farming activities are generally designed for easier application of chemical treatments and these appear to have met with some success, particularly, in the major carp farms of India and snakehead farms of Thailand. Both prophylactic and therapeutic treatments, generally involving the addition of quicklime (CaO), have been reported to have achieved satisfactory results; but, no large scale comparative assessment has been carried out involving a variety of affected species. Most of the compounds used are known to be ineffective against viruses, but even if the primary disease agent is a virus, they may be just as effective in limiting the severity of the ulcerative condition by treating secondary fungal or bacterial infections. The success of lime applications illustrates the need to improve environmental conditions that may prediscpose the fish to disease.

Liming

Lime is readily available from suppliers of construction materials and is a relatively simple and inexpensive means of enhancing water quality. This is largely achieved by increasing alkalinity and appears to be effective in helping to control losses in fish ponds. Agricultural lime ($CaCO_3$ or $CaMg(CO_3)_2$) stabilizes the pH of the water but has no sterilizing effect. Burnt lime (quicklime-CaO or slaked lime –Ca $(OH)_2$) could have a therapeutic effect against disease; but, has a very high pH and if too much is applied, it inhibits bacterial decomposition of organic matter; and, an overdose may even kill fish. Overdosing can easily be a problem in ponds which are naturally alkaline.

As a general husbandry procedure, ponds be ideally allowed to completely dry out; lime (preferably, slaked lime) spread evenly over the bottom; and, water supplied to a depth of 10 cm for 7-10 days before filling to the normal depth and stocking with fish. Emergency prophylactic and therapeutic applications can also be made directly to stocked ponds but treatment does vary considerably. Amounts of 150-600 kg/ha slaked lime in 1 m deep ponds at 2-4 week intervals have been recommended for treatments depending on the pH of the pond in the first place. If the alkalinity is between 150-200 mg/l $CaCO_3$, the pH would be considered fairly high and less lime would be required. Low levels could be applied for prophylaxis but Purkait (1990) noted that liming below 100 kg/ha fails to control the EUS.

Salt (NaCl)

Salt is a cheap, non-polluting and fairly effective agent, particularly if used in conjunction with lime. Salt is known to control some external parasites; but, more importantly, temporarily inhibits the activity of some bacteria and fungi and lowers the toxicity of nitrite and ammonia to fish. Tolerance of fish to salinity is species specific; but, some IMCs have been reported to be successfully treated with short dips of 3-4 per cent concentration; and, 1 per cent salt solutions have been used in unspecified fish ponds in Thailand.

Potassium Permanganate ($KMnO_4$)

This chemical has been recommended at concentrations of 5 mg/lit in bath treatments or applied at 3-5 kg/ha in ponds, to control secondary infections of diseased fish. However, Tachushong and Saitanu (1983) reported that, at 5 mg/lit., $KMnO_4$ took 120 minutes to inactivate *Aeromonas hydrophhila* in purified water. This could be due to the permanaganate ion reacting with organic matter and other reduced substances before it can react with the bacteria; and, most fish culture ponds are high in organic matter, such as phyto- and zooplankton, food residues and fish wastes. A positive consequence of breakdown of organic matter and deactivation of zooplankton is that, dissolved oxygen (DO) levels are enhanced due to a reduction in the Biochemical Oxygen Demand (BOD), although $KMnO_4$ can result in some depression in DO levels because oxygen is removed during oxidation of this organic matter. The chemical is, however, expensive and beyond the reach of many pond owners. Therefore, its use would, probably, be restricted to dipping valuable fish, such as, broodstock, or in hatchery environment, using clean water.

Bleaching Powder

As an alternative, or additional, prophylactic measure to liming, ponds can be disinfected with 50 mg/l bleaching powder (25-30 per cent chlorine) and left for 7-10 days before stocking. Chlorine at a sufficient concentration may act as a virucidal agent (Frerichs, 1990); but, such levels should never be used in ponds stocked with fish. In therapeutic treatments in India, bleaching powder applied at a concentration of 1 mg/l or 5-10 kg/ha has been reported to be effective in helping to heal early lesions.

Malachite Green

It is a very effective fungicide and ectopasiticide; but, may be toxic to some species of fish at high temperatures. Its use is restricted to the USA because it is suspected to be having teratogenic properties. Therefore, great care must be taken with its use on fish farms. Treatment dips of 1 mg/l eliminated fungal infections in carps from Bangladesh, allowing lesions to heal spontaneously (Roberts *et al.*, 1989); and, three daily spray applications of 0.1 mg/l concentration appeared to reduce mortalities in diseased *Trichogaster pectoralis* (Tangtrongpiros *et al.*, 1983). However, use of malachite green for commercial purposes may be restricted to hatcheries or smaller ponds where fish can be netted out for dipping.

Other

Formalin may also produce potentially toxic conditions to fish in warm water. Frerichs (1990) observed that formalin inactivated snakehead rhabdovirus but only at concentration which is quite unsuitable for fish life. It has been used at concentrations of 25-50 ppm, sometimes, in conjunction with Dipterex (0.01 mg/l); but, the main effect would probably have been in the control of unrelated parasitic infections.

Iodine has been used under experimental conditions, but its toxicity is greatly reduced by the presence of proteinaceous matter abundant in most fish ponds. Temeyavanich (1983) demonstrated that, catfish survived 12.5 mg/l iodine which killed *Aeromonas hydrophila* bacterial strains

The peroxide disinfectant Virkon S has been found to be effective in inactivating suspensions of snakehead rhabdovirus; but, *in vivo* tests have to be carried out (Frerichs, 1990).

'Traditional' treatments, such as, the application of crushed tamarind or banana leaves to infected ponds is said to have had some therapeutic effect but no scientific testing has been done.

Antibiotics

Antibiotics are a valuable but expensive means of controlling secondary bacterial infections, particularly, of *Aeromonas hydrophila*. Tetracycline has been found to be efficacious in oral treatments of *Puntius gonionotus* at 500 mg/500 g feed; but, in IMC broodstock, injection was found to be effective (Robers *et al.*, 1989). Expensive fish may benefit from an additional antibiotic bath. However, oral treatments are the easiest to apply and Jhingran (1990) recommended the use of erythromycin, nalidaxic acid, oxytetracycline or terramycine at doses of 60-100 mg per kg feed for seven days. Although antibiotic cocktails combining three drugs are reportedly more effective than a single antibiotic (Saitanu and Poonsuk, 1983) this be discouraged due to the increased risk of bacteria developing resistance to all these drugs. Poonsuk *et al.* (1983) had already recorded an increase in resistance to chloramphenicol, sulphadiazine, sulphathiazole and tetracycline in *A. hydrophila* strains isolated from farmed fish in Thailand. In addition, a sufficient withdrawl period be allowed before harvesting treated fish to prevent public health risks due to the presence of drug residues in marketed fish.

Monitoring

Fish Farmer

There are several steps which a farmer could take to monitor fish stock and the related environmental conditions to get a good idea of the state of health of the fish. Monitoring enables farmers to take immediate action to check any outbreak of disease in its earliest stages, if any changes are noticed.

Daily observations be made and attention given to any physical signs of injury or sickness, such as, body or gill colouration. Individual and group behaviour be monitored particularly in terms of swimming and feeding. The warning signs are evident when fish swims lethargically, close to the water surface, near the littoral zone of the pond, particularly around its corners. It is a good opportunity to observe the fish when it feeds. Loss of appetite, if noticed, could be due to the onset of the disease.

Intensive fish farms be advised to keep feed and stock records which are filled-in daily. A progressive loss of appetite can easily be determined by comparing records of the daily ration. Dead fish are a notorious source of the disease and be removed daily and mortality figures recorded. An increase in mortality is a clear indication that something is wrong. The dead fish could be examined to give a rough post-mortem diagnosis; fish farmers be familiar with the external symptoms indicative of EUS. Periodic sample weighing may be considered, in the culture of valuable fish with a long growth cycle, as a general indicator of the health and condition of the population.

Basic environmental monitoring is particularly relevant in the prediction of EUS episodes. This may range from simple observations of weather and rainfall patterns to recording potentially stressful water quality parameters, such as, temperature and pH. Particular attention be given to rainfall, as evidence suggests a higher risk of EUS when rain follows a long dry spell. DO levels may be significant and could be recorded before sunrise when readings are lowest; but, metres are usually expensive and difficult to calibrate. The depth of water in the pond be monitored during the dry season as low water levels could result in rapid diurnal fluctuations of other parameters.

Fish Health Worker

The need for a standard approach to EUS-monitoring throughout the region has been emphasized. A standardized monitoring from was prepared at the NACA Second Technical Workshop on EUS and the Environment. It calls for a basic site description including management practices of the rice-field or pond system, such as, application of fertilizer or pesticide; information on the occurrence of the disease itself and fish affected; water quality data; and, an economic impact assessment.

Histopathological examination of ulcerous lesions is a particularly useful diagnostic tool for investigating EUS; but, the ability to recognize characteristic features requires a great deal of experience.

Regional Monitoring Programme

Since 1987, NACA has implemented a Regional Research Programme on relations between EUS and the Environment. The research programme has established a region-wide monitoring system, for monitoring of the status of EUS in the region. Such a network provides the basis of an early warning system to the countries to alert them about the rapid spread of EUS.

Conclusions and Recommendations

Epizootic Ulcerative Syndrome (EUS) continues to cause significant losses to wild and cultured FW fish as it extends its range across the Indian sub-continent and recurs annually to a lesser degree in previously-affected areas of SE Asia. The serious socio-economic impact of EUS, and difficulties encountered in the diagnosis and control of EUS, have highlighted the urgent need for improved mechanisms for dealing with diseases in aquatic animals in developing countries of the Asia-Pacific region. Lessons learnt from EUS apply widely to dealing with other disease problems now threatening the development of aquaculture in the region. Many of the issues have been reviewed and summarized in a recent ADB/NACA joint study on fish diseases and fish health management in the Asia-Pacific region (ADB/NACA, 1991) which gives recommendations for the action required at the National and Regional level to upgrade capabilities to cope with the present and future problems of aquatic diseases. From the various reviews and the NACA Regional Research Programme on EUS, the following conclusions and recommendations could be made:

Rules and Regulations

Quarantine and Health Certification

The rapid rate at which the EUS has been spreading across the freshwaterbodies, regardless of country-boundaries, highlights the possibility that human agencies have been responsible for the spread of EUS, and the problem of current unrestricted transport of fish and their pathogens within and across National boundaries. It is possible that the trade in some EUS-susceptible species, such as, catfish, gouramies and the barbs which are of interest to aquarists, has been responsible for outbreaks of EUS in isolated areas (*e.g.*, Sri Lanka, the Philippines, etc., ?). Recommendations were made at the ADB-NACA Regional Workshop on Fish Disease and Fish Health Management for the establishment of centralized quarantine facilities and a system of health certification.. The importance of such protective systems, which are partly operational in some countries in the region, has been emphasized by the severe economic losses caused by EUS, and could, probably, be justified in economic terms alone.

Farm Registration

Related to quarantine, there is a growing need for regulations controlling domestic as well as international trade, to restrict the impact of infectious disease within countries or regions. Ideally, all farms and hatcheries should require a license and keep records of fish movements. The transport of diseased fish to unaffected areas could then be controlled; and, wherever necessary, be restricted. Commercial pressure from fish farmers for 'certified' disease-free fish could do much to pressure fingerlings and fry suppliers to improve the quality and disease status of their work, something that might remove the need for legislation, which may in any case be difficult to implement.

Chemotherapeutants

The indiscriminate use of chemotherapeutants, such as, antibiotics, may also serve to damage the aquaculture industry. Although most antibiotics are not directly harmful to human beings, they may cause disturbances in the intestinal flora of consumers or lead to the development of resistant bacterial strains (making human or fish diseases more difficult to treat). Few regulations are currently enforced and there are little or no guidelines concerning withdrawl periods after administration. Such information is needed to safeguard both the aquaculture industry and its workers, and the cosumers of aquaculture products.

Research

Aetiology

There is still insufficient knowledge regarding the aetiology of EUS, and studies are still needed, in investigating the role of oomycete fungi and viruses in the pathogenesis of the disease. More detailed descriptions of the histopathological characteristics of EUS, and similar ulcerative conditions, are also required.

Control

EUS episodes have illustrated the relevance of an integrated approach to fish health, utilizing a combination of all available management techniques. The application of chemotherapeutants, lime in particular, is a valuable part of these management techniques, but more detailed studies quantifying preventive and curative treatments are required. Such studies, combined with those on infectious agents, will provide the means for control of EUS in cultured fish populations, and possibly in wild fish populations in enclosed and semi-enclosed waters, *e.g.*, the 'Beels' in India.

Environment

The NACA Regional Research Programme, along with other studies, emphasized the importance of environmental factors in EUS, although the critical factors still remain elusive. Declining or unstable temperatures, low alkalinity and hardness and fluctuating pH (with possible high levels of aluminium and other elements, mobilized in low pH condition) appear to be significant; but, rigorous experimental work is required to determine their significance. Greater understanding of the importance of such factors might open the possibility for manipulation of the environment; or, alternatively, the avoidance of certain conditions developing, to control EUS in both wild and cultured fish populations. The continuous and region-wide monitoring programme of selected environmental parameters, as established under NACA, will continue to assist in elucidating these variables.

The Regional studies on EUS have also shown that, pollution of the aquatic environment, from organic and inorganic pollutants, is a growing and important constraint to fisheries and aquacultural development in the Asia-Pacific region. Our understanding of the complex relationship between EUS and the environment has been restricted by a chronic lack of data on the relationship between a fish and its environment. The practical benefits to be attained through improving the pond environment have been demonstrated on many occasions during the study of EUS, and a much greater understanding of the interactions between pollutants, fish production and disease would assist in developing strategies to both protect fish from the burgeoning losses, in both wild fish population and in aquaculture, due to pollution and disease.

Improved Facilities

EUS has also highlighted the need for improved fish disease diagnostic and research facilities in developing countries of the SE

Asian region. The economic and social benefits of developing the manpower and infrastructure to investigate such problems would outweigh the disadvantages and cost of such facilities (Note the millions of US$ lost due to EUS). To this end, the Royal Thai Government, with the support of ODA has established the 'Aquatic Animal Health Research Institute (AAHRI), which is supposed to provide facilities for sophisticated research on a regional and international basis.

The realization of the objectives indicated above for the prevention, control and eradication of EUS requires the adoption and implementation of appropriately planned policies by the Governments concerned. This, in turn, requires a high degree of organization and international co-operation when dealing with such a widespread problem as EUS, which clearly takes no heed of international boundaries.

Recommendations

There are several recommendations arising from the Reviews which closely follow those of recent NACA Second Technical Workshop on the Relationship between EUS and the environment held in August, 1990 (Phillips and Keddie, 1990). These Workshop recommendations were formulated by researchers from 11 countries in the region. As such, they represent distillation of priorities as seen by regional scientists. The main recommendations are given below:

1. The Workshop recognizes that EUS remains as a serious problem affecting the FW fisheries of Asia and is particularly serious in the Indian Sub-continent. The workshop recommends that, research be continued to determine the cause(s) and develop effective preventive and curative measures against EUS.
2. The Workshop, having completed a successful analysis of results from the Regional Programme, recommends the following priorities to be given in research:
   (*a*) Detailed studies of the pathogens involved
   (*b*) Detailed studies of the fish populations at affected and unaffected sites
   (*c*) Histopathological studies of the infected fishes

(d) Detailed studies of preventive and curative measures

(e) Detailed studies of climatic variations at infected sites

(f) Studies of selected environmental parameters.

3. The Workshop recognizes the threat to Aquaculture and Fisheries resulting from environmental pollution. The workshop recommends a detailed study of the effect of pollution on fisheries in the Asian region; as well as, to assess the impact on aquaculture, fisheries production and human health.

4. The Workshop recognizes the need to monitor and report on the spread of EUS in Asia; and, to establish a mechanism to disseminate information on EUS throughout the region. The Workshop recommends that a Central Co-ordinating office be established to provide this service; and, requests NACA to consider the establishment for such an early warning system and information service.

5. The Workshop recommends that the Regional Research Programmes be continued on the basis of the above recommendations.

6. The Workshop recognizes the need to develop manpower and infrastructure in the region to combat the on-going problem of EUS fish disease in the region; so that, the future programmes of aquaculture and fisheries are not much adversely affected by the growing problems of fish diseases, particularly, the EUS.

Definition of EUS

EUS, today, is a semi-global problem among the freshwater fishes. However, in view of its complex infectious aetiology, it is yet to be accurately defined (OIE, 1997).

Chapter 21

Biology of Fishes: Analysis of Life History Traits in Fishes

Food and Feeding Habits

Introduction

Assessment of the food and feeding habit of a fish enables us to determine its niche' in the ecosystem; and, thereby, its preferred food items. It tells us how much the food spectrum of a fish overlaps with that of the co-existing fishes. In other words, it gives us an idea how much competition the fish species faces for food from other species in the biotope.

According to Schäperclaus (1933), the natural food of fishes could be classified into three groups, *viz.*, (a)'Main food', *i.e.*, the natural food which the fish prefers under favourable conditions and on which it thrives the best; (b)'Occasional food', *i.e.*, the natural food which is said to be well-liked by the fish and is consumed as and when available; and, (c)'emergency food', *i.e.*, the food which is ingested by the fish when the preferred food items are not available and on which the fish is just able to survive.

Nikol'skii (1963) classified the food of fishes into four broad categories according to the relationship between the fishes and their food. These categories are: (a) Basic food, which the fish usually consumes, comprising the main part of the gut contents; (b) Secondary food, which is frequently found in the guts of fishes, but in small amount; (c) Incidental food, which is said to enter the gut of the fish only rarely; and, (d) Obligatory food, which the fish is said to consume in the absence of the basic food. Nikol'skii (1963), further, grouped fishes into the following groups: (i) Herbivorous and detritophagic: the fishes which feed on vegetable matter and detritus; (ii) Carnivores: the fishes which feed on animals; and, (iii) Predators: the fishes which prey upon. Nikol'skii (1963), still further, categorized fishes into three categories according to the extent of variation in the types of food consumed by them, *viz.*, (a) euryphagic:the fishes which feed on a variety of food items; (b) stenophagic: the fishes which feed on a selected few types of food; and, (c) monophagic: the fishes which feed on a single type of food.

Analysis of food components in the gut of a fish species collected from different ecosystems provides us with the information on how much that particular species is selective in food items or how much that fish species is flexible in feeding on different types of food items.

Study of feeding habits of cultivable fish species helps us in determining the desirable species combinations in culture systems with least interspecific competition for natural fish food. It also gives us vital clues in developing supplementary feed for the fish species.

Further, feeding habit is one of the primary criteria in enabling us to decide on transplantation of species to new ecosystems with least possible damage or competition with the indigenous native ichthyofauna.

In addition to the above, it is known that, natural food preference of many fish species changes as they grow through different ages and stages, *e.g.*, fry, fingerling and adult. This age-related information on feeding behaviour is very important in hatchery and nursery management practices.

Types of Fishes Based on Feeding Habits

As stated earlier, fishes could be classified broadly into four categories based on their feeding habits and predominant food items consumed by it:

Herbivore

Which primarily eat plant materials, *e.g.*, aquatic macrophytes (AM), filamentous or non-filamentous algae, periphyton, etc.

Carnivore

Which primarily eat animal materials, *e.g.*, molluscs, crustaceans, small fishes, etc. They could also be cannibals.

Omnivore

They are not very selective in their food items. Their food items could include phytoplankton, zooplankton, bottom dwellers, like mollusks, worms, insect larvae, etc. These fishes may also feed on small fishes, fish eggs and detritus depending on their availability.

Detritivore

These fishes feed mainly at the bottom of the water body on organically rich detritus, which could be composed of decayed plants, animals, plankton, etc. embedded in the soft mud.

In addition to the above, fishes could, further, be classified into three types on the basis of the layer of the water column they principally feed:

Surface Feeder

Fishes which primarily feed at the surface of the water body.

Column Feeder

Fishes which primarily feed at the mid-column region of the water body.

Bottom Feeder

Fishes which primarily feed at the bottom of the water body.

Some Important Considerations in Food Analysis

In most cases, while analyzing food items, one may not expect rigidity in food items in the stomach of the fishes examined. Unlike land animals, in water, the fish may not be able to be too choosy. In the process of water filtered through the gills, the fish may engulf many items which are available in the waterbody but are non-targeted food components. These are called 'accidental items'. Even items present in the stomach of a prey may also form a part of the stomach contents of a predator. Rational judgement based on a number of stomachs analysed, if possible, from a variety of environments, is the right way to decide the basic feeding habit of a fish species.

Simultaneous examination of the distribution of food elements in the environment and in the guts will enable us to decide on the selectivity or non-selectivity in the feeding habit of the fish.

It would also help us in determining the fish as bottom feeder, column feeder, surface feeder or a non-selective feeder.

It is essential to analyse food elements from the foregut as well as from the hind gut. This would make it possible to identify the food items which are digested in the gut and which are not digested. The latter ones do not constitute the preferred food elements and might have been engulfed along with other food items.

Methods of Analysis

Precautions During Sample Collection

1. Fish samples for gut content analysis be always collected from the wild rather than from culture systems; so that, an objective view of its natural food could be obtained. Sampling under culture systems are often necessary for specified objectives, *e.g.*, for determining the acceptability of artificial feed; or, to assess the health of a fish as revealed through active feeding.
2. It is better to get samples from cast net or drag net (rather than gill net) in order to get fresh and undigested food in the stomach. If gill nets are used for sample collection, it is advisable to get freshly gilled specimens through periodic checking of the nets.
3. Samples be obtained from more than one ecosystem to know the flexibility or rigidity in selection of food items.
4. Different age groups be represented in each sample, as far as possible, in order to establish the changes in feeding habits, if any, related to age, growth and life history stage.
5. A preliminary assessment by cutting open the stomachs of a few fishes collected in the morning and in the evening provides us with the first hand information on the feeding rhythm of the species in relation to time. The time at which the stomachs have more fresh food be noted. It is suggested to collect the fish samples during that time.

Sample Collection, Preliminary Observation and Labelling

Live fish may be killed with a sudden blow to the head. The total length (TL) and weight of the fish be noted.

The abdomen be cut open. By cutting closer to the pharynx and the vent, the entire gut be carefully removed using a pair of sharp forceps and a pair of scissors.

The fullness of the gut be observed and points be allotted for calculation of the 'Feeding Index' (vide the Model Table 2).

The gut be carefully uncoiled. The contents of the foregut and stomach be transferred slowly, by sqeezing, to a petri dish containing adequate quantity of 5 per cent formalin.

Similarly, the contents of the hind gut be also transferred, by squeezing, to another Petri dish containing 5 per cent formalin.

The gut content samples be put into separate tubes, sealed and labelled properly for analysis later; but, preferably, within a week.

The label be with the following information:

Sample No.:	Species:	Sex:
Date of collection:	Place of collection:	Time of collection:
TL of the fish (mm):	TW of the fish (g):	Portion of the gut (stomach/fore/hind gut):

NB: Labelling be legible and clear without causing any confusion later. Sample No. be invariably entered in duplicate. Simultaneously, all the relevant details of each sample be maintained in a Register. Only permanent markers (fine tip) or a sharp pencil be used. The label be protected with a transparent adhesive tape. This sample No. be not repeated during the entire process of study. In no case, 'labelling' be treated lightly.

The lips, cleft/gape of mouth, teeth, gill rakers, position of the vent, stomach, intestine be examined and the Gastrosomatic Index be calculated as given in Model Table 1 (Morphometric Assessment of Feeding Habits). Based on these morphological criteria, the preliminary inferences on the feeding habit be recorded.

Model Table 1: Morphometric Assessment of (Predominant) Feeding Habits

	Predominant Feeding Habits		
Mouth-Position	*Inferior*	*Terminal*	*Upward*
Mouth-gape	***Narrow and small*** Bottom feeders Grazers	***Medium and reaching up to orbit*** Omnivorous and nibblers Carnivorous	***Cleft beyond the orbit*** Plankton feeders
Lips	***Fringed and soft*** Bottom feeders/detritivores Grazers	***Non-fringed and stout*** Omnivores/carnivores	
Jaws	***Weak*** Herbivores/detritivores	***Moderate built*** Omnivores	***Strong and stout*** Carnivores
Teeth	***No canines, vomerine fine and poorly developed*** Herbvore/detritivore	***No stout canines, vomerine week*** Omnivore	***Strong canines, several rows of well developed vomerines*** Carnivores
Gill rakers	***Numerous and filamentous*** Planktivores	***Less and moderately long*** Omnivores	***A few, stout and sharp*** Carnivores

Contd...

Model Table 1–Cont...

	Predominant Feeding Habits		
Mouth-Position	*Inferior*	*Terminal*	*Upward*
Stomach	***Poorly developed, not distinguishable from the intestine very poor***	***Moderate musculature, continuous with the intestine but distinguishable from the intestine musculature***	***Strong musculature, clearly separable from intestine***
	Herbivore/detritivore	Omnivore/detritivore	Carnivore
Intestine and GSI RLG = Gut length/ Body length	***Thin, long and highly coiled gastrosomatic index more than 4***	***Moderately long, a few coils, gastrosomatic index 1-4***	***Not coiled, often straight, thick musculature GSI less than one***
	Herbivore	Omnivore	Carnivore
Position of the vent	***Posterior half of the body***	***Middle to posterior***	***Middle to first half***
	Herbivore/detritivore	Omnivore/detritivore	Carnivore

Model Table 2: Model Calculation of Feeding Index

The gut be observed after cutting open the fish and be graded as full, ¾th full, ½ full, ¼ th full and empty. The observations be recorded as follows:

Species: Date of Collection:

Sex: Size of Collection:

Sample No.	*TL/TW (mm/g)*	*Sex (M/F)*	*Stage of Maturity*	*Fullness of Stomach*	*Points Alloted*
1				1/2 F	10
2				1/4 F	5
3				F	20
4				1/2 F	10
5				1/2 F	10
6				1/4 F	5
7				3/4 F	10
8				F	20
9				E	0
10				E	0
Total points allotted					**95**

Feeding index: (100 × 95)/(10 × 20) = 47.5%. Maximum points for a full stomach is 20 in this case.

The Feeding Index (FI) be calculated for different months, different maturity stages and for different size groups. This has many other implications in the environmental and biochemical investigations.

Analysis of Stomach Contents

Some methods are available for the quantitative analysis of the gut contents. However, the methods are, in general, highly subjective. Judgment may vary from individual to individual, both on the methodology to be followed and also on the relative importance to be given to each food item. Some of the methods, generally followed, are given below:

1. Occurrence method
2. Number method
3. Points method

4. Index of Preponderence method
5. Volumetric method
6. Gravimetric method

The Occurrence method, the Number method and the Points method are briefly described below along with a glimpse of the Index of Preponderence method.

Occurrence Method

This method is based on the examination that a particular item is present or absent in the gut.

A few guts be analysed

To be ascertained, in how many guts a particular category of food item is present. The per cent of guts in which the item was present, is to be calculated. The relative importance is presented in terms of per cent.

In other words, if we put it in a formula:

$$\% \text{ Occurrence} = (n_1/N) \times 100,$$

where,

n_1 = The total No. of stomachs in which the given food item was present

N = Total No. of stomachs examined

The calculation is shown below in the Model Table 3.

Model Table 3: Estimation of Gut Content by Occurrence Method

Gut No.	*Food Items*				
	Algae	*Rotifer*	*Cladoura*	*Copepoda*	*Insect Larvae*
1	✓		✓		✓
2		✓		✓	
3			✓		
4	✓			✓	
5		✓			
6	✓		✓		
7	✓			✓	
ni	4	2	3	3	1
% Occurrence (ni/N) × 100	57.1	28.6	42.8	42.8	14.3

The only one standard applied in this case is the 'Physical presence' of an organism. The size, volume, weight, etc., of the item is not taken into consideration. Moreover, different food items get the same importance value irrespective of size or volume, *e.g.*, in a stomach, a rotifer and much bigger fish are treated as 1: 1 under this method. Hence, they are given equal value.

Number Method

In this method, the number of each food element in the sub-sample of the stomach content is counted and the per cent of each food item is determined from the total number of all the food items. Average per cent for each item from a number of sub-samples examined is considered as the relative importance of the item.

Here also, the size or volume is not taken into consideration. Hence, the problem which is encountered in occurrence method persists here too. But, since the smaller items are generally encountered in large numbers, the relative error is less in this case.

In other words,

$$\% \text{ abundance of each food item} = (n_1/N) \times 100,$$

where,

n_1 = Total number of each food item

N = Total number of all the food items

A calculation is shown below in Model Table 4.

Model Table 4: Percentage Abundance of Food Item in the Gut of a Fish by Number Method

Food Item	*Nos. Observed*				
	Sub-Sample I	*Sub-sample II*	*Sub-sample III*	*Total (ni)*	*% (100 × ni/N)*
Algae	70	60	30	160	74.1
Rotifera	12	18	3	33	15.3
Cladocera	1	9	6	16	7.4
Copepoda	3	2	-	5	2.3
Insect larvae	-	-	2	2	0.9
Total	**86**	**89**	**41**	**216 (N)**	

Points Method

In this method, the different food items in the stomach are counted and points are allotted to each category against a standard unit, *e.g.*, a copepod naulius larva is assigned 10 points by the investigator. Each unit of food item present in the gut is weighed against this unit and points are allotted based on eye estimation.

Atleast five sub-samples from a gut are analysed and points are allotted to each category of food element.

Points received for each category in the sub-samples are added together and is converted into per cent against the total points received for all the items

The per cent is considered as the relative importance of that food item

A calculation is shown in Model Table 5.

Model Table 5: Percentage Abundance of Each Food Item in the Gut of a Fish by Points Method

Food Item	*Points Received*						
	Sub Sample I	*Sub Sample II*	*Sub Sample III*	*Sub Sample IV*	*Sub Sample V*	*Total (ni)*	*% (100 x ni/N)*
Algae	1530	780	520	760	1240	4830	39.0
Rotifera	650	900	400	600	730	3280	26.5
Cladocera	475	320	150	340	480	1765	14.3
Copepoda	700	760	180	220	300	2160	17.5
Insect larvae	50	130	60	-	100	340	2.7
Total	3405	2890	1310	1920	2850	2315	100.0

Index of Preponderence Method

This method provides a measurable basis for grading the food elements as a combined picture of both occurrence as well as bulk; or, in other words, it combines the indices of both occurrence and the volume of each food item. The Index is constructed through three steps:

1. Calculation of Index of occurrence
2. Calculation of Index of volume
3. Calculation of Index of preponderance.

The frequency of occurrence of each category of food is calculated as per cent against total number of occurrences of all items, *i.e.*, if out of N fish examined, n_i fish contains food item $_i$; then, the per cent of occurrence of food $_i$ may be given as $(n_i / \Sigma n_i) \times 100$.

This will give the Index of Occurrence or O_i).

Similarly, the volume of each kind of food item in all the stomachs are added together and given as the per cent of the total volume in all the guts together, *i.e.*, out of N stomachs examined, the total volume of food item $_i$ is calculated as Σv_i. The per cent of v_i is calculated as $(v_i / \Sigma v_i) \times 100$.

This will give the Index of volume of V_i.

The Index of Preponderance of food item $_I$ (I_i) is, thus, calculated as follows:

$$I_i = (V_i O_i / \Sigma V_i O_i) \times 100$$

In this method, the food preference of the fish is more decisively brought out by the consistent occurrence of a food in good volume than by any partial examination of volume alone or frequency alone.

This method, however, poses little difficulties in fishes which feed on smaller items which are difficult to separate and estimate the volume.

A calculaltion is shown in Model Table 6.

Assessment of Selective Feeding

Often a question arises, how far a fish is selective in its food items available in the water body in which it lives. As such, Ivlev (1961) provided a formula for arriving at the 'Selectivity Index (E)' with regard to different kinds of food items present in the environment:

$$E = (r_i - p_i) / (r_i + p_i)$$

where,

E = Selectivity Index of a given food item

r_i = per cent of the given food item in the gut contents

p_i = per cent of the given food item among the samples collected from the environment.

Model Table 6: Estimation of Food Occurrence (%) and Food Preference by Index of Preponderance Method

Proforma	*Gut Content Analysis Through Index of Preponderance Method*
Species:	*Sex:*
Date of Collection	*Fullness of Gut:*
Location of Collection	*Stage of Maturity:*
	No. of Guts Examined:

Sample Calculation of Index of Occurrence

Food Item	*Gut Number*					*No. of Guts Occurrence*	*Index of Occurrence (O_i) (ni x 100/sum n_i)*
	1	*2*	*3*	*4*	*5*		
Crustaceae	✓	✓	✓	✓	✓	5	27.8
Algae		✓		✓	✓	3	16.7
Plants			✓	✓		2	11.1
Rotifera	✓				✓	2	11.1
Insecta		✓		✓		2	11.1
Protozoa	✓	✓	✓		✓	4	22.2
No. of Occurrence	3	4	3	4	3	18	100.0

Sample Calculation of Index of Occurrence

Food Item	*Volume of Food Items in Each Gut*					*Total Volume*	*Index of Volume (Vi) (ni x 100/sum n_i)*
	1	*2*	*3*	*4*	*5*		
Crustaceae	2.10	3.20	1.1	0.80	2.00	8.10	75.0
Algae		0.80		0.30	0.40	1.50	13.9
Plants			0.20	0.15		0.35	3.2
Rotifera	0010				0.10	0.20	1.9
Insecta		0.30		0.15		0.45	4.2
Protozoa	0.05	0.10	0.05			0.20	1.9
Total	2.25	4.40	0.25	1.40	2.50	10.80	100.0

Contd...

Model Table 6–Contd...

	Sample Collection Index of Preponderance			
Food Item	*Index of Occurrence (O_i) (n_i × 100/sum ni)*	*Index of Volume (V_i) (v_i × 100/sum n_i)*	*$O_i \times V_i$*	*Index of Preponderance (I_i) $O_i \times V_i \times$ 100/sum $O_i \times V_i$*
Crustaceae	27.8	75.0	2085.0	84.7
Algae	16.7	13.8	230.5	9.4
Plants	11.1	3.2	35.5	1.4
Rotifera	11.1	1.9	21.1	0.9
Insecta	11.1	4.2	46.6	1.9
Protozoa	22.2	1.9	42.2	1.7
Total	100.0	100.0	2460.9	100.0

'E' value ranges between +1 to -1, where +1 denotes complete selectivity; -1 denotes no preference for the item; and, 0 (zero) absence of selectivity.

NB: A good number of guts are to be analysed. An equal number of samples from the environment are also to be analysed. Both planktonic and benthic organisms are to be analysed to arrive at a reliable conclusion.

It is difficult to assess and quantify items like detritus in a bottom sample. The method is more reliable when assessed for the plankton feeders or benthic animal feeders.

If samples are less in number, it may also be possible that, if an accidental item recorded in the gut (engulfed along with other food items) will not be represented at all in the samples from the environment. But that will also give the value of maximum selectivity.

For example: Let us take an example of a fish scale engulfed along with detritus. Here, it may be supposed as follows:

% occurrence of fish scale in the gut: 1.5 per cent

% occurrence of fish scale in the environment: 0

Therefore, $E = (1.5\text{-}0)/(1.5 + 0) = 1.5/1.5 = +1$

This denotes a complete selectivity where as the fish, perhaps, never wanted to eat it. The Investigator is required to apply his/her judgement in such cases.

Assessment of Feeding Interaction

The method generally followed to explain the competition for food between the species is as follows:

1. The relative importance of each category of food item in the stomach of all the selected species is assessed.
2. The values, specieswise, is recorded in a Table
3. Then, conclusions are drawn.

A hypothetical calculation is shown in Model Table 7.

Model Table 7: Assessment of Feeding Interaction

Food Item	*Relative Importance (%) in the Gut*			
	L. gonius	*C. idella*	*B. sarana*	*H. molitrix*
Algae	20	2	8	68
Rotifers	2		2	8
Cladocora	—		16	—
Copepoda	—		13	4
Insect larvae	—		13	4
Molluscs	—		8	—
Detritus	78	12	23	24
Aquatic plants	—	86	—	4

Further, this type of data could be used for quantification of overlapping food spectrum using Simpson's (1949) Similarity Index for comparison of two environments:

$$S = 2C/(A + B)$$

where,

A = No. of items present in Sample A

B = No. of items present in Sample B

C= Items present which are common to both Sample A and Sample B.

A hypothetical calculation of the 'S' values, which represent the extent of overlapping of food spectrum between pairs of above species, is given in Model Table 8.

Model Table 8: Overlapping of Fc od Spectrum Between Pairs of Fish Species

	C. idella	*P. sarana*	*H. molitrix*
L. gonius	(2 × 2)/(3+3) = 0.67	(2 × 3)/(3+4) - 0.60	(2 × 3)/(3+5) = 0.75
C. idella		(2 × 2)/(3+7) = 0.40	(2 × 3)/(3+5) = 0.75
P. sarana			(2 × 4)/(7+5) = 0.67

Some General Considerations

Selection of Method

The method to be adopted depends on the purpose and the extent to which the accuracy is desired.. For example, if it is required to assess the feeding habit of a single species, it would depend on the feeding of that species. But comparative analysis of quantification of overlapping food spectrum requires a uniform methodology for all the species. It is advisable to adhere to the relative importance of each food item in terms of per cent rather than actual number or points. Per cent is said to be a common reliable standard for the purpose of comparison.

Taxonomic Level of Food Identification

It is also little indefinite that, how far one should slide down along the taxonomic level in recording the food components. In general, a group-wise recording in a Tabular form could be enough. A fish is not expected to be selective for food item(s) at species level. Thus, classification with the captions like 'Alage, rotifers, crustaceans, insects, molluscs, fishes' could be enough for recording data in a generalized manner. However, further elaborative recording could certainly enable assessing a more precise feeding habit of the fish. For example, recording with the captions like, 'Chlorophyceae, Bacillariophyceae, Myxophyceae, Rotifera, Cladocera, Copepoda, Ostracoda, Insect larvae', etc. This second level of food component analysis is generally adopted since it is difficult to go beyond this level due to partially digested condition of food components; and, also, that, the food components are often found embedded in mucous, mud, etc.

Nevertheless, an investigator is suggested to record as low taxonomic level as possible, *e.g.*, upto genera/species level, if possible, which could give him/her an opportunity to decide the level of consideration at a later stage.

Study of the Same Species by Different Groups from Different Habitats

A group of investigators could be involved in study of the food habit of a given species from different habitats (ecosystems). As such, it is necessary to arrive at a common policy for evaluation before the initiation of the data collection. The chief considerations could be as follows:

The primary objective of the investigation

Methods of collection of samples

Minimum number of each species to be studied

The method to be followed for gut-content analysis

The taxonomic level at which the food components are to be recorded

Without such common approaches, the final analysis of the pooled data could lead to erroneous conclusions. Further, wherever possible, the investigators could exchange the sub-samples of the gut contents to compare the inferences arrived at by each investigator.

Chapter 22

Length-Weight Relationship and Condition Factor

Knowledge regarding the age composition of a fish population is essential for proper management of a fish farm. Age and growth are closely related. The length frequency distribution has been used to estimate the age of fish since the last part of 19th century. This method is based on the fact that, the lengths of fish of one age tend to form a normal distribution.

'Growth' may be defined as increase in size actually. Growth is the direct result of the chemicals, osmotic and other forces, by which, materials are introduced into an organism and transferred throughout the many parts of the body.

Increase in size are usually measured during successive intervals of time. The growth curve is actually made up of points defining dimensions at specific times, measuring time along the X-axis and the dimensions (L,W) along the Y-axis. Thus, length plotted in relation to time produces a vector diagram which is called the 'growth curve'.

The growth rate of fish varies far more than those of a warm-blooded animal. The general characters of the curve remains the same; except that, because fish continues to grow even in extreme old age; it draws towards its ultimate or limiting growth with exceeding slowness. This is called 'indeterminate growth'.

Absolute growth in mammals begins slowly, attains a maximum velocity somewhat early in the life; and, then slows down, approaching a point where growth may cease altogether. Such a type of growth is called 'determinate growth'.

Further, the rate of growth is directly affected by temperature and also by other physical conditions. 'Absolute growth' is the average total size at each age. It is usually plotted as a regression of 'L' on age or 'average L' for each age group. Relative growth may be defined as the per cent growth, in which, the increase in growth in each time interval is expressed as a per cent of the growth at the beginning of the time interval.

Length-Weight (L-W) Relationship

Studies on the relationship between length and weight of fishes constitute important aspect of their biology. A formulation of this relationship could either be of purely academic interest as far as the formula could be used to calculate a corresponding length or weight when only one of these parameters is known. It could also be of practical utility in measuring the yield of fish from water bodies. It could further be useful in regulating the fisheries through maintenance of desirable mesh size of the fishing gears which could provide for the escapement of smaller individuals and retention of commercial-sized fishes (Lagler, 1956). Furthermore, since length-weight (L-W) relationship of fish varies depending on the condition of life and environment, the variations in this relationship provide a measure of condition co-efficient of fish, and the suitability of the environment.

Length (L) and Weight (W) of fishes may be determined with accuracy. W of fish may be considered a function of the L. This relationship of the L and W follows approximately the Cube Law relationship expressed by the formula,

$$K = W/L^3$$

where,

W = Symbol of Weight

L = Symbol of Length.

If form and specific gravity were considered to be constant throughout the life, the relationship could be used to describe the general L-W relationship in populations of fishes; and, thus, could serve as the basis for the calculation of unknown weights of fish of known length or *vice versa*. However, a fish or any animal is continually prone to changes in its bodily proportions during its life. Thus, the simple cube law expression may not hold good throughout the life history and growth period of the fish. Since the instances where the cube law does apply throughout the growth of a fish are the exceptions, a more satisfactory formula or general equation for the expression of the L-W relationship has been developed which is given below:

$$W = c\,L^n$$

where,

W = Weight

L = Length

c and n are constants

This equation, when expressed in logarithmic form, becomes

$$\log W = \log c + n \log L$$

The values of c and n could be determined empirically by fitting a straight line to the logarithms of L and W; or, by computing them in the following normal equation:

In other words, the following formulae could be used to determine the exponents:

$$\log c = \frac{\Sigma \log W.\ \Sigma (\log L)^2 - \Sigma \log L\ (\Sigma \log L.\ \log W)}{N.\ \Sigma (\log L)^2 - (\Sigma \log L)^2}$$

$$\text{and } n = \frac{\Sigma \log W - N \log c}{\Sigma \log L}$$

Beckman (1948) had presented a thorough description of the L-W relationship together with some of the constants for Michigan fishes. Some workers have made use of this relationship by merely plotting the log of the W against log of the L. This gives a straight line regression which may be easily compared with curves of a similar nature.

Condition Factor (CF)

Another use of the cube law relationship, although an entirely different expression from the L-W relationship, is the so-called 'Condition Factor' or the 'Co-efficient of Condition'. CF of fish is expressed by relating the length of the fish to its weight. Values of CF are interpreted as 'expressing the relative well being' of the fishes.

CF has been recognized as two kinds after the two ichthyologists:

(a) Fulton's Condition Factor

It is expressed by the following formula:

$$K = (W \times 10^5)/L^3$$

where,

K = The condition factor or the co-efficient of condition

W = Weight in grams

L = Standard length in mm.

The above equation was originally proposed by Ralph Hile and included in the L-W relationship paper by Beckman (1948). The Co-efficient of Condition is used to indicate suitability of an environment or to compare fish from one lake or area with a general average for an entire region. As fish grows older, they tend to gain proportionately more in weight than in length; so that, the value of K (or C) will increase with age. Often, sexual dimorphism, or the time of the year the sample is taken, is taken into account. Maturing fish can not be compared with fish which has just spawned. Proper sampling could tend to minimize many of the above difficulties; but, comparison could be considered as valid only if fish of the same length, age and sex; taken as nearly as possible on the same date; are used.

(b) Le Cren's Condition Factor

It is expressed by the following formula:

$K = W/\hat{W}$; or, expressed in words

K = Observed weight in grams/Expected weight in grams

Here, K stands for Condition Factor or Co-efficient of Condition

According to cube law, the weight of the fish equals the cube of the length times a constant. Mathematically, $W = aL^3$, where, W= Weight, L= Length and 'a' is constant. Le Cren (1951) has reviewed this cube parabola. The general parabolic equation is mathematically expressed as:

$$W = cL^n$$

where,

W = Weight of the fish in grams

L = Length of the fish in mm

c = A constant

n = An exponent

The equation $W = cL^n$ is calculated in its logarithmic form as:

$$\text{Log } W = \log c + n \log L$$

The value of $\hat{W}$ (as mentioned above) is calculated by using the formula:

$$\text{Log } W = \log c + n \log L$$

Some of the Essential Terminologies

1. Gastrosomatic Index (GSI) or Hepatosomatic Index (HSI):

   (Weight of the stomach content (g)/Total weight of the fish (g)) × 100

2. Relative Gut Length (RLG):

   Total length of the gut (cm)/Total length of the body (cm)

3. Index of Fullness or Degree of Satiation

   (Total weight of entire gut content (g)/Total weight of the fish (g)) × 100

NB: Here, '/' stands for 'divided by'

Chapter 23

Maturity, Breeding and Fecundity

The sustenance of a population depends on the availability of adequate breeding population in an environment and its capacity to produce sufficient number of progeny. Information on breeding size/age and the numbers produced per brood are vital information for deciding on exploitation strategies with regard to natural populations.

The study of maturity and breeding is important as it provides the information on the following aspects:

The size/age at first maturity

Synchronisation of maturity in male and female

Seasonal and non-seasonal secondary sexual characters

Breeding season

Breeding behaviour

Frequency of breeding

No.of eggs/offspring produced in a season

Fecundity in relation to size/age

Environmental requirements

Parental care, etc.

The knowledge on the maturing time, breeding migration, breeding grounds and breeding aggregation are important for fishery regulations and conservation measures.

The maturity size primarily determines the minimum exploitation size to be allowed in open waters. Prevention of the formation of under-sized fishery is crucial for maintaining population size. It is expected that a fish be given a chance to breed, at least once; so that, it adds to the population. Exploitation of large number of juveniles would lead to reduction of breeding population; and, ultimately, results in the collapse of the fishery itself.

Under culture fishery, the precise information on maturity condition, breeding period, size-related information on the number of eggs produced, etc., are necessary for planning the breeding schedule. The age/size related fecundity gives information on the quantity of brood stock to be maintained; and, also, the quantity of infrastructure/personnel required for the breeding operation.

Fishes are known to exhibit extreme variations in all aspects of breeding. Hence, species-wise information is inevitable before venturing into seed production in aquaculture or conservation of natural fauna.

Information on habitat preferences for breeding and the breeding migration helps identifying habitats which require conservation.

Some of the Terminologies

Fecundity

Fecundity is considered as the number of eggs produced in one year. Fecundity is said to be high where the eggs are liberated into open marine waters; but, it is lower in freshwater forms; and, also, where there is parental care. Fecundity and egg size tend to be inversely related. 'Absolute fecundity' is defined as the number of eggs in the ovaries of an individual fish which are ripening for the ensuing spawning period. The term 'relative fecundity' expresses the number of eggs per unit length or weight of the fish. Many authors believe that fecundity increases with length, weight or age.

In short, 'fecundity' is defined as the number of matured ova in female gonad prior to spawning period. It is an important aspect of fish biology and ecology due to its direct relation to fish production and fisheries. Fecundity also has a considerable bearing on the reproductive potential of a population.

Estimation of Fecundity

In other words, 'fecundity' is the number of mature eggs produced by a single female in a single breeding season. It could also be related to the number of mature eggs produced by a single female in a year or even in its life. Under such cases, it is necessary to specify such information while reporting the fecundity, *i.e.*, annual fecundity, life-time fecundity, etc. Nevertheless, some authors consider 'fecundity' as the number of ova actually shed during a breeding period. However, where a prolonged breeding period is observed for the individuals (*e.g.*, a fortnight when eggs are released in batches) with wide gap between two successive gonad maturations which extends to a few months, the fecundity includes the total number of eggs which are expected to be shed during that prolonged period.

Fecundity may be calculated by the following formula:

$$F = aL^b$$

where,

F = Fecundity

L = Length

a = Constant

b = Constant

In other words, the following formula could also be used:

$$F = W * (n_1 + n_2 + n_3)/(w_1 + w_2 + w_3)$$

where,

F = Fecundity

n_1, n_2, n_3 = No.of ova in the three samples

w_1, w_2, w_3 = Weight of the ovary samples in grams

Method of Assessment of Fecundity

1. Fish specimens be collected and its length (mm) and weight (g) be recorded. A number of fish specimens of varying size be collected and cut open in order to assess the status of maturity during the breeding season.

2. It is advisable to select only mature fishes of IVth and Vth stages of maturity. If mature fishes are available in adequate number, specimens belonging to maturity stage V be only selected.
3. The gonad be removed completely and cautiously without losing any egg and be packed in polythene bag and be properly labelled with information on species, date of collection, place of collection (location), sex, length and weight of the fish and serial number of the sample. The gonad be preserved in 10 per cent formalin and brought to the laboratory and be allowed to harden for a couple of days.
4. After a couple of days, the gonad be taken out to a Petri dish or a beaker. The excess moisture be removed by using a tissue paper or filter paper. The weight of the gonad be determined with a sensitive electronic balance.
5. Small portions of the ovary (gonad) be taken (chipped) out (*e.g.*, 0.1-0.5 g for eggs of upto 2.0 mm in diameter; or, more, depending on the size of the ovary) from the anterior, middle and posterior portions of the ovary.
6. Excess of moisture be drained out using a filter paper or tissue paper.
7. These three small portions of the ovary be weighed with the help of an electronic balance.
8. The eggs, contained in the petridish (with little amount of water) or on a piece of filter paper, be separated, using the finger. However, if the eggs are found to be very closely adhered together, the eggs be shaked in a solution to make them little loose. This would help to separate them and to count. A slight de-moisturising in the oven could help easy separation of the ova in many cases.

The above procedure may be briefed as follows:

For each matured fish specimens, the L and W of the fish as well as the ovary in fresh condition are noted. The ovaries are dissected out and preserved in 5-10 per cent formalin. Samples of the ova from the anterior, middle and posterior regions of the preserved ovary of individual specimens are taken and mixed thoroughly. A random sample of one gram with two replicates is counted and the total number of eggs in each ovary is estimated.

As indicated above, Fecundity may be calculated by the following formula:

$$F = W * (n_1 + n_2 + n_3)/(w_1 + w_2 + w_3)$$

where,

F = Fecundity

n_1, n_2, n_3 = No.of ova in the three samples

w_1, w_2, w_3 = Weight of the ovary samples in grams

The log Fecundity against fish length (mm) or fish weight (g) be plotted.

The regression equation be found to estimate the fecundity at given size or weight of the fish:

$$\text{Log } F = \log a + b * \log L; \text{ or,}$$

$$\text{Log } F = \log a + b * \log W$$

where,

L = Length of the fish in cm/mm

W = Weight of the fish in g

Fecundity could also be expressed in relative terms, *e.g.*, per 100 g of body weight or per kg of body weight, etc.

Gonado-somatic Index (GnSI)

The GnSI assessed through the seasons provide the indication of the progression of maturity in the fish and the periodicity of breeding. The spawners could be:

An annual breeder breeding within short span of few days.

An annual breeder breeding with prolonged time-gap

Even though an individual is a short-term breeder, the breeding season for the population could be a prolonged one. Observation of the gonads of a number of fishes at fortnightly or monthly intervals round the year gives the right knowledge on the breeding behaviour of individual fish; as well as, of the population. Assessment of GnSI is one straight tool for getting such information. However, information on GnSI is to be supplemented with macro- or microscopic examination of the gonads for precise information.

Estimation of GnSI

It could be defined as the per cent of the gonad weight to the body weight after the gonad is removed.

$$GnSI = (GW/(TBW-GW)) \times 100$$

where,

GnSI = Gonado-somatic Index
GW = Gonad Weight in grams
TBW = Total Body Weight in grams

Data recording and calculation of GnSI be done in Model Table 9 given below.

Model Table 9: Recording of Data and Calculation of GnSI in Fishes

Proforma: *Calculation of Gonado somatic index/Size at first maturity*

Species: *Date of Collection:*

Location of Collection:

Fish No.	*Fish Length TL (mm)*	*Fish Weight FW (g)*	*Sex*	*Stage of Maturity*	*Gonad Weight GW (g)*	*Fish Wt. Less Gonad Wt. (FW-GW)*	*GSI – GW × 100/ (FW-GW)*
1							
2							
3							
4							
5							
6							
7							
8							
9							
10							
11							

Contd....

Model Table 9–Contd...

Fish No.	*Fish Length TL (mm)*	*Fish Weight FW (g)*	*Sex*	*Stage of Maturity*	*Gonad Weight GW (g)*	*Fish Wt. Less Gonad Wt. (FW-GW)*	*GSI – GW x 100/ (FW-GW)*
12							
13							
14							
15							
16							
17							
18							
19							
20							
21							
22							
23							
24							
25							
26							
27							
28							
29							
30							

Gear Used: *Mesh Size (stretched):*

Max. Size in Catch: *Minimum Size in the Catch:*

Other Remarks:

Data recording and calculation of Fecundity be done in Model Table 10 given below:

Model Table 10: Recording of Data and Calculation of Fecundity in Fishes

Proforma *Calculation of Fecundity*

Species:	*Fish Length (cm)*
Location:	*Fish Weight (g):*
Date of Collection:	*Gonad Weight (Gw) (g):*

Sample No.	Sample Wt. (g) Sw_i	No. of Mature Eggs n_i	Fecundity ($n_i \times GW/sw_i$) f_i	Average Fecundity (f1 + f2 + f3)/3	Fecundity per 100 g Body Wt.
1					
2					
3					

Species:	*Fish Length (cm)*
Location:	*Fish Weight (g):*
Date of Collection:	*Gonad Weight (g):*

Sample No.	Sample Wt. (g) Sw_i	No. of Mature Eggs n_i	Fecundity ($n_i \times GW/sw_i$) f_i	Average Fecundity (f1 + f2 + f3)/3	Fecundity per 100 g Body Wt.
1					
2					
3					

Species:	*Fish Length (cm)*
Location:	*Fish Weight (g):*
Date of Collection:	*Gonad Weight (g):*

Sample No.	Sample Wt. (g) Sw_i	No. of Mature Eggs n_i	Fecundity ($n_i \times GW/sw_i$) f_i	Average Fecundity (f1 + f2 + f3)/3	Fecundity per 100 g Body Wt.
1					
2					
3					

Chapter 24

Aquaculture

Cultural practices performed in water are broadly included in the term called 'Aquaculture'. 'Pisciculture', the culture of fishes, forms an important component of aquaculture. The following are the significant aspects of successful pisciculture:

1. Construction and Management of Pond
2. Induced breeding
3. Hatching.

An an account of the above-mentioned aspects is briefly given below:

Construction and Management of Ponds for Fish Culture

Fish is a dependable source of protein and an important item of diet of the people of India. The simple aspect of fishery biology is the logical unity of the theoretical and practical knowledge to develop efficient and rational methods of raising more fish

An obvious pre-requisite for fish culture is the availability of different types of ponds for rearing various stages of fish. Some of the important considerations of a fish farm construction are:

Topography

The ideal topography of a fish farm site is a gently-sloping terrain of a wide valley or a basin surrounded on three sides with high lands and with a narrow outlet on the fourth.

Soil Type

The bottom soil be free from vertical and lateral seepage. Such impervious soils are ordinarily heavy clay, silty clay, clay loam, etc. Porous (sandy) soils could, however, be made satisfactory by the application of proper sealants. Water retention in ponds could be good in somewhat percolative soils, if the water table of the area is high.

Water Supply

The usual sources of water supply for a pond are, generally, the reservoirs, streams, springs, canals, surface run-off, wells, tube wells, etc. Nevertheless, the rain water and the ground water are, by far, the most common sources of water for a fish pond in addition to the above.

Size of the Pond

It depends upon the purpose of the farm. If subsistence fishing only is aimed at; then, smaller sites of <0.04 ha may be used. For commercial and experimental fishing, fish farms comprising of nursery, rearing, stocking and breeding ponds are required. It has been estimated that a watershed of 4-20 ha is essential for one surface hectare of pond area.

Planning and Construction

Before starting construction works, detailed contour survey and preparation of layout diagrams and plans is necessary. The map should indicate the location of ponds, their embankments, drains, mechanical spillways, various farm building, laboratories, store house, watchmen's tower, residential quarters, etc. A brief account of the planning norm is given below:

Embankment

The slope of the embankment depends on the quality of the fill. If the soil is of good quality, the slope could be 2:1 on the inner side or water side; but, it should be 3:1 for loamy, silty or sandy soils.

Drain Pipe

These be installed at the lowest point of a pond; so that, all the water could be drained out as and when necessary. Ponds upto 1.2 ha could have a drain pipe of *c* 153-203 mm dia made of asbestos, cement, etc. The inlet pipes be atleast 152 mm above the water level of the pond when the latter is full; and, they be provided with some sort of a screen to prevent entry of fish into the pond as well as to prevent escape of resident fish.

Mechanical Spillway

It is essential for all stocking ponds with waterspread area of > 4.0 ha. These handle normal overflows as well as act as safety outlets during heavy rains preventing ponds from flooding.

Some More Considerations in the Construction of a Fish Farm

Before any construction starts, all the vegetation comprising of trees, bushes, etc., be cut and the topsoil containing plant roots be scrapped off from the site; else, they may cause leakages. Turfing the pond bottom with low, creeping grasses be done soon after the completion of the embankment to check water seepage and soil erosion. A fencing be erected around the fish pond in order to keep off the livestock. Rats and crabs, which damage the embankment by burrowing, be controlled through trapping. Control of prolonged turbidity in newly constructed fish ponds through the application of agricultural gypsum (hydrated $CaSO_4$), though not much practised in India, could be tried.

Preparation of Different Types of Ponds

Fish culture operations begin with the construction of ponds. Specific types of ponds are required for the culture of particular species of fish and their life history stages. Different types of ponds required for the culture of IMC in farms are: (*a*) Nursery, (*b*) Rearing and (*c*) Stocking.

Nursery Pond

It is a pond where the fry are nursed till they attain a length of *c* 20-25 mm. A 4.0 ha farm could be divided into: (a) Nursery ponds 0.2 ha, (b) rearing ponds 0.8 ha, and (c) stocking ponds 3.0 ha. If a nursery pond is stocked @ 10 million spawn/ha, and an 80 per cent survival upto fry stage is expected, the total fry yield would be 8 million. At a stocking rate of 2 million fry/ha in the rearing ponds, the area required to rear 8 million fry upto fingerling stage would be 4.0 ha. Further, with an envisaged 80 per cent survival upto fingerling stage, the rearing ponds would have *c* 6.4 fingerlings. Adopting the present rate of stocking the fingerlings @ 5000/ha for raising table-sized fish in stocking ponds, *c* 1280 ha of water area would be required to stock the resultant *c* 6.4 million fingerlings. Thus, the ratio of nursery to rearing to stocking ponds would now be 1: 4: 1280.

Certain specific steps for the preparation of nurseries include:

Control of Predatory and Weed Fishes

Predatory fishes, like *Channa* spp., *Anabas* sp., *Notopterus* sp., *Clarias* sp., *Mystus* spp., etc., compete with the IMCs for food and space; and, also, directly prey upon them. The weed fishes like *Puntius* spp., *Chanda* sp., etc, feed upon the zooplankton which are the main food of the carp spawn (Alikunhi, 1955; Chaudhuri, 1960). The customary method of removing unwanted fishes from nursery ponds is by repeated drag-netting. Dewatering, even by resorting to pumping, be done to completely remove all the unwanted fishes. Further, though not to be much encouraged, several kinds of chemicals, called fish toxicants, are in use in nurseries to control undesirable fishes.

Fertilisation

The main objective of nursery ponds is to augment the production of zooplankton, the natural food of carp spawn. For this purpose, the nursery ponds are limed just after the removal of predatory and weed fishes. Then, the nursery ponds be treated with organic manures, such as, cowdung or with inorganic fertilizers (NPK) or with both, one following the other. Spaced manuring, with an initial dose of 10,000 kg/ha of cowding 15 days before the anticipated date of stocking the pond, followed by a second dose of cowdung @ 5,000 kg/ha 7 days after stocking is said to maintain sustained production of zooplankton. In India, cowdung is preferred to inorganic fertilizers as the former results in quicker production of a large number of zooplankton for the carps. Algal bloom, if formed, could be controlled by covering the pond surface with *Lemna* sp.(duckweed).

Aquatic Insects and their Control

Nursery ponds which are designed to rear spawn of IMC are invariably populated with a large number of aquatic insects mostly belonging to the orders Coleoptera, Hemiptera and Odonata (Alikunhi, 1955; Chaudhuri, 1960). These insects include the diving beetles, the giant water bugs, the backswimmers, etc. They harm the fishes by secreting toxins or by piercing their body. Repeated seining operations in the pond with fine-meshed nets, before fish stocking, may not always totally eradicate the predatory insects.However, spraying of oil generally kills the surface insects and often gives encouraging results.

Fish Food Organisms and their Production

Carp spawn, immediately after stocking, need a plentiful supply of natural food in the form of animalcules. The spawn, being tiny and tender, can feed upon plankton of restricted size only. Hence, a qualitative and quantitative knowledge of plankton biomass in the nursery pond is essential in order to fix the time as well as the number of spawn for stocking. The plankton biomass is determined in the laboratory by standard limnological methods. However, a rough field method for the estimation of plankton had been deviced by Alikunhi (1956) for the fish farmers. In this method, about 54.5 lit of water taken from different sections of the nursery pond are filtered through an organdi or muslin ring net with a 2.54 cm dia glass specimen tube tied to the lower narrow end of the net. A pinch of powdered common salt is added to the water in the tube after the plankton is collected. The tube is detached from the net in order to kill the plankton. Within 15-20 minutes of adding the salt, most of the organisms settle to the bottom of the tube. If the column of plankton sediment is 6.4 to 8.5 mm high from the bottom of the tube and the plankton consist mostly of the water-fleas, the pond water may be considered to be sufficiently rich in plankton to stock spawn @ 4,94,240 to 7,41,300/ha.

Stocking

The appropriate time for stocking of spawn in nursery pond is when plankton biomass conforms to the density recommended for stocking. According to Alikunhi (1956), stocking rate could be varied from 12 lakhs/ha to 20 lakhs/ha depending upon the density of plankton.

Supplementary Feeding

Soon after being stocked, spawn start feeding voraciously on zooplankton. At this time, the feeding requirements of spawn are so large that, within 2-3 days of stocking, the food available in the nursery ponds become very low. Alikunhi (1956) had observed that, for the first two days after feeding commences, the spawn do not take artificial feed; but, subsist almost completely on natural food. After this period, artificial feed, along with natural food, enhances the growth and survival of spawn considerably. The commonly administered feed for the IMCs are the rice bran and oil cakes of groundnut, coconut, mustard, etc. Chakraborty *et al.* (1973), on comparing the usefulness of various feed mixtures *vis-à-vis* zooplankton in spawn-rearing, found that, zooplankton alone was the best food for all the three species, *viz.*, *Catla catla*, *Labeo rohita* and *Cirrhinus mrigala*. Of the various feeds, *viz.*, mustard oil cake-rice bran mixture, ground nut oil cake-wheat bran mixture, silkworm pupae (tassar silkworm), soyabean and prawn waste; soyabean gave the best results in *C. catla* and silkwork pupae and ground nut oil cake-wheat bran mixture in *C. mrigala* and *L.rohita*. Yeast was found to increase the growth most followed by Vitamin B-complex, ruminant stomach extract with $CoCl_2$ and Vitamin B-12. The criteria for a successful fish feed are:

1. Ready acceptability
2. Easy digestibility
3. High conversion value
4. Easy transportability
5. Abundant availability
6. Low cost

'Conversion rate' expresses a ratio between food consumed for increase per unit weight gained by the body discounting the food requirement by the fish for its maintenance and energy requirement (sustenance). Mathematically,

Conversion rate = Quality of feed (food consumed) / Weight increase (flesh)

Rearing Pond

The management of rearing ponds is aimed at culturing maximum number of IMC fry to fingerling stage in the shortest possible time. As the breeding season of carps lasts for 2-4 monsoon months, *viz.*, June to September, the rearing of their spawn to fingerling stage has to be completed during this period. The fry which attain a size range of 20-25 mm in nursery ponds, are netted out and released into rearing ponds which are kept pre-prepared more or less on the same lines as the nursery ponds.The rearing ponds are stocked with carp fry in suitable combinations and ratios; supplementary feed given, if necessary; and, finally harvested for putting the fingerlings in the stocking ponds.

Stocking Pond

The goal of the management of stocking pond is to attain high production of table-sized fish from fingerling stage in the shortest possible time. Carp fingerlings, which commonly attain a size range of 75-125 mm within 2 months in the rearing ponds, are netted out and released into the stocking ponds. The latter are generally large perennial water bodies varying from 0.2 to 2.0 ha in area with an average depth of 1.5 to 2.0 m.

Clearing of the pond of weeds and predatory and weed fishes is a pre-requisite for the preparation of a stocking pond. The methods to be employed in this regard is the same as that applied to nursery and rearing ponds. However, old, heavily-silted ponds be desilted before going for stocking.

In order to augment production of natural fish food organisms in small stocking ponds, it is necessary to manure them in suitable instalments with organic manure (cowdung 20,000-25,000 kg/ha/year) and inorganic fertilizers (mixture of $(NH_4)_2 SO_4$, single super-phosphate and Calcium ammonium nitrate in the ratio of 11: 5: 1 @ 1000-1500 kg/ha/year) which may be suitably varied depending upon the water and soil quality.

Among the other points to be considered for attaining high fish production in stocking ponds, are, species combination, their ratio and rate of stocking. Every pond, irrespective of in whatever manner it is managed, can support a fish biomass only upto a certain weight limit. This limit is called the 'carrying capacity' (Yashouv, 1959) or the maximum standing crop (Hickling, 1962; 1971); which, of a pond, is related to their natural productivity and management practices adopted including fertilization, artificial feeding, the species stocked and the stocking rate. Hepher (1962) had demonstrated that, two ponds, one fertilized and the other not, each stocked with 1250 fish/ha, the production in the former was 391 kg/ha/year as against 47 kg/ha/year in the latter; thus, indicating that fertilization has a profound impact on fish yield. According to Charpy (1959), the carrying capacity varies with different levels of feeding and feed types. It depends on the conversion rate of the fodder as well: richer feeds support larger fish crops. Experiments conducted by Hepher and Chervinski (1965) on the interaction between population density and effect of protein rich diet indicated higher carrying capacity of the pond in which fish were fed with protein-rich diet (>2,600 kg/ha) than those in which the feed was not enough rich in protein (1,900 kg/ha).

An increase in the carrying capacity of a pond is obtained by culturing different species of fish with complimentary feeding habits or different weight classes of the same species. In this system of culture, an augmentation in carrying capacity is obtained not by increasing fish resources; but, rather, by better utilization of the existing fish food.

The depth of a pond also influences the carrying capacity. A deeper pond generally supports more fish than a shallow one simply because the former has more living space.

Notwithstanding the above, it is wise to stock fish ponds at a rate well below the carrying capacity in order to allow the fish to grow upto or near the carrying capacity in the shortest possible time. One of the best courses of determining the carrying capacity of a pond is by tiral and error based on common sense and empirical experience. However, the number of fish to be stocked may be computed by feeding the data in the following formula:

Number of fishes to be stocked/unit area =

[(Total expected increase in weight)/(Expected increase of weight of individual fish)] + Mortality (not more than 10 per cent)

Thus, if, in a pond (1 ha in area), a total production of 2,270 kg is the target and only a single IMC species is to be stocked, whose normal growth rate in the first year is, say, 908 kg, then, the number of fingerlings of that species to be stocked in the stocking pond is given by:

$$(2{,}270 \times 1000)/908 = 2{,}500$$

Considering 10 per cent mortality, the corrected stocking rate of this species is 2750.

Induced Breeding

Carp culture has been an age-old practice in many Asian and European countries; and, also, more recently, in certain parts of Africa and America. In India, popularity of carps as a source of food and for culturing them in confined waters, dates back to 350 BC. Their main desirable qualities being quite fast growth rate, short food chain, non-predatory nature, good taste resulting in a great market value and highly profitable returns to the fish culturists.

The breeding requirements of the cultivable IMCs, are quite varying although all of them have almost the same breeding habit. In rivers, generally, flood and rain, current of water, increased DO, optimum water temperature, shallow inundated areas (spawning grounds) and spawning congregation are considered as important ecological inducements for natural spawning of the IMCs. With the onset of monsoon, when the rivers carry a lot of freshwater, these carps ascend the rivers and migrate to adjacent inundated shallow areas for spawning purposes. They exhibit a sort of sex play, which probably stimulate the pituitary gland to secrete hormones; thus, induce ovulation and spawning.

It had been a practice by the pisciculturists, since long, to collect the carp spawn from the rivers by crude devices and release them in ponds in order to raise a population of IMCs therein. This type of fish rearing had problems in the transportation of spawn besides other disadvantages. Further, the fish seed, thus collected, is usually a mixture of economic and uneconomic species; and, their sorting-out by a fish farmer could be far from satisfactory. As a result, there had been a general feeling to ensure a dependable source of quality fish seed. To solve this problem, concerted efforts had been made to create lotic environment in ponds, so that, the IMCs could be induced to breed in confined water. Two such methods emerged out as a result of the efforts put in this direction. These methods of 'induced breeding' are the:

1. 'Bundh' breedig method and the
2. The Hypophysation technique

By inducing IMCs to breed in confined water, these methods could overcome the necessity of running water which is considered to be indispensable for their spawning.

Induced Breeding in 'Bundhs'

The IMCs have been induced to breed in special types of perennial and seasonal ponds locally called 'Bundhs' in some parts of Bengal, Bihar and MP. Bundhs may be either 'perennial', called 'wet bundhs'; or, could be 'seasonal', called 'dry bundhs'. These bundhs receive considerable quantity of rain water with washings after a heavy shower from the extensive catchement areas. Their large, shallow marginal areas serve as spawning grounds for the fish (Chaudhuri, 1968).

Wet Bundhs

A typical wet bundh is a perennial pond situated in the slope of a vast catchement area of undulating terrain with proper embankments and having inlet towards the upland and an outlet on the opposite. It has an extensive shallow area which dries up during summer; but, the actual pond retains water throughout the year; and, here, an adequate stock of IMC is maintained. After a heavy monsoon shower, fresh rainwater with washings from the upland area rushes into the bundh through the inflowing streamlets known as the 'dhals' (as said in Midnapore in West Bengal). During the heavy floods, the excess water is drained through the outlet known as the 'Bulan'. The shallow areas of the bundh, called 'Moan', are the main spawning ground of the carps. The outlet is protected by a bamboo fencing called 'Chhera' and the flow of water through the outlet could be controlled by plugging the 'chhera with straw and mud (Mookerjee, 1945).

Dry Bundhs

Dry Bundh is a seasonal pond and is shallower than the wet bundh. With the first shower, when water accumulates, selected mature carps from the nearby ponds are released. Usually, with the next heavy shower, when the dry bundh is inundated, fishes get stimulated and breed.

Spawning Behaviour of Carps in Bundh Breeding

In bundhs, spawning occurs after heavy shower because of large volume of rain water and considerable increase in water level.

At first, the smaller fishes get excited and migrate to the shallow areas of the bundh. Later, the bigger fishes start spawning in the same area.. During the sex play, as observed in the bundhs, the spawners rush at a high speed. A number of males chase a lone female with violent splashing of water and only one of them finally enters into sex play with the female. Mating lasts for a few seconds and is repeated till the female is spent. After mating, the spawners become quiet for sometime and swim back to deeper areas of the main bundh (Dubey and Tuli, 1961; Alikunhi *et al.*, 1964).

Factors Essential for Bundh Breeding

Water temperature of 27-29°C at the time of spawning, high turbidity, low pH, low TA, little high conductivity of water and abrupt rise of water level have been recorded as conducive factors for spawning in bundhs (Chaudhuri, 1968)

Hypophysation

The significant role of pituitary gland in reproduction in animals had been appreciated since long back; but, the application of this technique in inducing fish to breed is rather recent. Houssay of Argentina was the pioneer in introducing the concept, in 1930, that fish pituitary could be used for spawning of fishes. In India, the first attempt to induce spawning of carps by administering mammalian pituitary extract was made in 1937 by Khan (1938); but, without success. Not much effort was put in this direction till around 1955. Interest was found to grow in this field when the *Esomus danricus* (the carp minnow) and *Puntius atherinoides* (the little cat fish) was induced bred by pituitary hormone injection at the Cuttuck Sub-station of the CICFRI during 1955 (Chaudhuri, 1960). By 1957, the difficult-to-spawn IMCs were successfully induced to spawn by injecting fish pituitary hormones (Chaudhuri and Alikunhi, 1957). Later, the exotic carps were also induced to breed through hypophysation (Alikunhi, 1963). Subsequently, large-scale breeding operations were undertaken thorugh hypophysation (Chaudhuri *et al.*, 1966; Singh *et al.*, 1970).

An account of the induced breeding in fishes through hypophysation technique is briefly given below:

Stocking and Upkeep of Brood Fish

An adequate stock of healthy brood fish, preferably raised in the farm itself, is a pre-requisite. The brood fish stocking pond could have an area of 0.2-0.5 ha and an average depth range of 1.2-1.5 m during summer. These ponds be suitably manured to produce natural fish food. IMCs could be stocked here @ 1,500 to 2,000 kg/ha. The brood fish, so stocked, be adequately fed with supplementary feed, *e.g.*, groundnut oil cake + rice polish or wheat bran in 1: 1 ratio, everyday at 1-2 per cent of the body weight to start with. Under optimum feeding conditions, the fishes reach proper maturity and produce a large number of eggs. Excess feeding could lead to deposition of fat in the body cavity which may adversely affect maturity. Both natural food and supplementary feeding are necessary for raising a good stock of brood fish. The selected brood fish be examined periodically for their condition and to monitor the progress of their maturity; as well as, to check for any parasitic infection(s). Filling of the brood fish pond with freshwater, re-circulation and aeration of the standing water periodically, preferably during the summer months, is highly desirable. Infected fishes, if any, be treated suitably. Bathing the affected fishes in a solution of 10 ppm of $KMnO_4$ for 10-15 minutes or till the fishes show signs of distress, could be useful. 2-4 year old carps weighing 2-5 kg and having optimum level of fecundity are suitable for hypophysation. In the event of transporation, the brood fishes are transported individually in separate hand nets (or bag nets with handle) or in fish carriers. Splashless tanks (Mammen, 1962) are improvised devices for trasnsport of brood fishes. Use of anaesthetics have also sometimes been found to be satisfactory. The brood fish, after transport and being before released into the pond, are disinfected with 10 ppm solution of $KMnO_4$.

The cultivable carps are seasonal breeders. Their gonads start maturing gradually with the advent of breeding season and with the attainment of proper age. Besides age, hormonal secretion and quantity and type of food, gonadal development depends upon several external factors, like temperature, light, rainfall. By altering the period of exposure of a fish to light, a strain of trout, normally breeding in December, could be made to produce viable offspring in August. The recent success of breeding the IMCs twice in the same season, suggests the importance of light. Further, sudden dilution of the river or pond water (Chaudhuri, 1967; Singh, 1969; Singh *et al.*, 1970) or their water being rendered turbid all on a sudden (Dey and Kar, 1986) has a profound impact on the full maturity and spawning of carps.

Determination of Sex

Accurate determination of sex is a pre-requisite in successful hypophysation process. Among the IMCs, the males develop denticulations on the dorsal side of the pectoral fin rays with advancing maturity and approach of the breeding season. The females, on the contrary, have a smooth to touch pectoral fin. The ripe males ooze milt when the belly is gently pressed. The ripe female also generally oozes eggs when the abdomen is lightly pressed. A Hickling-type catheter (Chaudhuri and Singh, 1964), if inserted into the abdomen of a ripe female through the vent, gives a hint about the maturity condition of the brood fish.

The age of the cultivable carp species could be calculated by counting the number of concentric rings in its scale.

Factors Responsible for Induced Breeding of Carps by Hypophysation

Rainy weather, accumulation of fresh, cool rain water in the pond and a water temperature ranging from 26-30°C are generally considered to be conducive for a successful hypophysation venture. The temperature, however, may vary from place to place.

Collection and preservation of pituitary glands, preparation of pituitary extracts, determination of dosage and methods of injection:

Potent pituitary glands be collected from gravid, mature or near-maturing donor carps, or, even from induced-bred carps (immediately after spawning) with least uncertainty. The cultivable carps, being seasonal breeders, the gonads and the gonadotropins attain maximum functional capacity only for a limited period during the breeding season. Thereafter, they regress to almost a state as in juveniles. Therefore, glands be not collected from the immature fishes and from the fishes which have already spawned. Glands collected from fishes preserved at 1-3°C for 24 hours are also generally effective. However, breeding season of the donor fish is the appropriate time for the collection of the glands. In short, glands be collected from females with bulging abdomen and swollen vent or from males with convoluted cream-coloured testes which ooze milt on application of little pressure on the abdomen.

Pituitary glands may be removed from a fish after severing the head from the body; or, even keeping the fish intact. In the latter case, a portion of the cranium roof is chopped off to expose the brain.

After clearing the fatty substances over and around the brain, and after detaching the olfactory or ophthalmic nerves, the brain is lifted up and laid on its back. The pituitary gland, which lies just below the brain, just posterior to the optic nerve crossing on the floor of the cranium in cultivable carps, is dislodged with the help of a needle without damaging it and taken out.

In order to get rid of uncertainty and in order to help programming ahead, the glands be collected quite in advance of the fish breeding season and be preserved in 'pituitary banks'. Preservation of the glands in absolute alcohol is quite popular in India and Brazil. Ethanol is a dehydrating and defattening agent. Freshly collected glands are immediately put into ethanol; and, later, within 24 hours, the glands are transferred twice to air tight vials containing fresh ethanol. Finally, these vials are stored in dark air-tight bottles containing ethyl alcohol at either room temperature inside a desiccator containing anhydrous $CaCl_2$ in a cool place or preferably the desiccator itself is kept inside a refrigerator.

Pituitary hormones are readily soluble in water and not in pure alcohol. Hence, the alcohol used, be 100 per cent pure. While in store, the alcohol be changed every three months and the vials be kept air-tight. Steps be taken against quite high ambient temperature which could lead to loss of potency. However, glands stored at a temperature of about 40°C were found to be quite viable (Chaudhuri and Singh, 1984). Freshly collected glands could also be preserved in acetone.

Accurate dose of the hormone is important. However, there are some problems in standardization of the pituitary dosage because the amount of gonadotropin to be injected depends upon the size and stage of maturity of both the donor and the recipient fishes, freshness and potency of the glands, period of preservation of the glands and the important ecological factors of the breeding environment at the time of hypophysation. In India, although considerable success in hypophysation has been achieved by considering individual pituitary gland as a 'unit', yet, the need to evolve a standardized dose for hormone injection was keenly felt, mainly because the size of the pituitary gland varied considerably with the size of species of the donor fish. It was, however, considered reasonable to relate the dose of the pituitary hormones with the weight of the gland and the weight of the recipient fish. Hence, after determining the weight of the pituitary gland accurately, the dose of the hormone is expressed as weight of the pituitary gland in mg/kg of the body weight of the recipient fish (mg/kg body weight).

The dehydrated and defattened glands, after being taken out of alcohol on a watch glass, are picked-up with a pair of fine forceps, rolled over and wiped with filter paper. Then, the glands are weighed individually on an electronic or a chemical balance. It is advisable to maintain a constant interval of two minutes for drying the glands on a piece of filter paper. After weighing, the individual glands are put back into the alcohol in separate properly-labelled vials. Some of the useful information about the glands, like weight, date and place of collection, etc., are maintained individually in a gland register.

In conducive weather, during the breeding season, a ripe *Catla catla, Labeo rohita* or a *Cirrhinus mrigala* is injected twice at an interval of six hours. The first dose is of 2-3 mg/kg body weight and the second dose is about 5-8 mg/kg body weight of the recipient female fish. Conversely, the male is administered with a dose of 2-3 mg/kg body weight of the recipient fish at the time of 2nd injection into the female. Higher dose (4-5 mg/kg body weight) may be prescribed to a male which does not yield free milt during testing.

Not much accurate information is available on the prescription of a particular concentration of the hormone or injection of a definite volume of the extract. However, a concentration of 2-4 mg of the gland in 0.1 ml of the extract, and a dilution of 0.2 ml/kg body weight of the recipient fish have been found to be effective (Chaudhuri, 1984). Usually, 0.2-0.6 ml of the extract is considered to be adequate for brood fishes ranging between 1-4 kg; and, it could be increased upto 1.0 ml for fishes weighing upto 10 kg. Extracts of known concentration are generally kept prepared before the breeding season, preserved in glycerine and stored. The extract is suitably diluted with distilled water before injection.

From practical point of view, the extract is prepared after the breeding sets are formed, weights of the recipient fishes are recorded and the dosages for injection are decided. For making an extract, required quantity of pituitary glands are taken out of absolute alcohol and placed on a piece of filter paper. When the alcohol has evaporated, the glands are put inside a tissue homogenizer of suitable size (5-20 ml capacity) and then macerated with a little amount of distilled water or physiological salt solution. The homogenized glands are then, further, diluted to make up the desired volume. The triturated (macerated) gland suspension is injectd either after or without centrifugation. For intramuscular injection, the gland suspension is mostly centrifuged, and, the supernatant liquid alone is used for injection. In a simplified field procedure, each gland is crushed in its entirety by mortar and pestle; and, the suspension could be successfully used without centrifugation. Sometimes, a hand-operated centrifuge could also be used for centrifugation.

For preparing a glycerine extract (Ibrahim and Chaudhuri, 1966), glands of known weight are thoroughly macerated inside a homogenizer. The desired concentration of the extract is then decided and accordingly 1/3 of the total volume is made up by adding distilled water, in which, the hormone extract is obtained after maceration. This solution is kept inside a refrigerator for 24 hours and later glycerine is added to make up the remaining volume, the ratio of water to glycerine being 1: 2. The hormone suspension is again stored inside a refrigerator for another 24 hours; and, after that, it is taken out and centrifuged. The supernatant solution is filtered and stored in air tight bottles inside a refrigerator or sealed in ampoules with proper record of their concentrations and dates of preparation. Usually, a concentration of 40 mg of pituitary/ml of extract is found to be convenient for use.

The pituitary hormone is administered either intra-peritonially (as in USA) of intramuscularly (as in India and Brazil). The latter is generally given in the back muscles or in the caudal peduncle of the recipient fish. While injecting, the needle is, at first, inserted under a scale parallel to the body of the fish; and, then, the muscle is pierced through quickly at an angle of 45°; thereby, causing no damage to any of the scales. The needle be never inserted through the scales at an angle initially. That would hinder its penetration through the muscles as the scales are arranged in an imbricate fashion. A 2.0 ml hypodermic syringe with locking arrangement and graduations at 0.1 ml interval is conveniently used for injection. Generally, No.22 needle is used for fishes weighing between 1 and 3 kg; and, No.19 for larger ones. Needle No. 24 be used for fishes weighing < 1.0 kg. In case more than one injection to be given, pricks are made on either side (left and right) of the caudal peduncle alternately.

In the process of injection, the recipient fish, are, at first, taken out from the breeding hapas one by one inside hand nets and are gently put on a foam rubber cushion of suitable size. Services of two persons are required for administering the hormone into the fish. One of them keeps the fish still by placing both of his hands on the head while the other holds the caudal peduncle with one hand, gently but firmly; and, with the other hand, the hormone is injected in the region of the caudal peduncle quite quickly. The injected fishes are then quickly put back into the breeding hapas and the mouth of the hapa is securedly closed to prevent escape of the injected brood fishes into the pond.

In India, the immediate environment for induced breeding is considered to be the breeding hapa, inside which, the injected brood fishes of both the sexes are released to spawn. The breeding hapas are fixed in ponds which have received sufficient rain water, do not have algal blooms of any sort and are devoid of crabs, etc. Hapas could also be fixed in suitable positions of rivers for getting a constant flow of aerated water through the hapas.

The breeding hapa is an inverted mosquito net-type cloth container with a cover. Width-wise, it has an opening on one side provided with cloth loops and button holes to enable proper closing. Through the loops, a thin bamboo stick is passed to close the opening securedly after releasing the injected brood fishes inside. The hapa is fixed with the help of bamboo poles of suitable thickness and length. Usually, strong and durable fine-meshed mosquito-type net is used for making the hapas; so that, the ovarian eggs do not pass out of them when injected female is released into the hapa. The hapas be tanned and washed thoroughly before use. Hapas of the dimension of 2.4 m × 1.2 m × 1.0 m could be used for fishes measuring between 1.5–4.0 kg. The depth for fixing the breeding hapas is chosen according to the suitability of the fishermen in handling them. However, when fixed, the bottom of the hapa may not touch the muddy pond bottom. Atleast 15-20 cm of the hapa could remain above the water level.The IMCs spawn naturally inside the breeding hapa after a few hours of final injection.

The injected IMCs spawn naturally inside the breeding hapa; and, the fertilized eggs (spawn) are obtained as the released eggs by the female and sperm by the male are synchronized. Ordinarily, the IMCs spawn within 6-8 hours after the final injection to males and females. Under favourable conditions, spawning may take place even within 3-4 hours. Experiences revealed that, in cases where a single dose of injection is considered adequate, this could be done late in the evening, so that, spawning could be over by early hours of morning. In the event of two injections, which is usually the case, the first injection could be administered at 4 pm and the second at 10 pm; so that, spawning is over by early morning. If the injected female does not spawn even within 8 hours, a third dose of higher concentration (though not generally preferred) could be injected.

The injected hormone influences the pituitary and sex play between the male and the female is sometimes noticed a few hours after injection and before they release the sperms and the ova into the water. Good fertilized eggs exhibit blastodiscs and sometimes later stages of development. In bad, unfertilized eggs, the development of blastodisc is arrested after a certain stage or it does not develop at all; and, the central yolk mass gets diffused and shows disintegration; thus, making the eggs appear opaque and whitish unlike the fertilized eggs which are transparent showing a constant development of the embryo and which maintain their regular shape. The fertilized eggs soon start swelling-up and look like small pearl-like bodies. They gradually get water-hardened. By examining these eggs, total number of good (where normal development is noticed) and bad eggs, the per cent of fertilization, which is the main index of success of the induced fish breeding operation, could be determined. The stage of development, at which, an elongated, crescent ('C')-shaped embryo is seen with the help of naked eye, is considered to be a reliable index of expected yield of hatchlings from the fertilized eggs.

In some of the major carps, stripping is done. The female is said to be ready for stripping when it releases eggs freely. The eggs are stripped into 500 ml capacity enamel tray or basin; and, later, sufficient amount of milt is poured on it. The eggs and milt are then thoroughly mixed wih a piece of feather. The contents are rotated by hand to ensure proper fertilization of the eggs. After thorough mixing for about 5 minutes, some amount of water is gradually poured into the tray and mixing of the contents is continued. After a few more minutes, more water is poured-in; and, gradually, the excess of milt, the blood clots and other extraneous matter are completely removed through washing. By this time, the swelling of the eggs starts and the eggs gradually get water-hardened.

Some of the possible factors responsible for failure of spawning in this technique are the following:

1. The females are not in their prime stage of maturity.
2. Weather condition could be unfavourable, such as, rise in water temperature much above the optimum range.
3. Occurrence of plenty of unfertilized eggs even when they are naturally released, which, in some cases, could be due to non-synchronisation of spawning behaviou of the male and the female or due to improper quality and inadequate quantity of the milt.

Conditions Favourable for Induced Breeding

Spawning is an innate urge and is a state of absolute co-ordination reached between the external and the internal environments of the fish through its pituitary. The pituitary hormones (endo- and exogenous), in association with conducive ecological conditions, influence the selected brood fishes reach full maturity quickly. Observations made on the meteorological conditions and physico-chemical parameters of water during spawning of fishes induced by pituitary hormones revealed (Chaudhuri and Singh, 1984) that:

1. Breeding could take place both in clear and turbid water
2. Accumulated fresh rain water promotes successful spawning. Nevertheless, sudden and considerable dilution of the pond water by rain, not only promotes natural spawning of carps; but, also, influences release of germ cells in hypophysation process.
3. Breeding could take place both in day and at night.
4. Spawning generally occurs when the temperature of water ranges from 24-31°C; the most optimum temperature in the plains of India being around 27°C.
5. Cool, rainy days are generally more conducive to spawning than hot, sultry days.
6. Flowing water gives better result, especially during warmer days because it is cool and more oxygenated than stagnant water.

Hatching Technique

An efficient hatching technique is an indispensable step to ensure successful hatching of eggs, increased survival of hatchlings and young fry.

Fertilised eggs are immediately transferred from the breeding hapa to the hatching environment when they are sufficiently water-hardened which is indicated by the twitching movement of the embryo within. For removing the eggs, the bottom corners of the breeding hapa are, at first, united from the bamboo poles and then slowly lifted from the back end to bring all the eggs towards the open end. Next, the brood fishes are skillfully pushed to the other end of the hapa or are carefully removed by a wide-meshed hand-net without much disturbing the eggs. The hapa with the eggs is then lifted above water and the eggs are transferred to enamel or plastic buckets by means of a mug. They are collected and are then released into the hatching hapa after noting the total quantity.

The female brood fishes, on being removed from the breeding hapa, are re-weighed. The difference between the initial and the final weights gives an idea about the weight of eggs laid. Then, the females are either sacrificed for obtaining pituitary glands or released back into brood fish tank after giving them a bath in 10 ppm of $KMnO_4$. Females, which have spawned early in the season, could be induced-bred again in the same season for the second time (Bhowmik, 1976). In the event of shortage of male fishes, some of the treated males could be re-used, if found suitable.

Regarding optimum time for removal of eggs, it is mostly done during the morning hours and when the fertilized eggs get fully hardened. Also, with the rise in water temperature, further development of the fertilized eggs go on progressing. As such, the time-gap between spawning and removal of fertilized eggs is generally maintained around 4-8 hours.

Success of breeding by hypophysation is assessed on the basis of fertilization of eggs; and, for this purpose, an estimation of the total number of eggs laid by the female is essential. Information on the per cent of successful fertilization gives an idea of the total number of young fry which will be available for stocking on the 3rd day after hatching of eggs. It also gives an idea of the number of nurseries required for receiving the fry.

The procedure for estimation of eggs is given below:

Bucketful of eggs, after their removal from the breeding hapa, are slowly poured over a rectangular piece of cloth held by two persons. When all the water has drained out, the eggs alone are measured by a mug or a beaker of known volume. From the number of mugs or beakerful of eggs measured, the total of eggs is determined. Two or three small beakerful of samples are collected and taken to the laboratory for study. The number of eggs in the samples are counted and the average number of eggs in a sample of known volume is determined. From this and from the total volume of eggs obtained, the total number of eggs laid could be calculated. While counting the sample, the unfertilized or bad eggs and good developing ones are counted separately and the per cent of good eggs is determined. It has been experienced that, in majority of the successful spawning cases, the per cent of fertilization obtained is > 70 per cent, the maximum could be 99.8 per cent.

Hatching Environment

Double Hatching Hapa

Equal volumes of eggs are distributed into hatching hapas fixed in ponds as soon as the eggs are measured and samples collected.

A set of hatching hapa consists of two hapas, one fitted inside the other. The outer hapa is made of thick-meshed good durable cloth or of nylon stitched into a rectangular trough without any lid or cover. Each of the eight corners is provided with a loop for fastening thin cotton ropes. In addition to these, four additional loops are provided to the four lower inner corners. The standard size of the outer hatching hapa is generally 1.8 m × 0.9 m × 0.9 m. The outer hatching hapa is tied to four bamboo poles in marginal waters of a pond at a depth of little more than 1.0 m. The hapa may ordinarily emerge out at least 25-30 cm above the water surface, so that, other fishes and predatory organisms may not get in.

The inner hatching hapa, on the other hand, is made of round-meshed mosquito net cloth of the dimension of 1.75 m × 0.75 m × 0.45 m and stitched exactly in the same shape as the outer hatching hapa; but, is provided with eight outer loops. The inner hapa is fitted inside the outer one once the latter is fixed. The bottom corners of the inner hapa are tied to the inner loops of the outer hapa and top corners of the bamboo poles. The inner hapa is fully stretched, so that, the bottom is kept perfectly flat forming a hanging tray. In order to facilitate proper stretching, thin flat bamboo strips may be fitted inside the inner hapa.

The eggs are spread uniformly all over the surface of the inner hapa. The number of eggs to be released in each inner hapa depends on the per cent of good eggs. Ordinarily, the number varies from 0.075 to 1.0 million eggs/hapa. In terms of volume of eggs, it may vary from 2,000 ml to 3,000ml, with an average of 2,500 ml. A number of hatching hapas are required even when a few pairs of fishes breed successfully. As such, a row of hatching hapas are fixed together.

Modern Glass Jar Hatchery

In most of the countries, where hypophysation in fishes is done on a large scale, various types of hatcheries supplied with clear, cool, oxygenated running water have been developed. In China, floating 'net boxes'; and, in Taiwan, 'circular ponds', both with running water supply, are in use. In Russia, glass jars with a regulated water supply, also called 'Weiss apparatus', form the main hatchery component. Based on the Russian design, a glass-jar hatchery has been established at the Cuttuck Research Station of the CIFRI (Bhowmik, 1974) with regulated water supply. The capacity of the hatching glass jars is 6.34 litres; and, in each jar, *c* 50,000 fertilised eggs are put for hatching at a time. The rate of flow of water is kept around 600-800 ml/minute for the eggs of IMC maintaining a gentle movement (also called 'boiling of eggs'). When the temperature of this hatchery water ranges from 29-33°C, hatching could be completed nearly 3 hours earlier than in pond. Later, the hatchlings flow out of the jars through the pipes and get collected inside 'spawnery', which consists of hatching hapas fixed inside cement cisterns. The spawn (young fry) is collected from the spawnery after two days as in the case of hapas fixed in ponds.

The fertilized eggs get water-hardened and the embryo may also show twitching movement inside the egg by the time they are released into the inner hatching hapa of the 2-hapa hatchery system fixed in a pond. Most of the fertilized eggs of IMC hatch out within 14-18 hours, generally, at a temperature range of 24-31°C. The hatchlings escape to the outer hapa through the holes of the inner hapa leaving behind egg cases and bad eggs on the latter. The inner hapa is removed carefully when all the eggs have hatched out, washed thoroughly, given a dip in dilute $KMnO_4$ for disinfection and dried for future use. The hatchlings are left undisturbed till the 3rd day after hatching when they are collected and released in nursery ponds. By that time, the yolk is completely absorbed and the young fry start feeding.

In the event of large-scale mortality which may occur at any stage from the time of spawning till the fry are stocked in the nursery ponds, precautionary measures are adopted at every step during the hatching operation. As a chek-list, a few causes of mortality of eggs and hatchlings are given below:

1. Sudden rise in water temperature and oxygen-content of water leading to supersaturation. In other situation, asphyxiation may also lead to large-scale mortality.
2. Abrupt hike in water level of the river or of the pond due to severe rain submerges the hapa; and, thus, eggs and hatchlings are washed out. This may also make way for carnivores to get into the hapas resulting in large-scale predation of the developing eggs and hatchlings.
3. Mortality also results due to profuse rainfall leading to change in hydrobiological condition of the water.
4. Intense algal blooms cause large-scale production of oxygen through vigorous photosynthesis. Unhappily, a good number of such bubbles stick to the eggs imparting them buoyancy which eventually cause them to float. The surface water temperature being sufficiently high, the eggs start disintegrating, the embryos get bent and finally die.
5. Pollution of water from an unexpected source, such as, washing of petrol cans or vessels containing toxic substances.
6. In addition to escape of eggs and hatchlings through holes in the hapas made by crabs, or, due to disintegration of the hapa cloth, mortality of hatchlings also occur due to fast gale and copepod attacks.
7. Although the reasons for sudden loss of developing eggs and dropping down of the embryos from the eggs have not yet been fully analysed, studies revealed that adverse

physico-chemical conditions of water, the genetic components of the germplasm and the actual inner state of ripeness or maturity of the eggs could serve as important factors for accounting of such unnatural phenomenon. Shedding of eggs, comparable to abortion, may also occur due to excess dose of pituitary hormone; or, due to weak or genetically defective spawn.

8. Biswas and Saha (1972) had found low redox potential, sudden fluctuations of temperature and pH of water followed by strong sun or heavy rains accelerating ionic diffusion through the eggs to be the possible reasons for the rupture and dropping of the embryo; and, thereby, leading to mass disappearance of their eggs and embryos from within the breeding and hatching hapas.

The hatching hapas may meet with different types of hazards. With an aim to mainly prevent the occurrence of such hazards, the following measures could be adopted which could be effective:

1. Hatching hapas be preferably fixed in large ponds with clear water and particularly with shallow marginal areas.
2. Ponds with algal blooms be avoided
3. River water is not well recommended for fixing the hatching hapas because of the presence of silt in it. There is danger of hatchlings being killed due to asphyxiation because of silt covering the outer walls of the hapa and thus checking the flow of water.
4. Waters infested with crabs, common carp, tilapia, etc., if not got rid of them, be avoided. Use of wire net around the hapas is considered to be an effective step in screening off the predators and vectors.
5. Comparatively fine-meshed cloth be used for the outer hatching hapa in order to prevent the entry of copepods into the hapas.
6. Constant monitoring is done on the water level of the pond below the hatching hapa which should always be about a foot below its upper surface. The hapas be adjusted suitably in the event of sudden rise or fall of the water level.

Quantitative Estimation of Spawn

The spawn are estimated quantitatively as follows:

On the 3rd day, after hatching, the young fry, usually termed as 'spawn', start feeding. Directive movement is observed even earlier. The spawn be collected for stocking. Each outer hatching hapa be carefully united by four persons, each of them holding a bottom corner with one hand and the upper corner with the other. The hapa is slowly lifted above water. Simultaneously, one person slowly washes the sides, so that, spawn do not get stuck-up in the hapa cloth. Finally, when the spawn are concentrated in a small quantity of water, they be collected in a mug and transferred to a bucket half-filled with clear pond water, filtered previously through thin (fine-meshed) cloth. Spawn from all the hapas, thus collected, be transferred to a fry-carrier over a thin piece of cloth for measuring. The spawn are measured in cups for estimating its quantity. Estimation of the number of spawn produced is necessary for making an assessment of the results and for stocking purposes. The spawn are usually measured with the help of small aluminium cup of known volume having small perforations at the bottom, so that, water may escape through these holes. By actual counting of a few samples, it has been found that, approx. 500-600 spawn are present per ml in the cup. An aluminium-made measuring cup of 35 ml capacity contains approx. 20,000 spawn or young fry. The number may vary according to the size of the spawn of different species of IMC. Debris and water are not taken into account while measuring the fish fry; else, these are sieved through a piece of round-meshed mosquito net, if at all found to be present among the fry. Thus, in short, the mass of spawn, after being collected and measured by a measuring cup, are immediately released into the 'hundi' containing filtered water for transportation to the nursery tanks.

Transport of Spawn

The fry, after collection and measurement, are transported and stocked in nursery ponds. If the nursery pond is in proximity, they are carried in ordinary metallic 'hundies' For distance travel, the fry are packed in polythene bags with oxygen-packing. The bags are, at first, checked against leakage; and, then, after washing, they are filled with 6-7 litres of clear agitated tap water and stored. A measured quantity of spawn is then released into the bags; and, oxygen, under pressure, is bubbled through the water from an oxygen cylinder.

Oxygen occupies roughly a little over 1/3 the volume of the plastic bag. On introduction, the mouth of the bag is immediately tied firmly in order to prevent escape of oxygen. Finally, the polythene bags are put inside metallic cans of 18 litre capacity for transportation. The number of spawn to be stocked per bag varies according to distance and temperature of water, *e.g.*, for a distance of 50-60 km (2-3 hours), the stocking rate may be 0.06-0.10 million per bag. For long distance transport, the number of spawn per bag be reduced.

Stocking of Spawn

Extreme care is taken in the preparation of a nursery pond because, on its appropriate preparation and scientific way of management, depends the real success in fish breeding and production of advanced fry and quality fish seeds from the young ones stocked. The nursery pond be, therefore, cleared of weeds, miscellaneous fishes, predatory insects and other enemies of young fry and manured properly so as to produce natural food, preferably zooplankton in large number, which, incidentally form the main food of the young fry. Artificial feed in the form of finely-powdered ground nut oil cake and rice polish or bran, at 1:1 ratio, may be given everyday after the fry are released into the pond. Eradication of backswimmers, which take a heavy toll of the young carp spawn, is done by spraying soap-oil emulsions on the surface of the water body, 12-24 hours before stocking, @ 51 kg/ha of cheap vegetable oil mixed with 1/3 of its weight (17 kg) or ordinary washing soap. The insects are killed within 15-20 minutes after application of the emulsion at a time when the wind is almost still.

Young fry are ordinarily stocked @ 1.0-2.5 million (10-25 lakhs)/ha. Supplementary feeding with oil cake and rice bran is done @ about 2.8 kg/million fry for the first five days and 4.2 kg and 5.6 kg respectively for the next and last five days. The nursery ponds are usually netted out after two weeks of stocking and the advanced fry are collected and stocked in rearing ponds in combination with compatible species for growing to the fingerling stage.

After the production of fry through hypophysation, usual practice is followed in nursing fry to about 2.5 mm in nursery ponds; growing them to fingerling size in rearing tanks; and, finally, raising them to table size in stocking tanks, in which, ultimately, they are stocked for an indefinite period.

Hormonal Regulation of Fish Oocyte Maturation

Works done by Bhattacharyya *et al.* (2003) revealed the following:

The cascades of events for the maturation of fish oocytes are interwoven in an intricate manner. No doubt, GnRH is the key player in the upstream regulation while Maturation Inducing Steroid (MIS) and Maturation Promoting Factor (MPF) are downstream regulators. The sequence of events may be the following: GnRH released from the brain transported via the circulation to the pituitary gonadotroph cells, GnRHR complex triggers the release of GtH from pituitary cells, GtH then induces the synthesis of MIS in ovarian follicular cells and then the MIS triggers the formation of activation of MPF in oocyte cytoplasm. In rohu, action of MIS and association of other hormones with it in promoting the oocyte maturational process has been shown in the present communication. What could not be seen is the activation of MPF which is a transient phenomenon; and, for this purpose, a teleost has been selected which can be kept in the laboratory aquarium in healthy condition for a long time and this is a freshwater perch (*Anabas testudineus*). mGnRH injection to this perch induces the activation of MPF at 6 h. MPF is a master regulator of phase transition, its cytoplasmic agent, responsible for promoting the G2 to M transition in eukaryotic cell cycle.

The current scenario of explosive growth of carp seed production in India has created a high demand for pituitary glands; thus, making them extremely expensive. Breeding of Indian carp through hypophysation also poses a number of problems, such as, due to bad storage, potency of GtH in pituitary is questionable; transport of the pituitary glands from the market to the remote village and township where number of hatcheries are located and pituitary collected from the market contain pathogenic organisms. All these including the labile nature of glycoprotein GtH are frequently causing failure in induced breeding programme of economically important carps of India. Alternative methods, *i.e.*, using GnRH, which will induce the release of endogenous pituitary GtH affecting the final maturation and release of germ cells resulting in spawning and fertilization appears 'o be more convenient, cheaper and surer process. For these advantages, Ovarprim, a salmon GnRH analogue with other components (a product of Syndel Laboratory, Canada, marketed in India by M/S Glaxo Laboratories, Ltd.) has been a popular item for induced breeding of carps in India. Ovaprim, however, is expensive to the rural fish seed producers, as it costs about Rs. 600.00/vial, which can be efficiently used for spawning of 5-7 fishes having average body weight of 2-2.5 kg. This was the background for conducting investigations in the field with rohu to find whether mGn-RHs can be a better replacement of Ovaprim. For this purpose, investigations made in the field always consists of mGnRHs and Ovaprim to determine the comparative advantages.

Bhattacharyya *et al.* (2003) worked out a combination of mGnRHs (mGnRH I plus II), Ca^{2+} and domperiodone and this produces a significantly better result in the spawning of rohu as compared to the Ovaprim. The experiment has been conducted in Summer (early May) and monsoon (late July). In both the seasons, starting from the viable egg production, rate of fertilization, rate of hatching to spawn production, at every step, mGnRHs have shown very clear advantages over Ovaprim. Surprisingly, the egg size in response to the mGnRH analogue and mGnRHs is a native form of two, an anionic and a cationic GnRH from murrel brain. Such a high activity from native GnRH has no earlier report. What is more exciting is the advantages of mGnRHs over salmon GnRH analogue (Ovaprim) which definitely indicate a great promise for mGnRH. They have determined the sequence of events starting from mGnRHs administration followed by an increase in plasma GtH which causes a sharp peak; and, this withdraws diplotene halt which is reflected by the migration of germinal vesicle, and its breakdown (GVBD), alignment of metaphase plate chromosomes, extrusion of polar body and then finally ovulation.

This knowledge of regulation has given us the ability to manipulate oocyte maturation for better success in induced breeding.

Chapter 25

Freshwater Fisheries Management

Fishes are primarily-adapted cold-blooded aquatic vertebrates with paired and unpaired fins for locomotion and gills for respiration. Fish serves as an item of staple food of man in many regions of the world. Freshwater Fishery Management (FWFM) may be defined as the manipulation of freshwater habitats and fish population to maintain the habitat and to increase fish yield for commercial and recreational uses. The Fishery Managers generally apply a variety of techniques in order to maintain, improve and increase fish population. Some of the aspects of FWFM are discussed below:

Habitat Improvement

Fishery habitats may be altered to increase the competitive advantage for desirable species. Also, the total carrying capacity of the habitat may be improved for the production of fish biomass. Owen (1985) had noted that, one positive habitat manipulation often leads to other positive effects.

Habitat improvement may be done by the following means:

Alteration of the Land

This may include stabilization of the shoreline of the habitat, improvement of covers for fish, creation of artificial spawning sites, etc. (Cooper, 1980).

Alteration of the Water

It may be done by controlling turbidity, aeration and fertility. Such practices may change the vegetation pattern, control spawning and affect predation (Noble, 1980).

Enhancing the oxygen level of the water is a technique often used by fishery managers. Typically, the warm water fishes require oxygen level of about 3 mg/lit. During winter, snow may form a blanket of frozen lake in the cold region; thereby, may ıeduce photosynthetic oxygen production. To eliminate this problem, Fishery Managers could increase oxygen production by various techniques.

Fishery Managers could also alter the fertility of water by increasing phytoplankton production or, by directly adding organic manure.

Alteration of Vegetation

Freshwater habitat may be improved by controlling the associated aquatic vegetation. The excessive vegetation may be detrimental to fishery with regard to the fact that, it may upset the predator-prey relationship. Further, overlying thick vegetation deprive the phytoplankton of the needed nutrients and eventual decomposition of the same vegetation may result in decrease in the level of dissolved oxygen. The aquatic macrophytes may be controlled by mechanical, chemical or biological means. The mechanical means involve water level manipulation; the chemical control includes spraying or dusting of the waterbodies with herbicides; while, the biological methods include control by introducing the herbivorous fishes. There are various suggestions of 'creative approach' of managing the aquatic vegetation by considering the entire lentic body and its bays as an 'aquascape' and to manage it by judicious scientific ways.

Habitat Construction

Freshwater Fishery Manager could construct habitats in the form of reservoirs, artificial lakes, farms, ponds, etc., following specific designs which could allow manipulation of water level.

Biotic Manipulation

In addition to manipulating habitat (habitat improvement and habitat construction), Fishery Managers may also manipulate the actual fish population by:

1. Predator control
2. Competitor control
3. Artificial propagation, which includes stocking, aquaculture and breeding of pollution-resistant fish

Controlling the Predators

Since predation helps to keep the fish population from outstripping the carrying capacity of the water body, one could easily see, how thinning the predator population would likely to increase the population of desirable species of fishes.

Controlling the Competitors

Removing competitors or the undesirable fishes is another type of biotic manipulation known as rough or trash fish control. Non-game species, which often are considered as useless, may be removed by the use of appropriate dosage of fish toxicants, *e.g.*, rotoneone. However, such practices may not be always effective because they are said to be only temporary measures.

Artificial Propagation

Fishery managers may also manipulate fish population by using various methods of artificial propagation.

Aquaculture, also sometimes known as fish culture, or fish farming in a specific sense, may be defined as raising of fish in captivity. It is an effective method of increasing fish population. There is no doubt that, the potential for aquaculture is good; its importance in providing food will increase as world population grows. In order to lower capital investment costs, researchers are investigating the traditional practices of aquaculture in the 3rd World Countries, which, significantly, emphasizes 'polyculture', *i.e.*, the growing of fish and agricultural crops together. Introduction of new fish species and hybridization may not always prove to be beneficial.

Fishing Regulations

A more traditional means by which the Fishery Managers have cared for their resource, is the practice of imposing restrictions on the anglers, *e.g.*, managers may close a specific area from fishing

('Closed area'); or, forbid fishing during a particular season when fish are more vulnerable ('closed season'); or, restrict the number ('creel limit'); or, restrict the size of the fish ('size limit').

Sometimes, people working in the field and the anglers unite in order to call for stricter regulations on a fish species.

Habitat Protection

Fishery Managers are slowly beginning to realize that, many of the traditional management techniques would not be required, if fishery habitats are protected in the first place. As such, Fishery Managers are starting to make a transition from a quickfix solution used in the lake stream, reservoir, pond, etc., to longer-term solution for finding ways and means to prevent problems from occurring in the first place.

Prospects for the Future

The freshwater fisheries have tremendous potentiality in Asia and India, in general; and, in Assam, in particular, if their management is done in an effective way. Over the past century, only a few fisheries management techniques have been evolved, such as, fish culture, stocking, predator control, etc.

It is obvious that, new techniques; specially, the right attitude towards management, need to be developed to meet the challenges of the future.

Chapter 26

Management and Conservation of Fish Habitat and Fishes: A Case Study in Sone Beel in Assam

In view of 'fish' being the principal aquatic animal resource and staple food of the region, and the burgeoning demand for good quality and higher quantities of fish for the health-conscious people, efforts are being made towards sustaining these enormous freshwater habitats, such as, the Sone Beel, and their bioresources. However, there are certain forms of environmental pressures which adversely affect fish, such as, habitat loss, water contamination, drought, competition, predation, fishing pressure and disease, which must be tackled to achieve sustainable growth of fish, which is a pre-requisite for formulating effective management strategies and conservation (Kar *et al.*, 1996).

Impact of Habitat Loss and Topographical Changes

The construction of embankments and roads along Sone Beel often destroys shoreline habitats, resulting in sedimentation which spoils large quantities of fish eggs and also increases turbidity. Under such circumstances, molluscs, such as, mussels can neither feed nor easily settle at the sediment-clogged wetland bottom; and, thus, could contribute less to the food chain. Furthermore, the IMCs and the Indian shad [*Hilsa* (*Tenualosa*) *ilisha*] have been losing habitats due to the construction of the 'blind dam' in the River Kachua and diversification of the waterways; thus, registering a sharp decline in their population in the Beel (Kar, 1990). Prior to 1951, the river Kachua used to flow wild and free from the Sone Beel before emptying itself into the River Kushiara. Both *Hilsa* (Dey and Kar, 1989 b) and the IMCs (Kar and Dey, 2000) used to migrate long distances from Bangladesh to enter directly into the breeding grounds in Sone Beel. The blind dam in Kachua (constructed in 1951) was replaced by a 'lockgate' in 1964; and, the main outflow has since been diverted through the River Khagra. In addition, diversification of both the major inflow and the outflow through a network of channels within the lake has recently become a common practice in Sone Beel, mainly for agricultural purposes. Consequently, unless fish ladders, fish elevators, etc., are provided in the dams and lockgates, the long term future of the depleting fish species is not optimistic because the migration paths have been seriously disrupted. The impact of offsite erosion on the Beel fisheries is also very serious (La Roe, 1986).

Contamination of the Beel Water

Sone Beel receives a large amount of organic load in the form of sewage disposal which occasionally results in depletion of the DO in some sections of the Beel (Dey and Kar, 1987).

Although pesticides are applied to the paddy cultivation at the draw-down level of the Beel, there is no potential threat of industrial pollution just now, in view of the absence of any big industry around the Beel. As such, incidents of 'direct fish kills' due to drastic reduction of DO level, shifts in fish species (Wheeler, 1979), and the occurrence of fish with cancers that have been recorded in freshwater bodies elsewhere (Klee, 1991) have not been seen in Sone Beel (Kar and Dey, 1987).

Effects of Drought

Sometimes Sone faces prolonged periods of drought. In early 1979, the water level of the Beel was so low that, the spawning ground of the larger sized fishes were practically non-existent. Also, occasional high water temperatures could reduce the production of minor carps (Dey and Kar, 1990).

Competition and Predation

Introduction of exotic fish may be done only after examining the possible impact they might have on the autochthonous fish species of the Beel. The introduction of *Cyprinus carpio* in Sone resulted in a substantive depletion of the population of endemic species by competition, predation or hyhridization (Moyle, 1976), although, such a practice has been suggested to be useful elsewhere.

Fishing Pressure

Overfishing has been exerting heavy pressure on the Beel's resources, with fishing operations carried out throughout the year. Both the quality of the fishing gears and the frequency of fishing operations be regulated.

Remedial Measures

Sone Beel and its channels and associated rivers historically have been used for recreational and commercial fishing, navigation, mass transport, colonization, agricultural irrigation, water supply, and waste disposal (Kar, 1990). Therefore, in order to ensure a balanced development of the biota, and maintain the maximum sustained yield, the Beel habitat might be altered to increase the competitive advangage of desirable species; and, more importantly, to increase the total carrying capacity for maximum fish biomass production. In order to achieve positive effects through habitat manipulation, possible measures include: alteration of land, water and vegetation; habitat construction; and, biotic manipulation (Owen, 1985).

Alteration of Land

This procedure covers many aspects. Primarily, the Beel shore and stream bank erosion could be stabilized through afforestation. Rock deflectors could be built to direct stream currents. Artificial spawning sites, which have been found to be useful elsewhere (Cooper, 1980), could be created with sand, gravel or nylon mats in those portions of the Beel which are heavily mud-laden and do not serve as suitable natural spawning grounds.

Alteration of Beel Water

This is a difficult task for an enormous Beel like Sone Beel. However, attempts could be made in this direction by controlling the water level, turbidity, aeration and fertility. Manipulation of the water level would be useful in controlling vegetation and spawning;

and, would, thus, affect predation. It could also be an effective way for eliminating undesirable species by lowering the water level to expose and isolate their eggs (Noble, 1980). Furthermore, when the prey species become over-abundant, lowering the water level could bring them out from the aquatic vegetation which serves as their cover; thus, making them more susceptible to predators.

Turbidity

Turbidity is a menacing problem in this region and Sone Beel is subject to prolonged turbidity, both during monsoon (Dey and Kar, 1987) and dry seasons (Kar *et al.*, 1994). The latter is due to vigorous fishing at low water levels. Reasonably clear water is a prerequisite for the healthy growth of biota, notably the fishes. Methods to reduce turbidity include, among others, the following: (i)By directly adding green or dry organic matter to the Beel; which, after decomposing, causes the particles in the water to precipitate; (ii) by establishing herbs on the Beel-bottom at DSL; and, (iii)adding alum or gypsum to the Beel water.

Nutrients

Nutrients are an important factor in fish health; but, Sone Beel is oligotrophic (Hutchnison, 1967) with a low nutrient staus (Kar, 1990) which could be due to the characteristics of the surrounding watershed (Klee, 1991). The deficit could be improved by directly adding organic fertilizers to the Beel water; thereby, boosting phytoplankton production. Inorganic agricultural fertilizers may also be used; but, with caution against too heavy doses which could lead to the fish being killed by phytoplankton bloom. Such measures have only met with limited success elsewhere (Klee, 1991).

Vegetation Management

Sone Beel has a rich flora of aquatic macrophytes (AM) (Dey and Kar, 1989 a) which provide a higher biomass during the dry season than during the monsoon. The AM are a source of oxygen, provide cover for the young fish from predators, and serve as a surface on which food organisms could attach themselves and grow. However, excess vegetational growth in the Beel could be harmful to the fishery, if it upsets the predator-prey relationship, and often leads to the stunning of fish growth due to over-population (Kar and Dey, 1992).

Concomitantly, a low phytoplankton biomass in the Beel could also be due to less nutrients being made available to them due to excessive vegetational growth (Dey and Kar, 1994).

Elimination of excess of AM is essential and could be done mechanically by manipulating the water level and by the use of various types of cutting machines. The harvested vegetation could be used as a soil additive in the form of compost (Henderson and Markland, 1987). Spraying or dusting of algicides, herbicides and pesticides on the Beel water could be tested to chemically control the growth of AM; but, repeated doses may prove disastrous by resulting in fish kills due to low DO and copper toxicity. Biological control of AM growth in Sone could be attempted through the introduction of herbivorous fishes which would control a variety of weeds without interfering with other fish species. The ideal species which would be closest in meeting these requirements are the grass carps and the tilapia.

Biotic Manipulation

Manipulating the fish population in the wetland could be done by controlling predators, competitors, and through artificial propagation. Sone Beel harbours a rich population of cormorants, herons and cranes which help keep the fish population from exceeding the carrying capacity of the Beel. Thinning their population would likely to increase the population of both desirable and undesirable species of fish.

Non-game fish species which are often considered useless (*e.g.*, *Chanda baculis*, *c. nama*, *Badis badis*, *Lepidocephalus guntea* and *Danio devario*) may be removed by large-scale seine net operations and the Beel re-stocked with more desirable species. However, seining may endanger rare native species; thus, reducing the Biodiversity and overall richness of the Beel ecosystem.

Artificial propagation involves the stocking, aquaculture and breeding of pollution-resistant fish. Sone Beel has 8 prominent fishing centres where intensive aquaculture could be practiced through polyculture and other techniques for raising fry and fingerlings of desired species in hatcheries for future stocking of the Beel.

Habitat Construction

In addition to improving the existing habitat of the Beel, ponds and small reservoirs could be constructed adjacent to the Beel, conforming to standard specifications with proper management procedures for maintaining the fish stocks (Owen, 1985). This could release fishing pressure in the Beel.

Aquaculture and Agriculture

In order to reduce capital investment, attempts are now being made to combine aquaculture with agriculture. An example of this being practiced in Sone Beel and its environs, particularly, at DSL, is the paddy-cum-fish culture, which results in high yields of both fish and rice.

Fish Disease

Despite the high biotic potential of the fishes of Sone Beel, there have been heavy mortalities, since 1988, affecting about 70 per cent of the fish population. The cause is a hitherto unknown fish epidemic, called the Epizootic Ulcerative Syndrome (EUS). The details of this disease has already been discussed earlier.

Conservation

A hardline approach to the management and conservation of the aquatic resources of Sone Beel is to impose bans or restrictions on angling activities, such as, closing a specific area to fishing, forbidding fishing during a particular season, restricting the number or size of fishes caught, restricting the types of fishing gears, promoting aquacultural practices and supporting State and National regulations protecting a particular endangered species.

While imposing restrictions, it is essential to be adaptable and take into consideration the full spectrum of the users of the resource. Concomitant to administrative restrictions, encouraging results could be obtained through mass education of the Beel-users (particularly, the fisherfolk) by the NGOs regarding rational harvest and conservation and management of the Beel resources.

The potential health hazard posed to the Beel fishes by EUS is to be tackled through regular monitoring of the Beel environment and fish health; and, by adopting quarantine measures.

Conclusion

It is now gradually being realized that, many of the management techniques would not be required, if fishery habitats were properly protected; and, that, priority be given to finding ways to prevent the problems from occurring in the first place, rather than relying on quick-fix solutions. Beel and stream improvement programmes could now begin to focus on the entire human-land relationship within a watershed. Financial incentives could be offered which would encourage farmers to practice good soil and water resource management. These are some of the notable problems which need to be addressed in order to find a lasting solution.

Chapter 27

An Introduction to Aquaculture (Fish) Biotechnology

The Science of Biotechnology has been making every efforts to create new opportunities for the welfare of the humanity. In the recent years, it has created unprecedented opportunities, not only for the manipulation of biological systems for the benefit of mankind but, also for undertaking researches to unravel the various unanswered questions in life processes. Consequently, it has become the world's fastest growing and most rapidly changing technology.

Biotechnology is the application of biological organisms, systems or processes to manufacture and serve industries. It comprises of a number of technologies based upon increasing understanding of biology at the cellular and molecular level. The recent advent of genetic engineering and recombinant DNA technology and techniques like monoclonal antibody (mAb) preparation, tissue culture, protoplast fusion, protein engineering, immobilized enzyme, cell catalysis, sensing with the aid of biological molecules, etc., have accelerated advancements in the field of biotechnology for domestic and industrial uses. The science of 'Biotechnology' has collaborations with different disciplines of science. In other words, Biotechnology is an interface of 'Basic' and ' Applied ' sciences where gradual and subtle transformation of science into technology could be witnessed.

With the advances in different aspects of Biosciences, like Biochemistry, Molecular biology, Genetic engineering, etc., man has been easily thinking about harvesting economic benefits through application of biological knowledge. Consequently, biotechnology was born to serve as a link between Biological Sciences and Physical Sciences in order to achieve technological advancements. The word 'Biotechnology' is an abbreviated form of 'Biological Technology' and has been in use since mid-1970s as an alternative word for 'Bioengineering'. 'Biotechnology' may, thus, today, be defined as application of scientific and engineering principles to the processing of materials by biological agents to provide goods and services. Since mid-1983, 'Biotechnology', along with 'Biodiversity', seem to have become passwords and have been playing significant roles through their long-term possibilities in altering the genetic heritage of agricultural plants, farm animals; and, so on.

During the initial phases of growth of Biotechnology, deLaTowe from France and Schwann and Kurtz from Germany had proposed that, the products of fermentation, *viz.*, ethanol and CO_2 are produced by the microbes. However, their view was untenable because they had considered 'fermentation' as a clearly chemical process. Later, Louis Pasteur, through a series of experiments, showed that, alcohol could be produced from grape yeast mixture but not from albumin-yeast mixture.Pasteur had also discovered that, tartaric acid formed two types of crystals and he had succeeded in separating them microscopically. Pasteur had, further,demonstrated that, each type of fermentation process is mediated by a specific microbe which gains energy during the process. This latter observation had enabled Edward Buchner of Germany to discover cell-free metabolism in 1897. In India, some preliminary biotechnological thoughts had been given in working with Ayurvedic medicines.

The surviving fish species are said to be most numerous of all the vertebrates; of which, 41.2 per cent inhabit the freshwater and 58.2 per cent live in the seas. The whole study of 'Fishes' include morphology, anatomy, physiology, Biodiversity and taxonomy, embryology, ecology and endocrinology. 'Ichthyology' is the branch of Science which deals with the study of 'fishes', including their commercial aspects, hydrobiology, limnology, behaviour, environment, population dynamics, use of fishing gears and crafts, methods of fish catching and detection, etc. Fishes also excel as a source of commerce and recreation. They form a rich source of food, particularly for the developing countries; and, are cultured both intensively and extensively to give the best values of their raising. In addition, they yield a number of by-products of high economic value.

'Fishery' is a multidisciplinary science It is connected with different branches of Science, like, nutrition, health, microbiology, biotechnology, etc. Fish catching from natural waterbodies, for the purpose of consumption, is an age-old practice. With increase in population, the demand for fish also increased many folds and this led to a situation where the ratio of fish production (availability) and consumption has become imbalanced. Therefore, India needs the technology to increase sustainable fish production in order to meet the demand. A number of technologies have been developed in this direction as a *modus operandi* to increase production.

The Science of 'Aquaculture' deals with the artificial culture and propagation of organisms living in water. Application of the knowledge of Biotechnology aims at maximizing the production in aquaculture. In the composite fish culture system, the selected fish species having different feeding habits and dwelling in different ecological niches, are cultured in the same pond, so that, there is no competition for food and space. Fishes could be developed without harming each other and could grow in the same area. This practice increases per unit production. A considerable amount of success is generally achieved in this kind of culture. An integrated approach of aquaculture(fisheries) with agriculture, horticulture, poultry, duckery and piggery could also be successful in better utilization of land and by-products for the benefit of different components of the integrated system.

Some of the considerations for success in Aquaculture are listed below:

1. Complete control of the reproductive cycle of the fish species in culture.
2. Excellent genetic background of the stock fishes.
3. Efficient prevention and detection of disease infection.
4. Thorough understanding of the optimal physiological, environmental and nutritional conditions for growth and development.
5. Sufficient supply of good quality water.
6. Application of innovative mangement techniques.

Further, for increasing fish production, a number of technologies could be applied which are given below:

1. Seed production through natural and synthetic hormones.
2. Use of inorganic and organic fertilizers in aquaculture for improvement of productivity and formulation of nutritionally balanced fish feed for higher yield per unit area.
3. Semi-intensive and intensive culture system with higher seed stocking by maintaining good water quality to obtain higher production from the existing area.
4. Composite and polyculture system of farming in which multi-species of fish/prawn grow together to utilize all the possible ecological niches of pond and integrated fish farming system integrated with poultry, duckery, piggery, horticulture, agriculture and animal husbandry.

Aquaculture and Biotechnology

Application of gene transfer technology in fishes have major effects on aquaculture, particularly, fishery. It will exert a major impact on the economics of food production in India and elsewhere. For aquaculture, a faster growing fish, which can utilize the feed more efficiently, would be of obvious utility; and, indeed may take care of the aspects of profit and loss for many currently financially marginal firms, especially in the colder climates.

Aquaculture operations are worldwide in distribution, with concentrations in Europe, China, Japan, South America, Canada, and the USA. Thus, the successful production of growth-enhanced recombinant strains of fish would have wide application and major economic potential. Transfer of growth genes into fish is being viewed by the developing countries as a major means of increasing available protein supply and is a major goal of the Biotechnology programme of China. Currently, interest in genetic engineering of fish is accelerating, with groups working in Ontario, Newfoundland, and British Columbia, Canada, Japan, Norway, France, China, England, India, etc. The above-mentioned technologies proved promising for increasing fish production. However, there is still need for improvement in technologies for further increase in fish production in order to meet the demands of rising world population.

Among the various aspects of Aquaculture Biotechnology, Tissue culture forms an integral part of its study and research to take care of production, disease diagnosis and treatment

Animal Cell and Tissue Culture

Tissue culture developed quite naturally from some of the techniques of embryology, which were in use during the last century. Wilhelm Roux's experiment on maintaining the medullary plate of a chick embryo in warm saline for a few days was performed in the year 1885 and is the first recorded instance of a successful explantation. Concomitantly, Arnold (1887) implanted fragments of alder pith into frogs. When these had become invaded by the leucocytes, he had removed them to a dish of warm saline and subsequently observed that they had migrated and survived for a short time. These two experiments were performed ahead of their time; and, the possibilities that, excised animal tissues could be kept alive even longer in favourable conditions, was not explored until 1898 when Junggren demonstrated by re-implantation that, human skin could survive *in vitro* for many days, if stored in ascitic fluid. Jolly (1903) had performed experiments which had marked the first detailed observations on cell survival and cell division *in vitro*. He had maintained leucocytes from salamander in hanging drops for upto a month. This study was followed three years later, in 1906, by a paper by Beebe and Ewing which could be considered as a record of genuine attempt at tissue culture.

The foundation of animal cell and tissue culture was laid at the beginning of the 20th century when cells could be shown to survive and divide *in vitro*. It was Ross Harrison (1907), who had demonstrated quite unequivocally the continuation of normal functions *in vitro* and also offered a reproducible technique which has been generally accepted as the true beginning of tissue culture. Harrisson's explantation of small pieces of tissue from the medullary tube region of frog embryos into clots of frog lymph, when kept in aseptic condition, the fragments survived for some weeks and axons (nerve fibres) grew out of the cells. This had helped to settle certain controversies about the origin of these structures and incidentally illuminated the potentiality of experimental methods using surviving tissues *in vitro*. The traditional techniques of tissue culture were rapidly established thereafter. Burrows, in collaboration with, Harrisson, introduced the use of plasma clot in place of a lymph clot. Shortly afterwards, Burrows and Carrel undertook investigations into the effects of tissue extracts on growth. Alexis Carrel (1912) used tissue and embryo extracts as culture media and discovered that, embryo extracts had a strong growth promoting effect on certain cells. The technique of growing tissues in plasma clots supplemented with embryo extracts, then, had become a standard practice. The greatest difficulty in performing tissue culture (TC), at that time, was the avoidance of bacterial contamination. Carrel, who had already won a Nobel Prize for his work in experimental surgery, was largely responsible for the development of aseptic methods during the next few years. Consequently, he was able to obtain consistently successful results and the measure of his technical genius was his feat of keeping a strain of cells in active multiplication for about 34 years by means of the very tedious method he had available and in the total absence of antibiotics. Carrel's work could be said to have demonstrated, beyond doubt, that animal cells could be grown indefinitely *in vitro*. For about 50 years, mainly tissue explants, rather than, cells were used for culture techniques. Later, after 1950s, mainly dispersed cell cultures were utilised. As in plants, similarly, for animals, if the explant maintains its structure and function in culture, it is described as an organotypic culture irrespective of the tissue from which it was derived. Such a culture is described as histotypic culture.

The perfection of our present method of cell culture owe a very great deal to the group at the National Cancer Institute (NCI) headed by Dr. Wilton Earle. This group was the first to grow cells on glass surface in large numbers. They were also the pioneer to grow cultures from single cells and the first to propagate cells intentionally on suspension which, later, became known as 'suspension culture'.

An entirely different approach to the cultivation of tissues *in vitro* was initiated by Dr. David Thomson in 1914 which was later developed by Dr TSP Strangerways and Honourfell. Instead of trying to make cells grow as rapidly as possible, their aim was to maintain small fragments of tissues in a state as closely as possible to the state *in vivo*. The organ culture technique, as it is called, is, in many ways, the most direct descendent of Harrisson's original experiment.

Baker, in association with Carrel, also undertook investigations into the composition of the medium and attempted to identify the important constituents present in it through analytical procedures.This type of work was later carried on by Fischer; and, still later, by a large body of workers, *viz.*, Parker, Healy, Morgan, White, Waymouth, Eagle, etc.; and, resulted in the development of the present-day 'media'.

It was only with the rapid developments of 1950s, that the applications of Tissue culture in the field of virology, etc., became widespread. In addition to demonstration of survival of vaccinia virus by Steinhardt, Israeli and Lambert in 1913 in explanted cornea, Maitland (1928) developed a very simple TC method for virus multiplication. This consisted of suspension of fragments of tissue in a fluid medium. He(1948) had also showed that, the poliovirus could be cultivated *in vitro* in Gey's Hela cell line. Originally, tissue culture was concerned almost entirely with growth of tissue *in vitro*, *i.e.*, explantation; but, since 1940s, there have been a resurgence interest in transplantation, *i.e.*, culture of animal tissues *in vivo*. Transplantation of various tissues was first carried out successfully in 1912 by Murphy who had developed the technique of chorio-allantoid membrane transplantation in hen's egg and had later demonstrated that, irradiation with X-rays could permit heterografting in rodents. A great variety of techniques have become available today as different investigators modified previous methods for their own experiments. Slide techniques have been used mainly for morphological studies. Test tube cultures have been extensively used in virus research. For large-scale works with media preparation, a variety of flasks have been used. Also, the flask methods of cell culture have led to different techniques, in which, cells are either grown on the surface of large flaks or in suspension agitated by a stirrer.

Animal tissue culture is a technique in which the tissue/cells are taken out from an animal's body and grown; and, maintained in the laboratory *in vitro* in a suitable medium.

Usefulness of Tissue Culture Methods

1. It is possible to regulate and interpret the behaviour of cells in culture.
2. It is possible to maintain homogeneity of cell types (achieved through serial passages)
3. Legal, moral and ethical questions of animal experimentation are avoided.

Disadvantages of Tissue Culture Methods

1. Controlled physicochemical environment (pH, temperature, osmotic pressure, O_2, CO_2, etc.) is a pre-requisite which is often cost-intensive.
2. Cells in Tissue culture often results in unstable aneuploid chromosome constitution which may not be always useful.

Primary Culture and Primary and Established Cell Lines

There are three main culture methods, *viz.*, tissue culture, organ culture and cell culture. In both tissue and cell cultures, very small fragments of the tissue are placed in the medium and allowed to develop. In organ culture, special measues are taken to prevent the tissue from becoming disorganized. In tissue culture, no such precautions are taken. Since many cells are highly motile, they begin to migrate from the tissue fragment which soon becomes disorganized. Further, in cell culture, the tissue is intentionally disorganized at the beginning by disrupting it into individual cells. This could be achieved, for example, by digestion with trypsin. Cultures of this kind are called 'Primary cultures'.

A piece of tissue from the organism is usually quite complex and may contain, besides the specialised cells of the tissue, connective tissue cells, a variety of blood cells and reticuloendothelial cells. When a culture is first set-up, these all survive for at least a day or two. In usual culture conditions, many cells begin to migrate almost immediately.

The first cells to do this are often the macrophages, followed by fibroblasts which migrate in a radial manner from the explant. Nevertheless, many of the specialised cells remain immobile in the connective tissue stroma. The subsequent fate of these cells vary enormously. Many of them are quite short-lived, *e.g.*, most of the blood cells die and disappear from the culture within two or three days. Other cells, such as, neurons and muscle cells, frequently persist in culture, sometimes for months without dividing at all and then eventually die too. Still other cells begin to divide rapidly and continue to do so for sometime. However, many of these also die after a period which varies from weeks to months. If the cells multiply repeatedly for a long time, they can often be 'passaged'. This is usually achieved by, at first, obtaining the cells in suspension and then inoculating this into a new culture vessel along with fresh medium. As soon as cells have been 'passaged' in this way, the culture is designated as 'Primary cell line'.

Notwithstanding the above, often, the primary cell lines go on dividing at quite high rate for a long time and can be passaged repeatedly indefinitely. Even after quite a large number of passages, some primary cell lines cease to proliferate and die out. Sometimes, cell lines could be cultured for such a long period of time that they apparently have developed the potential to be subcultured indefinitely *in vitro*. Such cell lines are called 'established cell lines'. Generally, a line is not designated as an established cell line unless it has been subcultured at least 70 times at intervals of 3 days between subcultures.

In some instances, the transition from primary cell line to established cell line is smooth and gradual. It is very difficult to say when the primary cell line ceases and established cell line begins. In other instances, however, an established cell line arises from a primary cell line by a dramatic event called 'cell alteration' or more commonly called 'cell transformation'. Transformation may arise in a primary cell line which grows very slowly. Quite suddenly, a few rapidly-growing colonies of altered cells appear. These very quickly outgrow the rest of the culture and the predominant cell type. Established cell lines arising in this way often differ in many respects from the primary cell lines from which they are derived. Primary cell lines preserve many of the characteristics of the cells from which they are derived; where as, established cell lines often diverge from these. All primary cell lines, at first, have normal number of chromosomes; whereas, established cell lines almost invariably have an unusual number.

In addition to the above, primary cell lines do not behave in quite the same way as established cell lines. When established cell lines have become stationary, the medium in which they have been grown, is usually inadequate for maintenance of a fresh inoculum of cells. However, it can often be observed that, cultures of primary cell lines stop growing before the medium is exhausted. The reason for this is a phenomenon called 'contact inhibition' which was first described by Abercrombie and Heaysman. They observed that, when certain kinds of normal cells, mainly fibroblasts, came into contact, they were immobilised almost immediately. This is a very common property of primary cell lines. Due to contact inhibition, not only the cell movement is affected; but, also, the cell division, DNA synthesis, RNA and protein synthesis are much reduced or eliminated when a cell sheet becomes confluent. Further, growth of cells slow down progressively as they approach confluence; or, in other words, as they approach a limiting density. A limiting density of cells may result from exhaustion of nutrients, limitation of nutrients by diffusion and similar general factors. However, cells regulate their density within an enclosed system by some kind of negative feedback. Whether this is due to specific 'chalones' (which are envisaged as hormone-like cell-type specific substances)or by other signals is not clear. The phenomenon of 'density inhibition' and 'contact inhibition' are much less prominent and may sometimes be entirely absent in established cell lines.

Instruments and Glasswares for Tissue Culture

Glasswares

Although cells have been grown on many substances, including cellophane, Perspex, polythene and other plastics, the material most commonly used is glass, because, glass is so readily available. But, glasses vary very considerably in their composition and some contain materials which are higly toxic to cells, such as, lead and arsenic. Since these substances are slowly brought into solution by the slightly alkaline media used in TC, toxicity to cells can sometimes be traced to their source. It is, as such, generally, opined that, only borosilicate glasses could be used for tissue culture because they are less soluble in alkali. However, both borosilicate and soda glasses have been found to be toxic on occasions while glasses of both the types have also been found to be highly satisfactory. Thus, it is best to find a brand of glass which generally has no toxic properties. Two types of glasses which appear to be quite satisfactory, are the borosilicate glass made by Pyrex and the soda glass used for most medical prescription bottles. It is worth mentioning here that, new glassware is often found to be rather unsatisfactory for culturing cells even when rigorous cleaning methods are used. After it has been used atleast once, it seems to give more satisfactory results. The exact reason for this is not very well known; but, it may well have to do with surface charge on the glass, which has been shown by Rappaport (Paul, 1975) to be important in permitting cells to attach.

Rappaport *et al.*, observed that, the extent to which glass permits cells to adhere, is related to the charge density on its surface. The net negative charge could be determined easily by measuring the amount of stain retained when the surface is covered with 0.005 per cent crystal violet for one minute and then rinsed with water. The retained crystal violet is dissolved in ethanol for colorimetric determination. They could find that, soda glasses had a higher charge density than borosilicate glasses; and, that, the charge density could be increased and stabilized by a process which involves treatment with 0.1 M EDTA or dilute 0.025 N NaOH at 122°C for 30 minutes followed by treatment with Na_2CO_3.

Plasticwares

Polystyrene vessels are widely used for cell culture today. Most of the commercially available polystyrene dishes are unsuitable because cells will not adhere to them. It is essential that, the surface of the plastic should carry a negative charge which will bind Na^+ ions. According to Rappaport, the solubility of a surface is directly related to the density of the sodium ions. Plastic vessels manufactured by M/S Falcon Plastics and sterilized through ionizing radiations are perhaps the best. Nevertheless, plasticwares manufactured by M/S Nunclon are almost as satisfactory for many purposes. However, ordinary grade polystyrene vessels could be made suitable by sulphonating the surface layer of polystyrene. This can be achieved by treating the dishes with conc.sulphuric acid for one hour; and, then, after washing in water, to be treated with 10 per cent Na_2CO_3 for 10 minutes. After washing, until neutral, these are as satisfactory as the other dishes. These dishes are not usually re-used; but, if desired, they could be cleaned wih detergent and sterilized with 70 per cent ethanol. The dishes are wrapped while still wet with alcohol and dried in an incubator.

Stoppers for Cell Culture Vessels

Rubber stoppers contain substances which are toxic to tissue culture cells. This seems to be due to chemicals used in processing. In order to remove the surface impurities, rubber stoppers be, at first, boiled in dilute alkali (5 per cent Na_2CO_3) and thoroughly rinsed with water. Thereafter, they could be cleaned in the same way as the glasswares. Silicon rubber stoppers have been introduced to provide a solution to the problem of toxicity. The same problem could also arise in the case of rubber liners in screw cap bottles. The toxicity of rubber stoppers is particularly marked when protein-free synthetic media are used; and, is much less a problem when there is a high concentration of protein in the medium.

Rubber Tubing

In most cases, where tubing is used, the medium is exposed to it; but, generally, for a brief period, so that, we do not face much problem. However, silicone rubber tubing is available cheaply today and is used where there is risk of toxicity. Silicon rubber has the additional advantage that it can be sterilized by dry heat.

Instruments

Surgical instruments be kept scrupulously clean. They be wiped and dried immediately after use. Special cleaning is not very much necessary for surgical instruments which are in routine use. However, new instruments, particularly, the detachable knives, usually have a protective coating, either of grease or of an anti-rust compound. They be, therefore, wiped with a swab soaked in carbon tetrachloride and then a wet cloth before being dried and sterilized.

Aseptic Techniques and Basic Quality Control Measures

Contamination of glasswares, plasticwares, media and other materials with bacteria or other contaminating particles is one of the greatest hazards in most tissue culture works. Also, noxious chemicals sometimes impair tissue culture works. Thus, special care be taken in maintaining the basic quality control in Tissue culture. Such measures are broadly called 'aseptic techniques' and involves cleaning and washing, packing and sterilization of the materials used in TC works. The following is a brief account of the procedure followed in aseptic techniques, in general, for maintaining the quality control of the materials used in TC works.

Cleaning Procedures

Glasswares

All glasswares be immersed in water immediately after use. In this way, proteinaceous deposits are prevented from being drying-out and this also facilitates cleaning. Initiation of cleaning of the used glasswares begins with removing cottonwool plugs, etc.; and, then, rinsing with water. It is then subjected to whatever cleaning method is employed. A great many procedures are recommended and these could be divided under four general groups:

1. Cleaning with detergents
2. Cleaning with alkalis
3. Cleaning with oxidizing acids
4. Cleaning with ultrasonics.

Cleaning with Detergents

Detergents have the advantage that no boiling is required. On the other hand, the disadvantage is that, they are difficult to drain out completely. Also, many detergents are toxic to cells; so that, care be taken in choosing one. A detergent which has been evolved especially for cleaning TC glasswares, is known as 'Microsolve'and manufactured by Microbiological Association of Bethesda, USA. This detergent is practically non-toxic. In India, Extran, a superior quality of 'Surf', etc.are generally used for cleaning TC glasswares.

All glasswares are, at first, rinsed and all the labels and pencil marks are removed. The easiest way to remove wax pencils and magic marker marks is to soak the glasswares in warm detergent for some hours; and, then, wipe off the marks with a cloth carrying some abrasive cleaning power. These are then transferred to a tank containing RBS 25 or Decon 75. After soaking overnight, the glasswares are removed and rinsed thoroughly with tap water followed by a single rinse with deionised or distilled water. If the glasswares have been rinsed this way, before putting them into detergent, this detergent could be used over and over again for some weeks before it becomes necessary to renew it.

Pipettes could be cleaned similarly. After soaking in detergent, they are thoroughly rinsed with water in a pipette-washer, drained and dried in a drying oven.

Cleaning with Alkalis

Different kinds of alkalis are used. However, the most popular ones are soft soap, sodium triphosphate, Na_2CO_3 and Na-metasilicate. There is little to chose among them; but, Na-metasilicate is probably the most commonly-used; because, it does not introduce any foreign ion into the glassware. The usual practice is to prepare the alkali solution in a concentration of about 100 times stronger than which would be ultimately required. The glasswares are boiled in water, to which, this stock solution is added. The glasswares are then treated with dilute acid and rinsed with water, dried and sterilized.

Cleaning with Oxidizing Agents

Sometimes, strongly oxidizing acids are used for cleaning glasswares. Generally, hot H_2SO_4, chromic acid and nitric acid are used for these purposes. However, their use involves certain amount of danger. In the case of chromic acid, a monolayer tends to remain on the glass even after very prolonged rinsing. Chromic acid is highly corrosive and it is necessary to use it very occasionally at a higher temperature with great amount of care. Nitirc acid is sometimes used for cleaning small objects like cover glasses.

Cleaning with Ultrasonics

Special apparatus have been developed for cleaning glasswares utilizing very high frequency of sonic oscillations. These are effective; but, have no advantage over other methods for most purposes. They could, however, be used for cleaning particular scientific instruments having inaccessible holes and crevices.

Prevention of Contamination

Most of the microbes have much more rapid growth rate than cells and they frequently produce toxins which are harmful to them. Hence, the most important part of the TC technique comprises the avoidance of contamination and the growth of the tissues in aseptic condition. Contamination of materials may be prevented in two ways:

1. Sterilisation: This term implies the removal of microbes already present.
2. Aseptic technique: This term implies the prevention of contamination of materials which are already sterile.

In TC works, the sources of contamination could be the following:

1. The apparatus
2. The culture medium
3. The tissue itself

4. The atmosphere in which the operator has been performing
5. The operator.

Before commencing the work, the apparatus is rendered microbe-free by sterilization. Likewise, the medium is usually sterilized before use; but, some of the components are prepared aseptically. The tissue is usually obtained in sterile condition by the application of aseptic principles; but, occasionally, it may be necessary to sterilize it. Contamination from other sources is prevented almost entirely by the application of aseptic techniques.

Sterilisation Procedures

It may be noted here that, exposure to given sterilizing condition will kill a definite proportion of contaminating organisms. Sterilisation may be achieved by:

1. Physical destruction of the microbes by:
   (*a*) Dry heat
   (*b*) Moist heat or autoclaving
   (*c*) Radiation
2. Chemical destruction of microbes
   (*a*) Use of antiseptics
   (*b*) Use of antibiotics
3. Physical removal of microbes
   (*a*) Filtration
   (*b*) Centrifugation
   (*c*) Washing

Most of the apparatus are sterilized by dry heat or by moist heat. In general, apparatus, which is not damaged by high temperature, is sterilized by dry heat. Moist heat is most efficient because of the high latent heat of steam. Autoclaving (sterilizing by steam under pressure) is generally done for solution, rubberwares, cloth, etc. Simple boiling is usually done for minor instruments.. Antiseptics are generally used for plasticwares but they should be relatively harmless. 70 per cent ethanol is a commonly-used antiseptic. UV radiation could also be used for sterilizing some kind of materials.

Solutions are generally sterilized thorough filtration. Some of the solutions, which are not damaged by heat, could be autoclaved. Solutions could also be sterilized through UV radiations. Antibiotics are generally added to growth media to prevent contamination.

Laboratory furniture are generally disinfected with antiseptics like spirit. A thin film of oil could be used on the furniture in order to prevent dust particles from being kicked-out. It is also necessary to sterilize the air within the working environment by the use of filters. However, the actual working area is within a Laminar air flow cabinet which maintains a clean sterile environment. In dry sterilization, a hot air oven is used which has the provision of hot air circulation. The non-polythene labwares are kept inside the hot air oven at 160°C for 120 minutes for dry sterilization. The glass articles are to be wrapped with aluminium foil before inserting into the hot air oven for sterilization. Moist sterilization is done in an autoclave/pressure cooker. Articles (including polythene made ones) are kept inside the autoclave at 10-15 lb pressure for 10-15 minutes depending on the nature of the articles and the purpose of the work. In general, a temperature of 115°C and a pressure of 120 lb is maintained in an autoclave for about 10-15 minutes. Solutions may be sterilized in bottles or flasks, stoppered with cotton wool and wrapped in gauze. Plasticwares are generally sterilized in UV light. Gamma-rays are extensively used today for sterilization of pre-packed materials. Antiseptics are used for swabbing down laboratory furniture; and, 70 per cent ethanol is said to be the safest. Chloroform is frequently added to solutions in order to preserve them. Plastic dishes are generally sterilized with alcohol. Antibiotics are generally used to to prevent accidental contaminations. Penicillin is the most commonly used antibiotic. It is generally added as Sodium Penicillin G to the media to a final concentration of 20-50 units/ml. Also, Streptomycin may be used in such a way, that, it gives a final concentration of 50 µg/ml. Fungal infections may be inhibited by using Mycostatin to a final concentration of 20 µg/ml. Kanamycin, at a concentration of 100-200 µg/ml is generally used for eliminating Mycoplasma (PPLO).

Filtration is an efficient way of achieving sterile condition. Among the many different kinds of filters available, the following types are in common use:

1. Sintered glass filters

2. Millipore filters
3. Asbestos pads
4. Kieselguhr and porcelain filters
5. Porous cellulose acetate or cellulose nitrate filters (Millipore, Gelman, Sartorius, Oxoid) have all but replaced other kinds of filters. Their high cost is offset by their outstanding advantages, *viz.*, they give a very high filtration rate and are completely non-toxic. These filters are disposable. For filtering large volume of solutions, many special filter holders are made. For filtering very small volume of materials, Millipore discs are effective when used with Swinney adaptor fitted to a Lue-lock syringe. For sterilizing solutions for TC purposes, the HA (0.05 μm) grade of Millipore filter or a finer grade is used. Biological materials are generally filtered under pressure.

The other physical means of sterilization are centrifugation and washing. Heavily contaminated films, *e.g.*, bacterial cysts, etc., could be clarified by centrifugation before filtering.

Storage of Sterile Materials

It is convenient to sterilize materials in large batches and use them gradually. Solutions are best kept in tightly-sealed containers called 'medical prescription bottles'. Alkaline solutions may best be stored in a glass bottle for a long time. Other sterile articles are to be carefully stored in a sterile enclosure.

Sterility Testing

Routine sterility testing be done for all the media and solutions because filters and filtration techniques may sometimes fail. As a common practice, a small amount of the prepared medium is generally inoculated into glucose and Thioglycollate broth and incubated both at 23°C and 30°C. If any microbial growth is observed in the test medium, the prepared medium is rejected. Special methods are to be used to test for PPLO. This is done by spreading 0.1 ml of the infected material on special PPLO agar plates. The plates are incubated in a CO_2 incubator at 37°C for about a week and then examined. Some mycoplasma may be visualized by staining the cells with Jenner Giemsa.

Aseptic Technique

The following is a brief account of the techniques of prevention of contamination of materials which are already sterile.

Contamination from Tissue

It may be avoided by collecting the tissue aseptically or by sterilizing it after removal. Most internal tissue of animals are already sterile. In the case of adult tissue, bacteria are occasionally found even when they are removed from the internal organs. This does not happen very frequently. However, if necessary, the tissues be sterilized in antibiotics. The tissue is generally washed in an antibiotic solution of little high concentration before being grown in a medium containing the usual concentrations. A useful medium is the 'Balanced Salt Solution (BSS)' containing 1000 units of Penicillin and 0.5 μg of streptomycin/ml. Neomycin (0.5 mg/ml) is also highly effective.

Contamination from Air

Contamination may also arise from air in the room in which the work is done; and, even from the operator. The following set of rules have been found to be adequate in order to get rid of contamination from air:

1. The room be maintained clean, tidy and dust-free through the use of air conditioners and air filters.
2. No material, known to be contaminated, be opened in it.
3. When aseptic work is being performed, no body may enter the room or open the door without obtaining permission from the operator.
4. All working areas be swabbed with 70 per cent ethanol before use.
5. After use, the tables be cleaned thoroughly with polysan/ spirit.
6. The pipettes and test tubes be removed from the rack immediately after use
7. Unused materials be removed for re-sterilisation
8. Used pipettes be kept in a jar of water. Other materials be preferably removed from the area after use.

9. The working area be clean and preferably within a laminar air flow cabinet or atleast inside a hood.
10. The operator be always clean with clean dress and having air showers before enterig the TC lab.

Contamination from the Operator

Contamination, due to the presence of the operator, may arise from 3 sources:

1. Through direct contamination of instruments and cultures from non-sterile objects during operation
2. Through bacteria carried in expired air
3. Organisms carried by dust raised by the operators' movements

Direct contamination may arise from:

1. Contact of apparatus, particularly pipettes, with non-sterile surroundings or with the hands.
2. The entry of organisms present at the tip of the container.

The best way to try to avoid contamination of articles is to perform manipulations repeatedly under the supervision of an experienced person. Contamination from the hands may be minimized by sterilizing them before commencing the work.

On opening cell culture vessels, there is always a great danger of contaminating the contents with organisms from the lids. This could largely be avoided by flaming the necks of all the vessels before removing the stoppers.

Media

The medium is, by far, the most important single factor for culturing cells and tissues. The function of a medium is:

1. To provide the physical conditions of pH, osmotic pressure, etc., required for survival.
2. To provide the complicated chemical substances required by the tissue which it can not synthesise by itself. These latter substances include amino acids, carbohydrates and vitamins. Some microbes can synthesise all these component chemicals from simple organic substances; but,

all tissues from multicellular organisms require readymade molecules to some degree. In general, plant tissues are much less demanding than the animal tissues.

Ideally, the medium could be an accurately-defined mixture of chemical substances. On the whole, animal cells need either a completely natural medium or a medium supplemented with certain natural products, usually the serum. Although improved synthetic media will continue to be produced, the natural media are generally the cheapest and most convenient to use for the maintenance of cells.

A graphical presentation of the animal cell culture media is given below:

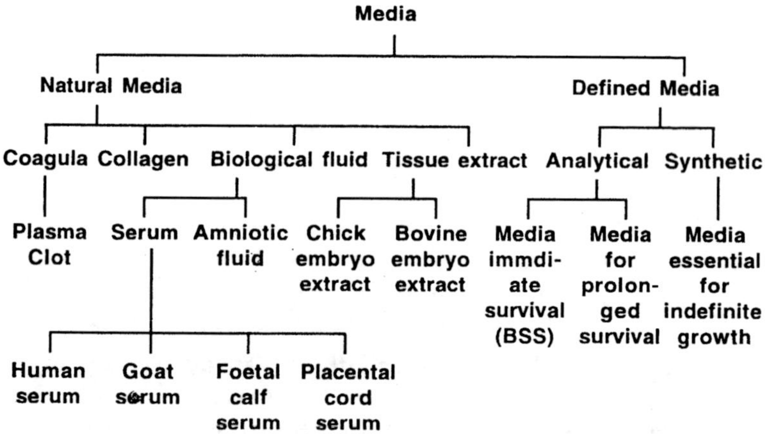

Plasma Clot

Here, the tissue fragments are grown in a coagulum. Ross Harrison grew a small piece of tissue from near the neural tube in adult frog limb. Later, Carrel and Burrows introduced plasma clots for the same purpose.

Biological Fluids

The most commonly used biological fluid is the 'serum'. Primates and human cells seem to survive better in human serum. Goat serum (GS) and the Foetal Calf Serum (FCS) are generally used in cell culture works. The active components of a serum have been identified by Healey and Parker as orosomucoid and α-2 macroglobulin. The latter could be replaced by non-protein molecules. Glycoproteins associated with Cohn's Fraction V are the important components of serum. However, some workers considered 'Fetuin' to be the important component; while, still others favour α-glycoproteins to be important factors of serum.

Serum is generally tested for toxicity. This is done by incorporating it into a growth medium and determining the plating efficiency of a standard cell strain. Some authors opine that the toxicity could be reduced by heat inactivation (heating to 65°C for 30 minutes)

Amniotic Fluid

The most commercially obtained is the bovine source. A pregnant uterus is obtained from the local slaughter house. It is carefully washed on the outside and then hung from a hook. An area near the base is swabbed with iodine. After this had dried, a tracer cannula, with the tube attached, is inserted into the uterus and the amniotic fluid is allowed to drain off into suitable vessel. After sterility testing, the material is ready for use.

Tissue Extracts

Almost all the media used for the cultivation of tissues contain some sort of tissue extract; particularly, embryo extract. However, some cells may be cultured quite satisfactorily in serum and balanced salt solution alone. The most important components of the embryo extract are the nutrients in a small molecular fraction; and, that, they could be effectively replaced by synthetic supplements which are usually mixtures of amino acids. Harris and Kutsky (1954) had demonstrated that, while the small molecular fractions constitute the essential components of the embryo extract, there is also a nucleoprotein fraction which greatly improves the growth of newly isolated cells. Embryo extract was first prepared by homogenizing whole 11-day old chick embryo with an equal volume of Hank's Balanced Salt Solution (HBSS). For the culture of tissues of cold-blooded animals, extracts of tadpole (for amphibia) and fish fry (for fish) are generally used. All tissue extracts tend to deteriorate rapidly at room temperature. Even at 4°C, they tend to lose a great deal of their growth promoting potency within a week. As such, tissue extracts be stored at -20°C.

Other Media of Biological Origin

The most useful supplements are the Lact. Albumin Hydrolysate (LH) used at a concentration of 0.1 to 0.5 per cent. This could replace the amino-acid component of defined media. Dehydrated yeast extract (0.1 per cent) could be used as a source of vitamins and co-enzymes.

Defined Media Used for Culturing Cells and Tissues

The TC media obtained from natural sources are widely used for many purposes, especially, for growing tissues freshy isolated from an organism. However, they suffer from certain amount of disadvantages. Since their composition is largely unknown, it is virtually impossible to reproduce conditions exactly from one experiment to another. Therefore, attempts have been made to develop media composed entirely of defined components. The problem was tackled in two ways, *viz.*,

1. The so-called analytical approach and
2. The so-called synthetic approach.

In the former, attempts are made to analyse the medium and identify the essential components. In the second method, metabolites, which are known to be essential and valuable for the survival and development of the intact organism, are combined to form media which are tested for their sterility in order to keep the cells alive. However, it may be said that, there will never be a universal medium for all cell types. A substance added to a medium may be beneficial or harmful. If the effect is beneficial, it may be that the substance is essential for the cells or it could have supplementary values. If it is harmful, it may be because it is intrinsically poisonous to the cells; or, because, it is present in an unsuitable concentration.

A brief account of the media used for culture of warm-blooded vertebrates is given below:

Here, the media could be classified into four groups:

1. Media essential for immediate survival (MEIS)
2. Media essential for prolonged survival (MEPS)
3. Media essential for indefinite growth (MEIG)

The components essential for the immediate survival of cells and tissues have been precisely defined. It is essential to control the osmotic pressure and pH; while, a source of energy and certain inorganic ions is necessay for all but the briefest period of survival. These requirements are met by a combination of salts and glucose. A simple medium of this kind is called a Balanced Salt Solution (BSS). Cells and tissues could survive for a brief period in BSS; but, for prolonged survival, other factors are essential. These factors, which are normally present in a whole serum, could be replaced by a synthetic mixture for survival for long period of time. The mammalian cells require, in addition to the BSS, all the essential amino acids, oxygen, vitamins and serum proteins. In the complete absence of proteins, the cells do not survive for a long period.

Recent years have witnessed rapid progress in the development of entirely synthetic media. At present, very few cell lines could be grown in media, from which, serum has been entirely omitted.

Balanced Salt Solutions (BSS)

All media used in TC have a basis of synthetic mixture of inorganic salts known as physiological or balanced salt solution. The function of this salt solution is to maintain the pH and osmotic pressure in the medium and also to provide an adequate concentration of essential inorganic ions. All BSS have been derived from the salt solution originally described by Ringer. The first one to be developed specifically for supporting the metabolism of mammalian cells was Tyrode solution. Most of them were devised with the aim of preventing Calcium deposition in the cultures; or, in order to obtain better buffering condition in the medium.

Salt solutions could be divided into two types, *viz.*,

1. Those intended to equilibrate with air
2. Those which are designed to equilibrate with a gas phase containing high tension of CO_2

According to Paul (1975), the most useful type of the former is Hanks BSS (HBSS) and one of the best examples of the latter is Earle's BSS (EBSS). Both these have the great advantage that, they are very easy to prepare by autoclaving of the different components. One of the best buffered salt solutions is undoubtedly EBSS but it requires a special gas mixture which is sometimes disadvantageous.The phosphate saline (PBS) of Dulbecco is used widely by virologists. The pH is adjusted to 7.2 with 5N NaOH and glucose must be present in the medium. Some attempts have been made to buffer the medium with buffering systems other than phosphate and bicarbonate. One of the best alternatives is the Tris. In many TC laboratories, cell strains like Hela, BHK21, etc., are kept in active proliferation for upto 7 years in media containing Tris as the buffer. Another kind of buffer, called HEPES, is today incorporated in many commercially available media.

Many modifications of Eagle's medium is used. Original version had rather low concentration of amino acids; and, most modifications are designed to correct this and permit higher cell yield. The RPMI medium is used for WBC culture. Trowell's T8 medium is used for organ culture, in which, there may be little overall increase in tissue mass during the culture period. Morgan produced a modification of 199 known as the M-150. The main differences are the use of Hanks BSS instead of Earle's BSS as a basis for the medium; addition of glucose, glutamine and antibiotics; replacemenet of xanthine by mono-Na-xanthine and Ba-salt of ATP by Na-salt; and, addition of bicarbonate immediately before sterilization. Many attempts have been made to simplify and identify the essential components in a medium. The most accurately defined medium of this kind is undoubtedly the one devised by Eagle (1955).Eagle's medium is said to be the simplest of the defined media in general use. It has been ascertained that, all its components are essential fo cell growth; and, that, most of the other components of most complicated media are not essential for general cell growth.

In making-up salt solutions, it is advisable to use double glass distilled water (DGDW); or, the water which has been deionised by passing through a mixed bed of resin and then distilled once in glass. Ordianry tap water generally contains a large number of ions, many of which are harmless; but, some of which, such as, lead, are toxic to cells. In addition, tap water may contain endotoxins released by the growth of gram negative bacteria. DGDW is not supposed to contain these impurities. Generally, DGDW is stored in polythene bottles to prevent accumulation of metallic salts from glass. Many metal ions may also be introduced as impurities of the salts used in solutions. Therefore, the salts must be of the highest analytical purity and stored in special bottles. Phenol red is added to many BSS as a pH indicator. It has been found to be non-toxic upto a 0.005 per cent, if serum is present in the medium.

Use of Stock Solutions

It is convenient to prepare stock solutions of BSS. A solution is made up with 10 times the concentration of all components except $NaHCO_3$. This is stored in a polythene bottle. If it is to be autoclaved subsequently, it could be preserved with little chloroform which will be driven off during sterilization. To re-constitute the BSS, distilled water (DW) is added to dilute the stock solution 10 times. To adjust the pH after sterilization, 1.4 per cent $NaHCO_3$ is used.

Cell Lines and their Maintenance

There exists differences between primary and established cell lines. Generally, after an initial phase of rapid growth, the cells slow down and stop growing. In many cases, they die out; but, in some cases, after a static period, there is a sudden outgrowth of new cells. Thereafter, they can frequently be carried indefinitely. The initial phase of rapid growth may last only a few days or as long as many months. During this phase, the line is referred to as primary cell line.

An account of the maintenance of a typical animal cell line is briefly given below:

Most cell lines are grown directly on the glass or plastic surfaces. They always grow uniformly as colonies, layers or suspensions. They are maintained in vessels having flat surface, on which, the cells could be grown, such as, Carrel flask, T flask, prescription bottles, Roux bottles, Erlenmeyer flasks, Petri dishes etc. In fact, any vessel with a flat bottom which could be conveniently sterilized and sealed, is suitable for the purpose.

The essential steps involved in maintaining cell lines are old medium removal, feeding with new medium and sub-culture (or transfer). Transfer requires that the cells be obtained in suspension.

Cells growing in a monolayer on a glass surface may be brought into suspension by:

1. By physical method, *viz.*, scraping from the glass
2. By chemical method: (i) by proteolytic enzyme, (ii) by chelating agents.

Some cell lines are very loosely adhered to the glass surface and may be suspended simply by agitating the flask gently. There are many cells which do not leave the glass surface so readily but which could be easily dislodged by scraping with a rubber/plastic policeman at the end of a glass rod. When the dislodged cells are sucked up and down once or twice in a pipette, a suspension of single cells is formed. However, many cell lines may not yield easily to physical removal but will round-up and leave the glass as single cells when treated with a dilute buffered solution of trypsin. Trypsin is commonly used at a concentration of 0.1–0.5 per cent. pH is an important factor during trypsinisation. Trypsin is virtually inactive below pH 7.0; while, above 8.0, the cells could be progressively damaged in its presence.

Having obtained the cells in suspension, it is necessary, at first, to estimate the number of cells present. This is usually done by making a cell count with the help of a haemocytometer.

Feeding of cells growing on glass or plastic surface is very simple. As a rule, it merely consists of removing the old medium and replacing it with a new. The old medium could either be decanted or pipetted-out. The medium could be completely exchanged at each feeding, which is usually done at an interval of 3-4 days. A mechanical dispenser greatly reduces the labour of adding medium to the vessels..

Animal Cell and Tissue Culture

Animal cell cultures have many applications including their use as model systems for biochemical, physiological and pharmacological studies and the production of growth factors, blood factors, monoclonal antibodies, interferons, enzymes, vaccines, and hormones.

The major problem associated with the isolation of free cells and cell aggregates from organs is that of releasing the cells from their supporting matrix without affecting the integrity of the cell membrane. Various methods have been employed to achieve this goal.

Established Cell Lines

Cells in culture display very much the same type of growth pattern as micro-organisms. In particular, they show the classical growth demonstrated by cultures of bacteria, yeast and protozoa. The culture exhibits following phases:

Lag Phase

This is the phase when practically no growth occurs and may last for some hours to a few days before growth commences.

Log (Logarithmic) Phase

During this phase, growth proceeds steadily with the population doubling every 15 to 20 hours in the case of fast growing cells.

Stationary Phase

At the end of log phase, maximum population is reached and the growth becomes stationary.

Chapter 28

Fish Genetics and Development of Transgenic Fishes

Agriculture followed by Animal husbandry is said to have been much benefited from the knowledge of Genetics and Biotechnology. Modern Biology seems to be having much impact, of late, on the aquaculture systems with regard to species ratio manipulations, stocking densities, health monitoring, aeration, replenishment of water, supplementary feeding, and, so on. However, the advantages of genetic potentials in aquaculture has been utilized quite slowly. Intensive and extensive polyculture, composite fish farming and Integrated fish farming are today considered as potential industries in aquaculture. Such techniques, grossly, have enhanced fish production level from a mere 0.6 mt to 12 mt (approx) per ha per year. Such a boost might have encouraged the fish culturing States to convert the agricultural fields into aquacultural ponds in order to possibly earn more profits.

Cellular Studies

Cellular studies in fishes was initiated around 1960s with one of the primary attempts to make comparative account of karyotypes in the parent species and their hybrids and to assess the status of fertility in the hybrids.

Genetic Manipulations

Also called 'chromosomal engineering', genetic manipulation was developed as early as 1980s. Suitable standard protocols have been developed to induce meiotic and mitotic gynogenesis and triploidy/tetraploidy in IMCs and exotic carps. Later, during the early 1990s, the Indo-Norwegian collaborative project on 'Selective breeding of Rohu was introduced in India with a broader genetic base.

Genetic Engineering

The gene transfer technology is now being used in aquaculture for developing transgenic individuals. The transgenic individuals, with the extra foreign gene received from other individual of the kind or otherwise, are expected to perform better, in terms of survival and growth, resistance to disease and unfavourable/adverse environmental conditions. Much work is being done abroad on transgenic fish technology with fishes, like Atlantic salmon, gold fish, channel catfish, common carp, etc. In India, attempts have been made to introduce foreign gene into the embryos of IMCs, like Rohu.

Biochemical and Molecular Genetics

These are also being put to use in modern aquaculture, particularly for characterization of stocks and for identification of fish populations through isozyme and DNA fingerprinting studies respectively. The International Network on Genetics in Aquaculture (INGA) was established in 1993-94 with the active participation of International Centre for Living Aquatic Resources (ICLARM) (now called World Fish Centre).

Hybridisation Studies

In a broad sense, it is called heterospecific insemination. It aims at producing offspring of superior quality than the parent species based on the transfer of desirable traits from one species to another. Fish hybrids are generally intermediate in position with regard to expression of phenotypic characters pertaining to the parents. However, they may show advantageous qualities like faster growth rate, better food conversion, higher survival, resistance against unfavourable environment and diseases, etc. In India, several inter-generic and inter-specific hybrids of IMCs have been produced through hybridization during the late 1950s. At CIFE, some amount of studies have been going-on on the selective breeding in fishes

Interspecific hybrids are obtained, generally, by crossing four species, *viz.*, *Labeo rohita*, *L. calbasu*, *L. gonius* and *L. bata*. Among the inter-specific cross of *Labeo rohita* and *L. calbasu*, the reciprocal cross could produce highly viable hybrids. The growth rate of these hybrids are generally more than that of *L. calbasu*.

In India, genetic upgradation/improvement work had its initiation during the late 1950s through simple inter-specific and inter-generic hybridization works. It may be noted here that, the inheritance of the positive traits appear to be partial, *i.e.*, the hybrid progeny could acquire a body depth wider than in rohu and mrigal but smaller than in catla.

Hybrids have also been produced by involving the following exotic carps in the crossing process:

Silver carp × Grass carp

Grass carp × Silver carp

Bighead carp × Silver carp

Bighead carp × Grass carp

Hybrids have also been produced by involving the IMCs with the exotic carps in the crossing exercise.

Natural Hybridization

It occurs more likely between an abundant species and an introduced or search species than it is between two abundant species. The probability of incidence of hybridization is increased when the spawning areas in nature are greatly limited and the brooders are, thus, forced into close proximity. Such common hybrids are rohu-catla, catla-rohu and mrigal-rohu.

Important Fish Hybrids

Catla × Rohu and Rohu × Catla Hybrids

1. These hybrids are mainly phytophagous in food habit but they have adaptability towards artificial feed.
2. The hybrids grow faster than rohu.
3. The hybrids generally show more compatibility with rohu than with catla

4. Attempts could be made to replace either rohu or catla with these hybrids in composite fish culture.

Rohu × Calbasu and Calbasu × Rohu Hybrid

1. It attains maturity within two years
2. Head is smaller and growth rate is better than in calbasu.

Catla × Calbasu and Calbasu × Catla Hybrid

1. Growth rate is faster than that of calbasu.
2. It feeds on a different kinds of food and also accepts artificial feed readily.

Application of Hybridisation in Fisheries

1. Races of the species of the same genus or two different genera are crossed to combine their desirable qualities and evolve a new strain or a hybrid.
2. The hybrids generally possess faster growth rate, higher rate of food conversion, better utilization of food from the environment, more viability, greater resistance against disease, better taste of their flesh and higher nutritional quality.

The development of transgenic fish has enabled us to answer to many questions in the areas of oncogenesis, developmental biology, gene regulation, etc.in fishes. This seems to be an attractive technology for quick genetic improvement in fishes. Much of the results depend on the fulfillment of the following:

1. Proper integration of exogenous genes into host DNA which could be inherited in a Mendelian fashion.
2. Characterisation of genes regulating synthesis of species-specific growth hormones in fishes along with their promoters.
3. Formulation of a suitable protocol for gene-targeting experiments in fishes.
4. Development of efficient gene delivery system.
5. Acquiring of knowledge with regard to DNA pathway from cytoplasm to nucleus.

Enzymes in Genetic Engineering

One of the principal aims of Modern Biotechnology is to work through manipulation of DNA which forms part of nucleic acid enzymology. There was no method available for cutting a DNA molecule into discrete fragments before 1970. Discovery of DNA metabolizing enzymes had enabled scientists to comprehend and initiate serious works in the field of genetic engineering. Further, the Recombinant DNA technology has enabled man to cut DNA molecules at defined sequences. More precisely, it was based upon the discovery of type II restriction enzymes. Further, the development of PCR (Polymerase Chain Reaction) as one of the simple but spectacular techniques has permitted amplification of specific DNA sequences.

Restriction Endonucleases

This was coined by Lederberg and Meselson (1964) to describe the nuclease enzymes which destroy (restrict) any foreign DNA entering the host cell. These enzymes have been classified into three types, *viz.*, Type I, Type II and Type III. In gene manipulation technology, restriction endonucleases are popularly called molecular scissors.

Chapter 29

Fish Genetics

'Genome' is the gene string containing the familial relationships of the individuals. The individuals comprise a species. It refers to the organism's complete set of chromosomes. The term 'genomics' is derived from genome. It indicates the act of mapping, sequencing and analyzing the genome.

Gene Mapping

It refers to the analysis of loci on the genome indicating the linear order of different genes on the chromosomes. Broadly, gene maps are of two types:

Physical Maps

These are based on the attempts to locate the loci on the chromosomes and could be accomplished mainly by the methods, such as, somatic cell hybrid panels, *in situ* hybridization and comparative mapping

Genetic Maps

These are constructed by studying the meiotic recombination between two or more loci through linkage analysis.

DNA Markers

Detection of major genes influencing the desirable traits is said to have been made possible because of the availability of DNA markers. DNA marker maps are the infrastructures for the identification and mapping of any gene. As such, development of these markers in a species could be of interest in aquaculture.

Advantages of DNA Markers in Fish Biotechnology

1. They help in the preparation of Genetic linkage map for identification of single gene/locus and quantitative trait locus (QTL). They are responsible for regulating traits of economic importance like growth, fecundity, age, maturity, disease resistance, etc.
2. They help in mapping of genes regulating diseases.
3. They help in disease diagnostic processes.
4. The markers help in analysis of the Transgenics.
5. They assist in Enhancing production of cultivable fish species thorugh reliable selection programme.
6. They also regulate controlled reproduction involving GnRH/GnRH-Receptor-dependent maturation and spawning.
7. The markers play a significant role in Fish health management involving therapeutic proteins and vaccines against infections and diseases, fish transgenics and immunogenetics.

Chapter 30

Androgenesis and Gynogenesis in Fishes

Fish productivity could be augmented thorugh genomic manipulation which has potential application in fish breeding. Alteration of the original genome structure is the main purpose of Genome manipulation and leads to gynogenesis or polyploidy (triploidy/tetraploidy). It provides a new genetic status to an individual for better performance. Gynogenesis could be artificially induced by eliminating/denaturing the genetic material (DNA) of the milt through irradiation either by UV or Gamma rays and activating the eggs with irradiated milt. Diploidy could be restored by giving thermal or hydrostatic pressure shock.

Aims of Gynogenesis

1. To make highly homozygous inbred lines in much shorter time than through conventional inbreeding process
2. Heterosis effect could be achieved by topcrossing the gynogen lines with the heterozygous stock.

Procedure

1. Ovaprim or pituitary hormone is administered into mature female brood fishes. Males are taken preferably from another species. It could help in distinguishing the gynogens from that of hybrid.

2. Male milt is generally stripped before an hour of the expected ovulation time of the female fishes.
3. The milt is then mixed with freshly prepared Hank's solution in the ratio of milt one part and Hank's four parts.
4. Milt is then segregated into two parts. One half is kept for the control (normal) at about 4°C. The second half is poured into small petridish to make 1-2 mm thick column.
5. The Petri-dish is now kept on an ice tray and the assembly is inserted into the UV chamber and placed on the mechnical shaker.
6. Irradiation of milt process is continued for 17-20 minutes in the IMCs.
7. Milt is stored at 4-5°C after completion of irradiation.
8. Stripping of the female is done and the eggs are collected in basins. The eggs are splitted into four groups. After thawing the milt properly, two groups of eggs are mixed with irradiated milt. One group of eggs is fertilized with normal milt of the same species and the fourth group with the normal milt of the species whose male specimen is used for the experiment.
9. Few drops of water is added to egg and milt and mixed them together before activation/fertilization is done
10. One of the groups of irradiated milt-activated eggs are given thermal/pressure shock to restore diploidy.
11. The other groups of irradiated milt-activated eggs are allowed to develop without any shock treatment.
12. Eggs fertilized with normal milt are also allowed to develop normally to determine the quality of eggs used in the experiment.

After gynogenesis developed over the gynogen lines, they are reared separately. The rearing process requires special care because of their homozygosity. The resulting gynogenetic offspring may all develop into females. Methyltestosterone (MTS) is used for sex reversal. The highly homozygous gynogens could be top-crossed with the heterozygous stock to realize heterosis effect.

Androgenesis

It is a process to obtain all male progeny. The induction of androgenesis is relatively uncommon in nature. As such, it is difficult to induce artificially. However, androgenesis, occurring spontaneously, was reported when female common carp was crossed with male grass carp. It is induced by eliminating maternal genome in a similar way as is done for gynogenesis. Insemination of normal sperm is conducted with irradiated eggs followed by shock treatment to restore the diploid nature.

Importance of Gynogens and Androgens in Aquaculture

1. These are shorter routes for developing new fishes.
2. These are also efficient procedures for building up homozygous inbred lines in a faster way.
3. These could also lead to production of monosex population.

Chapter 31

Transgenic Fish

A transgene used in producing transgenic fish for basic research or for application is a recombinant gene which produces a genetic product at an appropriate level in the desired tissue(s) at the desired time(s). Therefore, the prototype of a transgene is usually constructed in a plasmid to contain an appropriate promoter-enhancer element and a structural gene sequence.

Gene-transfer studies have been conducted in many different fish species, including channel catfish, common carp, goldfish, etc.

Classical systems of genetic selection have possibly led to the development of strains of many commercially important species of fish which have improved growth characteristics. The ideal fusion genes, for the purpose of development of transgenic fishes, are composed of fish-derived promoter/enhancer elements linked to a piscine growth gene. A transgenic fish contains only extra DNA sequences which are originally derived from the same species, *i.e.*, fish without any heterologous genes and no viral or mammalian promoters.

Some of the general steps for developing transgenic fishes are given below:

1. The fish species be chosen, depending on the nature of the studies.
2. A specific gene construct be prepared

3. The gene construct could contain the structural gene encoding a gene product of interest and the regulatory elements which regulate the expression of the gene in a temporal, spatial, and developmental manner.
4. The gene construct to be introduced into the developing embryos for the production of transgene be integrated strongly into the genome of every cell.
5. A screening method be adopted for identification of transgenic individuals.

Aquaculture operations are worldwide in distribution. Successful production of growth-enhanced recombinant strains of fish could have wide application and major economic potential. Transfer of growth genes into fish is being viewed by developing countries as a major means of increasing available protein supply.

Recently, growth hormone (GH), cDNA and genomic DNAs have been isolated. Significant enhancement of growth of fish could be achieved through exogenous application of biosynthetic GH. Fish GH gene could be readily transferred into embryos of many fish species and integrated into the host genome. Although few, there are documentations indicating transfer of foreign GH gene into the target fish species; its integration into fish genome; and, its subsequent transfer genetically into future generations; thereby, demonstrating expression of foreign genes in transgenic fish. Expression of the foreign GH gene may result in enhancement of growth rates of both P and F_1 generations.

Cost Effectiveness of GH Factor Introduction

The following points are noteworthy in favour of GH factor introduction:

1. High cost in producing large-scale purified biosynthetic GH.
2. Treatment of the individual fish with the hormone is labour-intensive
3. GH-uptake into fish from an exogenous source is inefficient.
4. Transgenic fish strains, once generated, would have their own means of producing and delivering the hormone; and,

they could transmit their enhanced growth characteristics to their offspring.

Fish Growth Characteristics for Improved Aquaculture

The transgenic fishes are said to show characteristics for improved aquaculture:

1. Better growth rate and early maturation.
2. Enhanced somatic growth rate providing larger body size for better market value.
3. Possess improved feed conversion efficiencies.

Future Prospect

Transgenic fish technology has a great potential in revolutionizing the aquaculture industry. Improved transgenic fish strains could be produced for aquaculture by introducing desirable genetic traits into fish. The traits may include higher growth rate, improved food conversion efficiency, resistance to diseases, tolerance to low oxygen concentration, etc.

Future Possible Thrust Area in Fish Gene Transfer Technology

The transgenic technology offers a direct and selective method for introducing a well-defined DNA encoding a gene product for a specific biological function into the genome of the recipient. However, the following aspects could be taken care of with regard to application of transgenic fish technology in aquaculture:

1. Proper identification of genes for desirable traits to develop aquaculture.
2. Development of targeted gene-transfer technologies
3. Proper identification of suitable promoters to direct the expression of transgenes at optimal levels during the desired developmental stages.
4. Assessment of safety of the transgenic fish against environmental impacts.

Biotechnological techniques could play a profound role in health management in aquaculture through the application of hybridomas, monoclonal antibodies, vaccines, etc.

References

ADB/NACA (19910). Fish Health Management system in Asia-Pacific. Report on a Regional Study and Workshop on Fish Disease and Fish Health Management. ADB Agriculture Department Report Series No.1. Network of Aquaculture Centres in Asia-Pacific (Bangkok).

Agarwala, H.N. (1994). Endangered Sport Fishes of Assam: pp. 209-212. In: Threatened Fishes of India, pp.384, Natcon Publication No. 4 (UP).

Agersborg, H.P.K. (1933). Salient Problems in the artificial rearing of salmonid fishes, with special reference to intestinal fungisitosis and the cause of White Spot Disease. Trans. Am. Fish. Soc., 63: 240-250.

Ahmad, N. (1954). Hilsa fishery of East Bengal. J. Asiat. Soc. (Sci.), 20: 55-58.

Ahmed, M. and Rab, M.A. (1995). Factors affecting outbreak of Epizootic Ulcerative Syndrome in farmed fish in Bangladesh. Journal of Fish Diseases, 18: 263-271.

Ahne, W.; Joergensen, P.E.V.; Olsen, N.H.; and, Wattaravijarn, W. (1988). Serological Examination of a Rhabdovirus isolated from snakehead (*Ophicephalus striatus*) in Thailand with ulcerative syndrome. J. Applied Incthyology, 4: 194-196.

Alikunhi, K.H. (1956). Fish Culture techniques in India. Proc. Fish. Development in India: 63-73.

Alikunhi, K.H. (1957). Fish Culture in India. F_m Bull.: Indian Coun.Agric. Res., 20: 144 p.

Alikunhi, K.H.; Dutta, A.; Singh, V.D.; and Dubey, G.P. (1964). Observations on the breeding of carps in Bundhs, near Nowgong, Madhya Pradesh, during July-August, 1964. Bull. Cent. Inst. Fish. Educ.(Bombay).

Amend, D.F.; Yasuteake, W.T.; and, Mead, R.W. (1969). A haemopoietic virus disease of rainbow trout and sockeye salmon. Trans. Am. Fish. Soc., 98: 796-804.

Amend, D.F. and Chambers, V.C. (1970). Morphology of certain viruses of salmonid fishes, I. *In vitro* studies of some viruses causing haematopoietic necrosis. J. Fish. Res. Bd. Can., 27: 1285-1295.

Angermeier, P.L.; and Karr, J.R. (1983). Fish Communities along environmental gradients in a system of tropical streams. Environmental Biology of Fishes, 9(2): 117-135.

Angeermeier, P.L. and Winston, M.R. (1998). Local vs Regional influences on local diversity in stream fish communities of Virginia. Ecology, 79 (3): 911-927.

Annandale, N. (1918). Fauna of the Inle Lak. Rec.Indian Mus., 14 (1 and 2): 4-28.

Anon (1964). Report on the Fish spawn prospecting investigations. I. Uttar Pradesh and Gujarat. Bull. Cent. Inl.Fish. Res. Inst., Barrackpore, 4: 1-191.

Anon (2000). Artificial propagation of hilsa, *Tenualosa ilisha* through hatchery system: CIFRI's efforts. Fishing Chimes, 19 (10-11): 79-81.

APHA (1995). Standard Methods for the Examination of Water and Wastewater, xxxix + 1193, American Public Health Association (USA).

Armontrout, N.B. (1990). Aquatic Inventory: pp. 32. Bureau of Land Management, Eugene District (USA).

Arnold, J.(1887). Ueber Theilungsvargänge an den Wanderzellen, ihre progressiven und retrogressiven Metamorphosen. Arch. Mikr. Anat., 30: 205.

Arunachalam, M.; Madhusudan Nair, K.C.; Vijberg, J.; and, Kortmulder, K. (1988). Food and Habitat usage of cyprinid fish assemblage in stream pools of a South Indian river. Rep. 19: pp. 89. Limnological Institute of Royal Academy of Sciences (The Netherlands).

Balasuriya, L.K.S.W. (1994). Epizootic Ulcerative Syndrome in fish in Sri Lanka: Country Status Report. In: Roberts, R.J.; Campbell, B.; and, Mac Raem I.H. (Eds.) Proceedings of the ODA Regional Seminar on Epizootic Ulcerative Syndrome, January, 1994, pp.39-47. Aquatic Animal Health Research Institute (Bangkok).

Banarescu, P and Nalbant, T.T. (1982). New data about the Malayan Indo-Chinese affinities of the aquatic fauna of the Western Ghat, South India. Rev. Roum. Biol. Anim., 27: 23-27.

Banerjea, S.M. (1967). Water quality and soil condition of the fish ponds in some States of India in relation to fish production. Indian J. Fish, 14 (1-2):115-144.

Barua, G. (1990). Bangladesh Report: pp. 2-4. In: Regional Research Programme on Relationships between Epizootic Ulcerative Syndrome in Fish and the Environment, 13-26 August, 1990, NACA (Bangkok).

Barua, G (1994). The Status of Epizootic Ulcerative Syndrome of Fish of Bangladesh: pp.13-20. In: Proceedings of the ODA Regional Seminar on Epizootic Ulcerative Syndrome, 25-27 January, 1994 (Eds.) Roberts, R.J.; Campbell, B.; and, Mac Rae, I.H., Aquatic Animal Health Research Institute (Bangkok).

Basu, S.P. (1950). Myxosporidian parasites of Indian carps, *Catla catla (Ham.), Labeo rohita (Ham.), Cirrhina mrigal (Ham.)* and their seasonal variations. Proc. Indian Sci. Congr., 37: 237 p.

Battish, S.K. (1992). Freshwater Zooplankton of India, vi + 233, Oxford and IBH Publishing Co., Ltd. (New Delhi).

Beckman, W.C. (1948). The Length-Weight relationship factors for correlation between standard and total lengths and co-efficient of conditions for seven Michigan fishes. Trans. Am. Fish. Soc. 75: 237-256.

Benke, A.C.; Van Ardsall, T.C. (Jr.); Gillespie, D.M.; and, Parish, F.K. (1984). Invertebrate productivity in a sub-tropical backwater river: The importance of habitat and life history. Ecol. Monogr., 54: 25-63.

Bennet, G.W. (1962). Management of artificial lakes and ponds, pp.1-283, Reinhold (New York).

Bensam,P. (1987). On two post-larval stages of the shad, *Hilsa kelee* (Cuvier). Indian J. Fish., 34: 105-108.

Bentham, G. and Hooker, J.D. (1862). Genera Plantarum, Vol. I-II, London.

Bhanot, K.K. (1973). Observations on the spawning of *Hilsa ilisha* (Hamilton) in the Hooghly estuary. J. Inland Fish. Soc. India, 5: 162-170.

Bhat, A. (2002). A study of the diversity and ecology of the freshwater fishes of four river systems of Uttar Kannada district, Karnataka, India, xiv + 195, PhD thesis, Indian Institute of Science (Bangalore).

Bhattacharyya, S.; Dasgupta, S.; Basu, D.; and, Ravi Kumar, L. (2003). Hormonal Regulation of Fish oocyte maturation, pp. 69-72. In: Welfare Biology in the New Millennium (Eds) Kar, D.; Dey, S.C.; and, Datta, N.C., xx + 97, Allied Publishers Pvt. Ltd. (Bangalore).

Bhaumik, U.; Pandit, P.K.; and, Chatterjee, J.G. (1990). Impact of Epizootic Ulcerative Syndrome on the Society. National Workshop on Ulcerative Disease Syndrome in Fish, 6-7 March, 1990 (Calcutta).

Biswas, K. and Calder, C.C. (1955). Handbook of Common Water and Marsh Plants of India and Burma (1936), pp. 216, Health Bull., 24. Malaria Bureau of 11, Govt. of India Press (Calcutta).

Bhowmick, R.M. (1976). Techniques of Induced Breeding of Indian and Exotic carps, hatchery techniques and operational cost involved, pp. 11, Fish Farmers' Development Agency Officials' Training, Central Inland Fisheries Research Institute, Barrackpore (India).

Biswas K.P. and Saha, G.N. (1972). A new observation on the large-scale disintegration of eggs of Indian major carps during development. Fish Technol., 9 (1): 383-384.

Biswas, K and Calder, C.C. (1984). Handbook of Common Water and Marsh Plants of India and Burma, pp. 216, Bishen Singh Mahendra Pal Singh, Dehra Dun (India).

Blandford, W.T. (1901). The Distribution of Vertebrate animals in India, Ceylon and Burma. Proc. Roy. Soc., London (b), 194: 335-436.

Boerma, G.H. (1970). Additional notes on *Phoma herbarum*. Persoonia, 5: 15-48.

Bondad-Reantaso, M.G.; Lumanlan-Mayo, S.C.; Natividad, J.M., and, Phillips, M.J. (1990). Environmental monitoring of the Epizootic Ulcerative Syndrome (EUS) in fish from Munoz, Nurva Ecija, The Philippines. In: Symposium on Diseases in Asian Aquaculture, 26-29 Nov 1990, Bali, Indonesia, Asian Fisheries Soc., Manila (The Philippines).

Bondad-Reantaso, M.G.; Paclibare, J,O; Lumanlan-Mayo, S.C.; and, Catap, E.S. (1994). EUS outbreak in the Philippines: A Country Report: pp. 61-87. In: Proceedings of the ODA Regional Seminar on Epizootic Ulcerative Syndrome, 25-27 January, 194, (Eds) Roberts, R.J.; Campbell, B.; and, Mac Rae, I.H., Aquatic Animal Health Research Institute (Bangkok).

Boonyaratpalin, S.; Supamataya, K.; Duangsawad, M.; Reungprach, H.; and, Tayapachara, N. (1983). Relationship between pesticide residues and disease: pp. 226-252. Practical Report of the Ulcerative Fish Disease Committee, 1982-83 (Bangkok).

Boonyaratpalin, S. (1985). Fish Disease Outbreak in Burma. FAO Special Report, TCP/BUR/4402: 7 pp.

Bootsma, R. (1971). Hydrocephalus and Red Disease in pike-fry (*Esox lucius* L.). J. Fish Biology, 3: 417-419.

Bootsma, R..; dekinkelin, P.; and, Le Perre, M. (1975). Transmission Experiments with pike fry (*Esox lucius* L) Rhabdovirus. J. Fish Biol., 7: 269-276.

Brandt, A.V. (1980). Fish Catching Methods of the World, pp. 418, Fishing News Books Ltd. (UK).

Bray, R.H. and Kurtz., L.T. (1945). Determination of Total Organic and available forms of Phosphate in soil. Soil Sci., 59: 39-45.

Brett, J.R. (1958). Implications and Assessments of environmental stress in the investigation of fish power problems. H.R. MacMillan Lectures on Fisheries, University of British Columbia.

Cairns, J.; McCormick, P.V; and, Niederlehner, B.R. (1993). A proposed framework for developing indicators of ecosystem health. Hydrobiologia, 263: 1-44.

Callinan, R.B.; Fraser, G.C.; and, Virgona, J.L. (1989). Pathology of Red Spot Disease in Sea Mullet, *Mugil cephalus* L. from Eastern Australia. Journal of Fish Diseases, 12: 467-479.

Callinan, R.B. (1994 a). A Comparative Review of *Aphanomyces* sp. associated with Epizootic Ulcerative Syndrome, Red Spot Disease and Mycotic Granulomatosis: pp.248-252. In: Proc. ODA Regional Seminar on Epizootic Ulcerative Syndrome,25-27 Jan, 1994 (Eds.) Roberts, R.J.; Campbell, B.; and MacRae, I.H. Aquatic Animal Health Research Institute (Bangkok).

Callinan, R.B. (1994 b). Red Spot Disease-Epizootic Ulcerative Syndrome in Australia: pp: 82-88. In: Proc. ODA Regional Seminar on Epizootic Ulcerative Syndrome, 25-27 Jan., 1994 (Eds.) Roberts,R.J.; Campbell, B.; and, MacRae, I.H. Aquatic Animal Health Research Institute (Bangkok).

Callinan, R.B.; Paclibare, J.O.; Bondad-Reantaso, M.G.; Chin, J.C.; and, Gogolewski, R.P. (1995). *Aphanomyces* sp. associated with Epizootic Ulcerative Syndrome (EUS) in the Philippines and Red Spot Disease (RSD) in Australia: Preliminary Comparative Studies: pp.233-238. In: Symposium on Diseases in Asian Aquaculture, 26-29 Nov 1990, Bali, Indonesia, Asian Fisheries Soc., Manila (The Philippines).

Callinan, R.B.; Sammut, J.; and, Fraser, G.C. (1996). Epizootic Ulcerative Syndrome (Red Spot Disease) in estuarine fish: Confirmation that exposure to acid sulphate soil runoff and an invasive fungus, *Aphanomyces* sp., are causative factors. Proceedings of the National Conference on Acid Sulphate Soils. Robert J. Smith and Associates and ASSMAC, Australia.

Callinan, R.B. (1997). Pathogenesis of Red Spot Disease (Epizootic Ulcerative Syndrome) in Estuarine Fish in Eastern Australia and the Philippines, pp. 232, University of Queensland (Australia).

CAMP (1998). Report of the Workshop on 'Conservation, Assessment and Management Plan (CAMP) for Freshwater Fishes of India'. Zoo Outreach Organization and NBFGR, pp. 156, 22-26 Sep., 1997 (Lucknow).

Carmichael, J.W. (1966). Cerebral mycetoma of trout due to a Philophora-like fungus, *Sabouraudia*, 6: 120-123.

Cartwright, G.A.; Chen, D.; Henna, P.J.; Gudkovs, N.; and, Tajima, K. (1994). Immunodiagnosis of virulent strains of *Aeromonas hydrophila* associated with Epizootic Ulcerative Syndrome (EUS) using monoclonal antibody. Journal of Fish Diseases, 17: 123-133.

Casperi, W.A. (1910). Deposits of the Scottish Freshwater Lochs. In: Murray and Pullar's Bathymetrical Survey of the Scottish Lochs, 1: 261-274.

Caullery, M. and Mesnil, F. (1905). *Sur les Haplosporidies parasites de poisons marins.* C.r. Se'anc. Soc. Biol., 58: 610-612.

Cavalier-Smith, T. (1987). The origin of fungi and pseudo-fungi: pp. 339-353. In: Evolutionary Biology of the Fungi (Eds.) Rayner, A.D.M.; Brasier, C.M.; and, Moore, D., Cambridge University Press (UK).

Chacko, P.I. and Ganapati, S.V. (1949). On the Bionomics of *Hilsa ilisha* (Ham.). Proc. 38th Indian Sci. Congr., Poona, 1951, Pt.III, Abstracts: p. 227.

Chacko, P.I. (1954). Past, Present and Future of the hilsa fisheries in Madras State. J. Asiat. Soc. (Sci.), 20: 55-58.

Chacko, P.I.; Kurien, C.E.; and, Thyagarajan, S. (1954). A Survey of the Fisheries of Cauvery River. Contr. Freshw. Fish Biol. Stn., Madras, 12: 1-19.

Chakraborty, M.M. (1939). Studies on Myxosporidia from the Fishes of Bengal with a note on the Myxosporidian infection in Aquaria fishes. Arch. Protistenk, 92: 169-178.

Chakraborty, R.D.; Sen, P.R.; Chatterjee, D.K.; and, Kotwal, G.V. (1973). Observations on the relative usefulness of different feed for carp spawn and fry. J. Inland Fish. Soc.India, 5:182-188.

Chakravary, A.N. and Dastidar, S.G. (1991). Repeated isolation of chemoautotrophic nocardioform bacteria from fish suffering from Epizootic Ulcerative Syndrome. Indian J. Experimental Biology, 29: 623-627.

Charpy, B. (1965). *La Croissabce des Tilapia et la production de gross poisons.* 2nd Symposium on African Hydrobiol., and Inland Fisheries, Brazzaville.

Chattopadhyay, U.K.; Pal, D.; Das, M.S.; Das, S.; and, Pal, R.N. (1990). Microbiological Investigations into Epizootic Ulcerative Syndrome (EUS) in Fishes. The National Workshop on Ulcerative Disease Syndrome in Fish, 6-7 March, 1990 (Calcutta).

Chaudhuri, B.L. (1916). Fauna of the Chilka Lake: Fish, Part I. Mem. Indian Mus., 5 (4):403-439.

Chaudhuri, H. and Alikunhi, K.H. (1957). Observations on the spawning in Indian carps by hormone injection. Curr. Sci., 26: 381-382.

Chaudhuri, H. (1960). Experiments on induced spawning of Indian carps with pituitary injections. Indian J. Fish, 7: 20-48.

Chaudhuri, H.; Singh, S.B.; Sukumaran, K.K. (1966). Experiments on large-scale production of fish seed of the Chinese grass carp, *Ctenopharyngodon idellus* (C and V) and the silver carp, *Hypophthalmichthys molitrix* (C and V) by induced breeding in ponds in India. Proc. Indian Acad. Sci., (B), 63 (2): 80-95.

Chaudhuri, H. (1967). Breeding and selection of Cultivated Warmwater Fishes in Asia and the Far East. FAO Fish Report, 44 (4): 30-36.

Chaudhuri, H. (1968). Breeding and Selection of Cultivated warm water Fishes in Asia and the Far East: A Review. FAO Fish Report, 4 (44): 30-66.

Chaudhuri, S. and Singh, S.B. (1984). Induced Breeding of Carps, pp. 82, Indian Council of Agricultural Research (New Delhi).

Chauhan, B.S. (1953).Studies on the Trematode fauna of India: Parts 1-4.. Rec. Indian Mus., 51 (2): 113-393.

Chauhan, B.S. and Ramakrishna, G. (1958). On the occurrence of fish mortality due to helminthic infestation by cestode cysts in a stocking tank at Nagpur (India). Indian J. Helminth., 10 (1):53-55.

Chinabut, S. (1990). The Histopathology of EUS in Asia: pp.75. In: Proc. Regional Research Programme on Relationship between Epizootic Ulcerative Syndrome in Fish and the Environment: A Report on the 2nd Technical Workshop, 13-26 Aug, 1990, (Eds.) Phillips, M.J. and Keddie H.G.: Network of Aquaculture Centres in Asia-Pacific (Bangkok).

Chinabut, S.; Roberts, R.J.; Willoughby, L.G.; and, Pearson, M.D. (1995). Histopathology of snakehead, *Channa striatus* (Bloch), experimentally infected with the specific *Aphanomyces* associated with Epizootic Ulcerative Syndrome (EUS) at different temperatures. Journal of Fish Diseases, 18: 41-47.

Chondar, S.L. and Rao, P.S. (1996). Epizootic Ulcerative Syndrome disease to fish and its control: A Review. World Aquaculture, 1996, Queen Sirikrit National Convention Centre (Bangkok).

Chulalongkorn University (1987). Proc. Of the 2nd. Seminar on Living Aquatic Resources, 17-18 Dec., 1987, Chulalongkorn University (Bangkok).

Chumkasein, P. (1983). *Aeromonas hydrophila* in man with special reference to diarrhoeal cases during the freshwater fish epidemic: 1982-83, pp. 547-549. In: The Symposium on Freshwater Fishes Epidemic, 1982-83, 23-24 June, 1983, Chulalongkorn University (Bangkok).

Chumley, J. (1910). Bibliography of Limnological literature: pp. 659-673. In: Murray and Pullar's Bathymatrical Survey of the Scottish Lochs, 1 (List of Publications by Forel, pp. 685-689).

Coates, D.; Nunn, M.J.; and, Uwate, K.R. (1989). Epizootic Ulcerative Disease of Freshwater Fish in Papua New Guinea, Science in New Guinea, 15: 1-11.

Cook, R.E. (1977). Raymond Lindeman and the Trophic dynamic concept in Ecology. Science, 198: 22-26.

Cooper, B.L. (1980). Fisheries Management in Streams: pp. 297-322. In: Fisheries Management (Eds.) Lackey, R.T. and Nielsen L.A., Blackwell Scientific (Boston).

Cordington, K. DeB (1939). Notes on Indian Mahseer. Journal of Bombay Natural History Society, 46: 336-344.

Costa, H.H. and Wijeyratne, M.J.S. (1989). Epidemiology of Epizootic Ulcerative Syndrome occurring for the first time among fishes in Sri Lanka. Journal of Applied Ichthyology, 1: 48-52.

Coxton, F.E. and Cowden, D.J. (1950). Applied General Statistics, pp. 944, Sir Isaac Pitman and Sons (London).

Cruickshank, R. (1972). Medical Microbiology, pp.1070, Churchill Livingstone and ELBS (England).

Cruz-Lacierda E.R.and Shariff, M. (1995). Experimental transmission of Epizootic Ulcerative Syndrome (EUS) in snakehead (*Ophicephalus striatus*): pp. 327-336. In: Diseases in Asian Aquaculture, II (Eds.) Shariff,M.; Arthur, J.R.; and, Subasinghe, R.P., Fish Health Section, Asian Fisheries Society (Manila).

Darlington, P.J. (Jr.) (1957). Zoogeography: The Geographical Distribution of Animals, pp. 675, John Wiley and Sons (New York).

Das, M.K. and Das, R.K. (1993). A Review of the Fish Disease, Epizootic Ulcerative Syndrome in India. Environment and Ecology, 11 (1): 134-145.

Das, M.K.; Pal, R.N.; Ghosh, A.K.; Das, R.K.; Joshi, H.C.; Mukhopadhyaya, M.K.; and Hajra, A. (1990). Epizootic Ulcerative Syndrome: A comparative account. Proc. National Workshop on Ulcerative Disease Syndrome in Fish, 6-7 March, 1990 (Calcutta).

Datta Munshi, J.S. (1960). The Structure of the gills of certain freshwater teleosts. Indian J. Zoology, 1: 1-40.

Datta Munshi, J.S. and Srivastava, M.P. (1998). Natural History of Fishes and Systematics of Freshwater Fishes of India, pp. xviii + 403, Narendra Publishing House (Delhi).

Day, F. (1873). Report on the Freshwater Fish and Fisheries of India and Burma, pp.22, 23, 35, 36 (Kolkata).

Day, F. (1878). The Fishes of India, being a Natural History of the Fishes known to inhabit the Seas and Freshwaters of India, Burma and Ceylon, xx + 778, 195 pls, Text and Atlas in 4 parts, W_M Dawson and Sons Ltd. (London).

Day, F. (1885). Relationship of the Indian and African Freshwater Fish Fauna. J. Linn. Soc. (Zool.), 18: 308-317.

Day, F. (1889). The Fauna of British India, including Ceylon and Burma: Fishes, Vols. I and II: pp. 548 and 509.

De, D.K. and Datta, N.C. (1990). Studies on certain aspects of morphophysiology of Indian shad, hilsa, *Tenualosa ilisha* (Hamilton) in relation to food and feeding habits. Indian J. Fish., 37:189-198.

De, D.K. (1991). Hilsa fisheries in the context of Farakka barrage. In: Course material on Open water Fishery Resource Assessment and Management for FAO Trainees held at CICFRI, Barrackpore, 6 Nov–2 Dec, 1991: 1-8 pp.

De, D.K.; Sinha, M.; and, Ghosh, A. (1994). Impact of Farakka barrage on the spawning of *Tenualosa ilisha* in the Hooghly estuary. J. Inland Fish. Soc. India, 26: 121-124.

De, D.K.(1996). Present status of hilsa fisheries in India and its strategies for development. Sem on Fisheries Development and its impact on Environment, Department of Zoology, Ashutosh College and The Zoological Society, Calcutta, 28 Nov 1996.

De, D.K. and Sinha, M. (1997). Studies on the immigration of hilsa, *Tenualosa ilisha* by tagging experiments. National Seminar on Changing Perspectives in Inland Fisheries. Inland Fish. Soc. India and Central Inland Capture Fisheries Research Institute, 16-17 March, 1997 (Barrackpore).

De, D.K.; Sinha, M.; and, Banerjee, K.S. (1998). Floy, T-Bar, Anchor tags for tagging hilsa, *Tenualosa ilisha* (Hamilton). Indian J. Fish., 45: 99-101.

De, D.K.(1999). Hydrological changes, Fish population trends and production potential of estuarine ecosystem with special emphasis on hilsa. In: Open water Fisheries Technologies and Extension Methods. Bull. Cent. Inland Capture Fisheries Research Inst., Barrackpore, 96: 33-39.

De, D.K.; Moitra, P.M.; Mandal, N.C.; Sen, H.K. (1999). Status of Fisheries in the Freshwater Stretch of Hooghly estuary, West Bengal and the required Conservation. J. Inland Fish. Soc. India and Central Inland Capture Fisheries Research Inst., Barrackpore, 22-23 Dec., 1999: p. 12.

Dekinklein, P.; Bootsma, R.; and Galimard, B. (1973). Isolation and identification of the causative agent of 'Red Disease of Pike' (*Esox lucius* L.). Nature, 241: 465-467.

Desai, V.R. (1994). Endangered, Vulnerable and Rare Fishes of River Mahanadi, pp.63-68. In: Threatened Fishes of India, pp.384, Natcon Publication No.4 (UP).

Dey, S.C. (1973). Studies on the Distribution and Taxonomy of the Ichthyofauna of the hill streams of Kamrup-Khasi-Garo Regions of Assam with special reference to the Functional morphology of some rheophillic Fishes, D.Sc. Thesis, xi + 299, University of Calcutta (India).

Dey, N.C. and Dey, T.K. (1980). Practical Pathology and Microbiology, pp. 188, Allied Agency (Calcutta).

Dey, S.C. (1981). Studies on the hydrobiological conditions of some commercially important lakes (Beels) of Kamrup district of Assam and their bearing on Fish production, pp. 177, 18 Figs., 8 pls., Final Technical Report, North-Eastern Council, Govt. of India (Shillong).

Dey, S.C. and Kar, D. (1987). Physico-chemical Complexes of Water and Soil in Sone, an Ichthyologically potential tectonic lake of Assam. J. Assam Sci. Soc., 30 (1): 1-11.

Dey, S.C. and Kar, D. (1989 a). Aquatic Macrophytes of Lake Sone in Assam. Environment and Ecology, 7 (1): 253-254.

Dey, S.C. and Kar, D. (1989 b). An account of *Hilsa ilisha* (Hamilton) of Lake Sone in the Karimganj district of Assam. Bangladesh J. Zoology, 17 (1): 69-73.

Dey, S.C. and Kar, D. (1989 c). Fishermen of Lake Sone in Assam: Their socio-economic status. Science and Culture, 55: 395-398.

Dey, S.C. and Kar, D. (1990). Fish yield trend in Sone, a tectonic lake of Assam. Matsya, 15-16: 39-43.

Dey, S.C. and Kar, D. (1994). Phytoplankton dynamics of Lake Sone in Assam. Bull. Life Sciences, IV: 49-54.

DFID (1998). A Report on the 2nd Mission to investigate Epizootic Ulcerative Syndrome (EUS) in Pakistan, 19-30 April, 1998.

Dhar, N. (2004). Certain aspects of Limnology and Fishery of Baskandi Lake in the Cachar district of Assam, pp. 165, PhD Thesis, University of Gauhati (Assam).

Dhar, N.; Goswami, M.M.; and, Kar, D. (2004 a). Aquatic Macrophytes in Baskandi Lake, an Oxbow Lake of Barak Basin, Assam. Environment and Ecology, 22 (Spl.4):606-608.

Dhar, N.; Goswami, M.M.; and, Kar, D. (2004 b). Cooperative Movement: Need for the hour in Capture Fishery sector: pp. 113-116. In: Cooperative Movement in Assam: Problems and Prospects (Eds.) Goswami, P.J. and Acharjee, P.R., pp. xiv + 124,

G.C. College Employees' Credit and Thrift Cooperative Society Ltd. (Silchar).

Dhu, S. (1923). The anglers in India on the mighty Mahseer, pp.180, Government Press (Allahabad).

Dickman, S.R. and Bray, R.H. (1940). Colorimetric determination of Phosphate. Indus and Engg. Chem. (Anal.), 12: 665-668.

Doha, S. and Hye, M.A. (1971). Fecundity of Padma river hilsa, *Hilsa ilisha* (Hamilton). Pakistan J. Science, 22: 176-184.

Doty, M.S. and Slater, D.W. (1946). A new species of *Heterosporium* pathogenic on young Chinook salmon. Am. Midl. Nat., 36: 663-665.

Dubey, G.P. (1994). Endangered, Vulnerable and Rare Fishes of West Coast River System of India: pp. 77-95. In: Threatened Fishes of India, pp. 384, Natcon. Publication No. 4 (UP).

Dubey, G.P. and Tuli, R.P. (1961). Observations on the breeding of the major carps in Madhya Pradesh. J. Bombay Nat. Hist. Soc., 58 (1): 81-91.

Dudgeon, D. (1995). The Ecology of rivers and streams in Tropical Asia: pp. 615-657. In: Ecosystems of the World 22: River and Stream Ecosystems (Eds.) Cushing, C.E.; Cummins, K.W.; and, Minshall, G.W., Elsevier Press (UK).

Dunn, I.C. (1982). Hilsa resources Development and Management in Bangladesh. FI: DP/BGD/72/016: Field Document: FAO Rome: 47 pp

Dutt, S. (1966) The Indian shad, *Hilsa ilisha* (Hamilton) in the sea. Curr.Sci., 35:339-330.

Eagle, H. (1955). The specific amino acid requirements of mammalian cells (Strain L) in tissue culture. J. Biol. Chem., 214: 839.

Egusa, S. and Masuda,N. (1971). A new fungal disease of *Plecoglossus altivelis*. Fish Pathology, 6: 41-46.

Exley, C and Phillips, M.J. (1988). Acid rain: implications for the farming of salmonids. In: Recent Advances in Aquaculture, vol. 3 (Eds.) Muir, JF and Roberts, R.J., Croom Helm (London).

FAO (1962). *Rapport au Government de la Re' publique du Niger sur la situation et' evolution de la pe^che au Niger: Base' sur le travail de M. Jaques M.A. Daget', Rapp.FAO/PEAT,1525*: 27 pp.

FAO (1963). Modern Fishing Gears of the World. Proc. 2nd. World Fishing Gear Congress (London).

FAO (1974). Design of Fisheries Statistical Surveys: Inland Waters. Food and Agricultural Organisation. Fisheries Tech. Pap., 133: 1-122.

FAO (1986). Report of the Expert Consultation on Ulcerative Fish Diseases in the Asia-Pacific Region (TCP/RAS/4508), Bangkok, 5-9 Aug 1986. FAO Regional office for Asia and the Pacific (Bangkok).

Fernando, C.H. and Furtado, J.J. (1975). Reservoir Fishery Resources of South-East Asia. Bull. Fish. Res. Stn. Sri Lanka, 26 (1-2): 83-95.

Fijan, N.N. (1968). The Survival of *Chondrococcus columnaris* in the waters of different quality. Bull. off. Int.Epizoot., 69: 1159-1166.

Fijan, N.N. (1969). Systematic mycosis in channel cat fish. Bull. Wildl. Dis. Ass., 5: 109-110.

Fijan, N.N. (1972).Infectious dropsy in carps: a disease complex: pp: 39-51 In: Diseases of Fish. Proc. Symp. No. 30, Zoological Society, London, May, 1971, (Ed.) Mawdesley-Thomas, L.E., Academic Press and The Zoological Society (London).

Fijan, N.N.; Petrinec, Z; Sulimsnovic, D.; and, Zwillenberg, L.O. (1971). Isolation of viral causative agent from the acute form of infectious dropsy in carp. Vet. Arch., 41: 125-135.

Finlow, R.S.(1920). Annual Report of the Department of Fisheries, Bengal, Bihar and Orissa for the year ending 31 March, 1921: p.3.

Forbes, S.A. (1893). A Preliminary Report on the Aquatic invertebrate fauna of the Yellowstone National Park, Wyoming; and, of the Flathead Region of Montana. Bull. U.S. Fish Commission for 1891: pp. 207-258.

Forbes, E.B. (1897). A contribution to a knowledge of North American freshwater Cyclopidae. Bull. Illinois State Lab. Nat. Hist., 15: 27-83.

Forel, F.A. (1892-1904). *Le Le'man, Monographie limnologique, Lausanne, 3 vols.*

Forel, F.A. (1895). *Le Le'man. Monographie limnologique, vol. II. Lausanne, pp. 651.*

Forel, F.A. (1901). *Handbuch der Seenkunde. Allgemeine. Limnologie, Stuttgart, pp. 249.*

Forman, R.T. and Godran, M (1986). Landscape Ecology, John Wiley and Sons (New York).

Fraser, G.C.; Callinan, R.B.; and, Calder, L.M. (1992). *Aphanomyces* species associated with Red Spot Disease:an ulcerative disease of estuarine fish of Eastern Australia. Journal of Fish Diseases, 15: 173-181.

Frerichs, G.N. (1990). Efficacy of chemical disinfectants against snakehead Rhabdovirus. J. Appl. Ichthyol., 6: 117-123.

Frerichs, G.N.; Millar, S.D.; and, Roberts, R.J. (1986). Ulcerative Disease Rhabdovirus in Fish in South-East Asia. Nature, 322: 216 p.

Frerichs, G.N.; Hill, B.J.; and, Way, K. (1989). Ulcerative Disease Rhabdovirus: cell line susceptibility and serological comparison with other Rhabdoviruses. Journal of Fish Diseases, 12: 51-56.

Fritsch, F.E. (1965). The structure and reproduction of Algae, vol. II: Phaeophyceae and Myxophyceae. Cambridge University Press (UK).

Gaarder, T. and Gran, H.H. (1927). Investigations of the production of plankton in the Oslo Fjord. Cons. Perm. Int. P.Expl. Mer. Rapp. Proc. Verb. Reun., 42: 1-48.

Gadgil, M. and Reddy, B.M. (1989). Fisherfolk of Karnataka Coast: Ecological overview: Technical Report, pp. 113, Centre for Ecological Sciences, Indian Institute of Science (Bangalore).

Gadgil, M. and Kar, D. *et al.* (2000). Conserving India's Biodiversity: Let People Speak: pp.583-589. In: Setting Biodiversity Conservation, pp. 707, World Wide Fund (WWF) for Nature-India (New Delhi).

Ghittino, P. (1965). Viral Haemorrhagic Septicemia (VHS) in rainbow trout in Italy. Am. NY Acad. Sci., USA, 126: 468-478.

Ghose, A. and De, D.K. (1999). Some aspects of dietary shifting in *Tenualosa ilisha* (Hamilton) in the freshwater zone of Hooghly estuary. Proc. Conf., 22-23 Dec 1999, Inland Fish. Soc. India and Central Inland Fish. Res. Inst. (Barrackpore).

Ghosh, S.K. and Lipton, A.P. (1982). Ichthyofauna of the NEH Region with special reference to their economic importance. ICAR Research Complex, NEH Region, Shillong, Spl. Bull. 1: 119-126.

Gopalakrishnan, V. (1963). Controlling pests and Diseases of Cultured Fishes. Indian Livestock, 1 (1): 51-54.

Gopalakrishnan, V. (1964). Recent Developments in the prevention and control of parasites of fishes cultured in Indian waters. Proc. Zool. Soc. Calcutta, 17: 95-100.

Gopalakrishnan, V (1968). Diseases and Parasites of Fishes in warm-water ponds of Asia and the Far East. FAO Fish Rep., 5. (44): 319-344.

Gopalakrishnan, V. and Gupta, P.D. (1960). An eye disease which causes mortality of the *Catla catla* (Ham-Buch). Curr. Sci., 29 (6): 240 p.

Goswami, U.C.; Datta, A.; Sarma, D.K.; and, Sarma, G. (1988). Studies on certain aspects of prevention and treatment of fish suffering from Epizootic Ulcerative Syndrome, pp. 16-17. In: Proc. Symp, on Recent Outbreak of Fish Diseases in North-Eastern India, 30 Dec., 1988, Guwahati (Assam).

Govt. of Assam (1980). Statistical Handbook of Assam. Directorate of Economics and Statistics, Govt. of Assam (Guwahati).

Guizhen, M.J. (1990). China Report: pp. 11-13. In: Regional Research Programme on Relationships between Epizootic Ulcerative Syndrome in Fish and the Environment, 13-26 Aug 1990, NACA (Bangkok).

Gunther, A. (1880). An Introduction to the Study of Fishes, pp. 720, Adams and Charles Black (Edinburgh)

Gupta, S.P. (1961). A reference list of trematode parasites of freshwater fishes of India with a discussion on their systematic position. Indian J. Helminth., 13 (1): 35-60.

Gupta, P. and Khan, H.A. (1994). Mahseer Fisheries in India, pp. 181-190. In: Threatened Fishes of India, pp. 384, Natcon Publication No. 4 (UP).

Haines, A.K. (1983). Fish Fauna and Ecology: pp.367-384. In: The Purari tropical environment of high rainfall river basin (Ed)Petr, T, Dr.W.Junk Publishers (The Netherlands).

Hardy, R.W. (1990). Fish Health Management and Nutrition in the Asia-Pacific region, pp:425-434. In: Fish Health Management in Asia-Pacific. Report on a Regional Study and Workshop on Fish Disease and Fish Health Management. ADB Agricultural Department Report Series No.1. Network of Aquacultural Centres in Asia-Pacific (Bangkok).

Hamilton, B. (1822). Account of the Fishes found in the River Ganges and its tributaries, pp. 405, Edinburgh (UK).

Harris, M. and Kutsky, R.J. (1954). Synergism of nucleoprotein and dialysate growth factors in chick embryo extract. Exp. Cell Res., 6: 327

Hasan, R and Qasim, S.Z. (1962). *Trypanososma punctati* n.sp. from the *Ophicephalus punctatus* Bloch, common freshwater murrel of India. Z. Parasitenk, 22 (2): 118-122.

Hatai, K. (1980). Studies on the pathogenic agents of saprolegniansis in freshwater fishes. Special Report of Nagasaki Prefectural Inst. of Fisheries, No. 8, Matsugae-Cho, Nagasaki, Japan.

Hatai, K (1994). Mycotic Granulomatosis in ayu (*Plecoglossus altivelis*) due to *Aphanomyces piscicida.* In: Proc. ODA Regional Sem. On Epizootic Ulcerative Syndrome, 25-27 Jan1994, (Eds.) Roberts, R.J.; Campbell, B.; and, MacRae, I.H., Aquatic Animal Health Research Institute (Bangkok).

Hatai, K.; Egusa, S.; Takashashi, S.; and, Ooe, K. (1977). Study on the pathogenic fungus of Mycotic Granulomatosis: I: Isolation and pathogenicity of the fungus from cultured ayu infected with the disease. Fish Pathology, 11 (2): 129-133.

Hatai, K.; Nakamura, K,; Rha, S.A.; Yusa, K.; and, Wada, S. (1994). *Aphanmyces* infection in dwarf gourami (*Colisa lalia*). Fish Pathology, 29 (2): 95-99.

Heckel, J. (1844). *Fische Kuchemrs in von Hugle's Kachir and Reich der sienk,* 4 (2): 351-384.

Hedrick, R.P.; Eaton, W.D.; Fryer, J.L.; Groberg, W.G. (Jr.); and, Boonyaratpallin, S. (1986). Characteristics of a Birna virus isolated from cultured sand goby, *Oxyeleotris marmoratus*. Diseases of Aquatic Organisms, 1: 219-225.

Henderson, S. and Markland, H.R. (1987). Decaying Lakes: The origins and control of Cultural Eutrophication: Detailed Analysis. John Wiley and Sons (New York).

Hephar, B. (1962). Ten years of research in fish pond fertilization in Israel,I: The effect of fertilization on fish yields. Bamidegeh, 14 (2): 29-38.

Hepher, B.and Chervinski, J. (1965). Studies on carp nutrition: the influence of protein-rich diet on growth. Bamidgeh, 17 (2): 31-46.

Hickling, C.F. (1962). Fish Culture, 2nd. Ed., pp.295, Faber and Faber (London).

Hickling, C.F. (1971).Fish Culture, 2nd Ed., pp.317, Faber and Faber (London).

Hofer, B.(1904).*Handbuch der Fischrankhereiten*. Munich: *Allg. Fischerie-Zeitung*.

Hooker, J.D. (1872). Flora of British India, vols. I-VI, London.

Hora, S.L. (1921 a). Fish and Fisheries of Manipur with some observations on those of the Naga Hills. Rec. Indian Mus., 22: 165-214.

Hora, S.L. (1921 b). Indian Cyprinoid fishes belonging to the Genus *Garra* with notes on the related species from other countries. Rec. Indian Mús., 22: 633-687.

Hora, S.L. (1930). Ecology, Bionomics and Evolution of torrential fauna with special reference to the organs of attachment. Phil. Trans. Roy. Soc. London (B), 218: 171-282.

Hora, S.L. (1936). Nature of substratum as an important factor in the ecology of torrential fauna. Proc. Nat. Inst. Sci. India, 2: 45-47.

Hora, S.L. (1937). Geographical distribution of Indian Freshwater Fishes and its bearing on the probable land connections beween India and the adjacent countries. Curr. Sci., 7: 351-356.

Hora, S.L.(1938). A Preliminary note on the spawning grounds and Bionomics of the so-called Indian shad, *Hilsa ilisha* (Hamilton) in the Ganges. Rec. Indian Mus., 40: 147-158.

Hora, S.L. (1939). The Game Fishes of India, VIII: The Mahseers of the large-scaled barbs of India, I: The putitor mahseer *Barbus (Tor) putitora* (Hamilton). J. Bombay nat. Hist. Soc., 41: 272-285.

Hora, S.L. (1940). Dams and the problem of migratory fishes. Curr. Sci., 9: 406-407.

Hora, S.L. (1941). The Game Fishes of India, XIII: The Mahseers or the large-scaled barbs of India, 6: The 'jungha' of the Assamese *Barbus (Tor) progenius* McClelland. J. Bombay Nat. Hist. Soc., 42: 526-532.

Hora, S.L. (1943). The Game Fishes of India XVI: The Mahseers of the large-scaled Barbs of India, 9: Further observations on Mahseers from the Deccan. J. Bombay nat.Hist. Soc., 44: 1-8.

Hora, S.L. (1951 a). Knowledge of the ancient Hindus concerning Fish and Fisheries of India, 3: Matsyarinoda or a Chapter on angling in the Manasollasa by King Somesvara (1127 AD). J. Asiatic Soc., Letters, 17: 145-169

Hora, S.L. (1951 b). Fish Geography of India. J. Zool. Soc. India, 3 (1): 183-187.

Hora, S.L. (1951 c). Some observations on the Palaeogeography of the Garo-Rajmahal gap as evidenced by the distribution of the Malayan fauna and flora to Peninsular India. Proc. Nat. Inst.,Sci. India, 17: 437-444.

Hora, S.L. (1953). Fish Distribution and Central Asian Orography. Curr. Sci., 22 (4): 93-94.

Hora, S.L. and Law, N.C. (1941). Respiratory adaptations of the South Indian Homalopterid Fishes. J. Asiatic Soc. Bengal, 8: 39-46.

Hora, S.L.;Malik, G.M.; and, Khajuria, H. (1955). Some interesting features of the Aquatic Fauna of Kashmir valley. J. Bombay nat.Hist. Soc., 53 (1):140-143.

Hora, S.L. and Menon, A.G.K. (1952). Distribution of Indian Fishes of the past and their bearing on the geography of India.1: The extinct freshwater Dipnoan and Ganoid Fishes of India. Everyday Sci., 1: 26-37.

Hora, S.L. and Mukherjee, D.D. (1935). Fishes collected by the Third Netherland Karakoram Expedition Wiss. Eergeb. Niederl. Exped. Karakoram, 1: 426-445.

Hora, S.L. and Silas, E.G. (1952). Evolution and Distribution of glyptosternoid fishes of the Family Sisoridae (Order: Siluroidea). Proc. Nat. Inst. Sci. India, 18 (4): 309-322.

Hora, S.L. and Pillay, T.V.R. (1962). Handbook on Fish Culture in the Indo-Pacific region. FAO Fish Biol. Tech. Pap. 14: 204 pp.

Hoshima, T.; Sano, T.; and, Sunamayo, M. (1960). Studies on the Saprolegniasis of eel. J. Tokyo University Fish., 47: 59-79.

Hugueny, B.; and Paugy, D. (1995). Unsaturated fish communities in African rivers. American Naturalist, 146:162-169.

Hutchinson, G.E. (1939). Ecological observations on the fishes of Kashmir and Indian Tibet. Ecol. Monogr., 9 (2): 147-181.

Hutchinson, G.E. (1957). A Treatise on Limnology, Vol. I. Geography, Physics and Chemistry, ix + 1015, John Wiley and Sons Inc. (New York).

Hutchinson, G.E. (1967). A Treatise on Limnology, vol. II: Introduction to Lake Biology and the Limnoplankton, pp.1115, John Wiley and Sons Inc. (New York).

Huxley, T.H. (1882). *Saprolegnia* in relation to salmon disease. Q. Fl. Microsc. Sci., 22: 311.

Ibrahim, K.H. and Chaudhuri, H. (1966). Preservation of fish pituitary extract in glycerine for induced breeding of fish. Indian J. Exp. Biol., 4 (4): 249-250.

ICSF (1992). Enigma of EUS Consultation on Epizootic Ulcerative Syndrome *vis-à-vis* the Environment and the People, 25-26 May 1992 Trivandrum (Kerala). International Collective in Support of Fish-workers, 40 pp, Madras (India).

IUCN (1988). Red List of Threatened Animals, IUCN, Gland (Switzerland).

Jain, S.L. (1959). Some observations on the monogenetic trematodes from the gill filaments of some Indian Freshwater Fishes. Curr. Sci., 28 (8): 332-333.

Jackson, M.L. (1973). Soil Chemical Analysis, xiv + 498, Prentice Hall of India Pvt. Ltd. (New Delhi).

Jackson, D.A. and Harvey, H.H. (1989). Biogeographic associations in fish assemblages: Local *vs* Regional processes. Ecology, 70: 1472-1484.

Jayaram, K.C. (1981). The Freshwater Fishes of India, Pakistan, Bangladesh, Burma, Sri Lanka: a Handbook, xxii +475, Zoological Survey of India (Calcutta).

Jayaram, K.C.(1999). The Freshwater Fishes of the Indian Region, xvii +551, Narendra Publishing House (Delhi).

Jayaram, K.C. (2003). Ecostatus and Conservation Strategies for Mahseer fishes of India with special reference to Deccan species: pp. 3 12. In: Welfare Biology in the New Millennium, (Eds.) Kar, D.; Dey, S.C. and Datta, N.C., pp. xx + 97, Allied Publishers Pvt.Ltd. (Bangalore).

Jensen, M.H. (1963). Preparation of fish tissue culture for Virus research. Bull. of int. Epizoot., 59: 131-134.

Johnston, M.R.L. and Davies,A.J. (1973). A Pirhemocyton-like parasite of the blenny, *Blennius pholis* L. (Teleostei: Blenniidae) and its relationship to immuno-plasma Neumann, 1909. Int. J. Parasit., 3: 235-241.

Jolly,J. (1903). *Sur la dure'e de la vie et de la multiplication des cellules animals en dehors de I'organisme.* C.R. Soc. Biol. (Paris), 55: 1266.

Jones, S. and Sujansingani, K.H. (1951). The hilsa fishery of the Chilka Lake. J. Bombay nat. Hist. Soc., 50: 264-280.

Joshi, C.B.; Sehgal, K.H.; and, Shyam Sunder (1978). Observations on the Fishery resources of the hills of Jammu Province with special reference to Mahseer and other commercially important species. Indian J. Fish., 25 (1-2): 197-206.

Joshi, K.D. and Kapoor, D. (1994). Endangered Mahseer of Kumaon Hills: pp. 169-171. In: Threatened Fishes of India, pp. 384, Natcon Publication, 4 (UP).

Jothy, A.A. (1981). Preliminary Report on the outbreak of Wabak Kudison in Freshwater fish in paddy-growing areas in Peninsular Malaysia. Report to the Ministry of Agriculture, Malaysia, Dec., 1981: pp. 17.

Junge, F. (1885). *Der Dorfteich als Lebensgemeinchaft. Kiel* (Germany).

Kar, D. (1984). Limnology and Fisheries of Lake Sone in the Cachar district of Assam (India), viii + 201, PhD Thesis, University of Gauhati (Assam).

Kar, D. (1990). Limnology and Fisheries of Lake Sone in the Cachar district of Assam (India), Matsya, 15-16: 209-213.

Kar, D. (1996). Biodiversity Conservation Prioritisation Project (BCPP) in India. Proc. International Project Formulation Workshop of BCPP, World Wide Fund (WWF) for Nature-India, 1 (New Delhi).

Kar, D. (1998). Biodiversity Conservation Prioritisation Project (BCPP) in India. Proc. International Project Finalisation Symposium of BCPP, World Wide Fund (WWF) for Nature-India, 1: New Delhi.

Kar, D. (1999 a). Preliminary Study of Limnology and Aquatic biota of Rudra Sagar wetland in South Tripura. Proc. Regional Seminar on Biodiversity, Guwahati, Assam, 1.

Kar, D. (1999 b). Microbiological and Environmental Studies in relation to Fish Disease in India. Gordon Research Conference, Connecticut, USA.

Kar, D. (2000 a). Present status of Fish Biodiversity in South Assam and Tripura: pp.80-82. In: Fish Biodiversity of North-East India (Eds.) Ponniah, A.G. and Sarkar, U.K., NBFGR-NATP Publication No. 2: pp.228 (Lucknow).

Kar, D. (2000 b). Socio-economic Development of the Fisherwomen through Aquaculture with emphasis on Integrated Farming in the villages around Chatla Haor Wetland in Silchar, Assam. Seminar presented on DBT-sponsored Awarness Workshop on Biotechnology-based Programmes for Women and Rural Development, NEHU, Shillong, 1: pp. 13-14.

Kar, D. (2001). Species composition and distribution of Fishes in the rivers in Barak valley region of Assam and the Principal rivers in Mizoram and Tripura in relation to their Habitat parameters. Proc. National Workshop, NATP-ICAR Project Mid-term Review, 1: Central Marine Fisheries Research Institute, Cochin, 25 p.

Kar, D. (2002).Present status of Biodiversity of Fishes of Barak valley region of Assam with a note on their Management and Conservation, pp. 3-10. Proc. UGc-Sponsored State-level Seminar on Biodiversity of Assam, Session Chairman's Lecture, 30 Jan 2000 (Eds.) Bhattacharjee, M.K.; Dattachoudhury, M.; and, Mazumder,P.B., Karimganj College, Assam University (Assam).

Kar, D. (2003 a). Fishes of Barak drainage, Mizoram and Tripura: pp. 203-211. In: Environment, Pollution and Management (Eds.) Kumar, A.; Bohra, C.; and, Singh, L.K., pp. xii + 604, APH Publishing Corporation (New Delhi).

Kar, D. (2003 b). An account of the Fish Biodiversity in South Assam, Mizoram and Tripura along with a brief account of Epizootic Ulcerative Fish Disease Syndrome in freshwater fishes. UGC-sponsored Invited Lecture in Dept. of Environmental Engg., Guru Jambeswar University, Hissar (Haryana).

Kar, D. (2003 c). Peoples' Perspective on Fish Conservation in the Water bodies of South Assam, Mizoram and Tripura: pp. 325-328. In: Participatory Approach for Fish Biodiversity Conservation in North-East India (Eds.) Mahanta, P.C. and Tyagi, L.K., v + 412, National Bureau of Fish Genetic Resources (ICAR) (Lucknow).

Kar, D. (2004). A Glimpse into the Fish Bioresources of North-East India with a note on their Management, Conservation and Biotechnological Potential. Proc. (Invited Lecture) DBT-sponsored National Symposium on Biodiversity Conservation and Sustainable Utilisation of Environmental Resources: Tripura University, 10-11 Jan., 2004: 9 p.

Kar, D. (2004 a). Inventorying of Fish Biodiversity in North-East India with a note on their Conservation. Proc. National Conference on Fish and their Environment, 9-11 Feb 2004, B.A. Marathwada University, Aurangabad, 1.

Kar, D. (2004 b). Conservation and Ichthyodiversity in the River Systems of Barak drainage, Mizoram and Tripura. Proc. National Symp. Management of Aquatic Resources for Biodiversity Maintenance and Conservation, 2-4 Feb 2004, J.N. Vyas University, Jodhpur, 1.

Kar, D. (2005 a). Fish Genetic Resources and Habitat Diversity of the Barak drainage: pp. 68-76. In: Aquatic Ecosystems, Conservation, Restoration and Management (Eds.) Ramachandra, T.V.; Ahalya, N.; and, Rajsekara Murthy, C., xiii + 396, Capital Publishing Company (Bangalore).

Kar, D. (2005 b). Fish Fauna of River Barak, of Mizoram and of Tripura with a note on Conservation. J. Freshwater Biol., 16.

Kar, D. (2005 c). Fish Diversity in the Major Rivers in Southern Assam, Mizoram and Tripura: pp.679-691. Proc. 2nd International Symposium on GIS and Spatial Analyses in Fisheries and Aquatic Sciences, 2-6 Sep 2002, University of Sussex, Brighton (UK), (Eds.), Vol.2, Nishida, T.; Kailola, P.J.; and, Hollingworth, C.E. Fisheries and Aquatic GIS Research Group, Kawagoe, Saitama (Japan).

Kar, D. (2005 d). Sustainability issues of Inland Fish Biodiversity and Fisheries in Barak drainage (Assam), in Mizoram and in Tripura. Proc. International Symposium on Improved Sustainability of Fish Production systems and Appropriate Technologies for Utilisation, Cochin University of Science and Technology, School of Industrial Fish and Fisheries, 1.

Kar, D. and Dey, S.C. (1982 a). *Hilsa ilisha* (Hamilton) from Lake Sone in Cachar, Assam. Proc. Indian Sci. Congr., 69: 77 pp.

Kar, D. and Dey, S.C. (1982 b). An account of *Hilsa ilisha* (Hamilton)of Sone Beel (Cachar district, Assam, India). Proc. All-India Sem. Ichthyol., 2: 3 pp.

Kar, D. and Dey, S.C. (1986). An account of ichthyospecies of Lake Sone in Barak valley of Assam. Proc. All-India Sem. Ichthyology, 2: 3 p.

Kar, D. and Dey, S.C. (1987). An account of the Fish and Fisheries of Lake Sone in the Barak valley of Assam (India). Proc. Workshop Development of Beel Fisheries in Assam, 1: 13 pp.

Kar, D.and Dey, S.C. (1988). An account of Fish yield from Lake Sone in the Barak valley of Assam. Proc. Indian Sci. Congr., 75: 49 pp.

Kar, D. and Dey, S. (1988).Preliminary Electron Microscopic studies on Diseased fish tissues from Barak valley of Assam. Proc. Annual Conference of Electron Microscopic Society of India, 18: 88 pp.

Kar, D. and Dey, S.C. (1988 a). A critical account of the Recent Fish Disease in the Barak valley of Assam. Proc. Regional Symp. On Recent outbreak of Fish Diseases in North-East India, 1: 8 pp.

Kar, D. and Dey, S.C. (1988 b). Impact of Recent Fish Epidemics on the Fishing Communities of Cachar district of Assam. Proc. Regional Symp. On Recent outbreak of Fish Diseases in North-East India, 1: 8 pp.

Kar, D.and Dey, S.C. (1990 a). Fish Disease Syndrome: A Preliminary Study from Assam. Bangladesh J. Zoology, 18: 115-118.

Kar, D. and Dey, S.C. (1990 b). A Preliminary study of Diseased Fishes from Cachar district of Assam. Matsya, 15-16: 155-161.

Kar, D. and Dey, S.C. (1990 c). Epizootic Ulcerative Syndrome in Fishes of Assam. J. Assam Sci. Soc., 32 (2): 29-31.

Kar, D. and Dey, S.C. (1991). Gill nets in Lake Sone of Assam with their Economics and Impact on Fishery. J. Applied Zool. Res., 2 (2): 76-79.

Kar, D. and Dey, S.C. (1993). Variegated Encircling gears in Lake Sone of Assam. J. Appl.Zool. Res., 4 (2):171-175.

Kar, D. and Dey, S.C. (1993). Interrelationship and Dynamics of Fish Population of Lake Sone in Assam. Environment and Ecology, 11 (3): 718-719.

Kar, D. and Dey, S.C. (1996). Scooping Gears of Lake Sone in Assam. J. Applied Zool. Res., 7(1): 65-68.

Kar, D. and Dey, S.C. (2000). Yield and Conservation of Indian major carps of Lake Sone in Assam. Environment and Ecology, 18 (4): 1036-1038.

Kar, D. and Dey, S.C. (2002). On the occurrence of Advanced fry of *Hilsa (Tenualosa) ilisha* (Hamilton-Buchanan) in Chatla Haor Seasonal wetland of Assam. Proc. Zool. Soc. Calcutta, 55 (2): 15-19.

Kar, D.and Barbhuiya, M.H. (2000 a). Length-weight Relationship and Condition Factor in *Gudusia chapra* (Ham-Buch) and *Botia dario* (Ham-Buch) from Chatla Haor (flood plain wetland) in Cachar district of Assam. Environment and Ecology, 18 (1): 227-229.

Kar, D. and Barbhuiya,M.H. (2000 b). Ichthyodiversity of Chatla Haor: A Floodplain wetland in Barak valley Region of Assam: pp.3-6. In: Advances in Zoology and Environmental Degradation and Biodiversity, pp. 279 (Ed.) Pandey, B.N. and Singh, B.K., Daya Publishing House (New Delhi).

Kar, D. and Barbhuiya, M.H. (2001). Ecology of Aquatic Macrophytes of Chatla Haor, a floodplain wetland in Cachar district of Assam. Environment and Ecology, 19 (1): 231-233.

Kar, D. and Barbhuiya, M.H. (2002). Macrophytic diversity in certain Wetlands of Barak valley region of Assam: pp. 86-89. In: Restoration of Lakes and Wetlands (Eds.)Ramachandra, T.V.; Rajasekara Murthy, N.; and, Ahalya, N., pp. xxii + 400, Allied Publishers (P) Ltd. (Bangalore).

Kar, D. and Barbhuiya, M.H. (2004). Abundance and Diversity of Zooplankton in Chatla Haor, a floodplain wetland in Cachar district of Assam. Environment and Ecology, 22 (1): 247-248.

Kar, D. and Barbhuiya, M.H.(2005). Length-weight Relationship and Relative Condition Factor in *Hilsa ilisha* (Hamilton) of Barak drainage in Assam. Indian J. Environment and Ecoplanning, 10 (1).

Kar, D. and Mazumdar, J. (2004). Biodiversity, Disease and Conservation of Fish Bioresources: A perspective. Proc. National Workshop on Sci. and Tech. for Regional Development: Case for North-East India, 3-6 Feb 2004, Tezpur University and I.I.T. Guwahati: 34 pp.

Kar, D. and Das, M.(2004 a). Preliminary Histochemical studies with diseased fishes suffering from Epizootic Ulcerative Syndrome in Assam. Proc. National Symposium for Management of Aquatic Resources for Biodiversity Maintenance and Conservation, 1: 27 pp.

Kar, D. and Das, M. (2004 b). Preliminary Enzymological studies with diseased fishes suffering from Epizootic Ulcerative Syndrome in Fishes. Proc. National Conference on Fish and their Environment, 1: 25 pp.

Kar, D.; Dey, S.C.; Michael, R.G.; Kar, S.; and, Changkija, S. (1990 a). Studies on Fish Epidemics from Assam, India. J. Indian Fisheries Association, 20: 73-75.

Kar, D.; Bhattacharjee, S.; Kar, S.; and, Dey, S.C. (1990 b). An account of 'EUS' in Fishes of Cachar district of Assam with special emphasis on Microbiological studies. Souv. Annual Conference of Indian Association of Pathologists and Microbiologists of India (NE Chapter), 8: 9-11.

Kar, D.; Dey, S.C.; Kar, S.; Bhattacharjee, N.; and, Roy, A. (1993). Virus-like particles in Epizootic Ulcerative Syndrome of Fish. Proc. International Symp. On Virus-Cell Interaction: Cellular and Molecular Responses, 1: 34 pp.

Kar, D.; Dey, S.C.; Kar, S.; Roy, A.; Michael, R.G.; Bhattacharjee, S.; and, Changkija, S. (1994). A Candidate virus in Epizootic Ulcerative Syndrome of Fish. Proc. National Symp. of the Indian Virological Society, 1: 27 pp.

Kar, D.; Roy, A.; Dey, S.C.; Menon, A.G.K.; and, Kar, S. (1995 a). Epizootic Ulcerative Syndrome in Fishes of India. World Congress of In Vitro Biology, *In Vitro*, 31 (3): 7 pp.

Kar, D.; Kar, S.; Roy, A.; and, Dey, S.C. (1995 b). Viral Disease Syndrome in Fishes of North-East India. Proc. International Symp. of International Centre for Genetic Engg. and Biotechnology (ICGEB) and the Univ. of California at Irvine, 1: 14 pp.

Kar, D.; Dey, S.C.; and, Kar, S. (1995 c). A Viral Disease among the Fishes of North-East India. Annual Congress on Man and Environment, National Environment Sci. Acad. and National Institute of Oceanography, 10: 62 pp.

Kar, D.; Dey, S.C.; Kar, S.; Michael, R.G.; and, Gadgil, M. (1996). Ichthyoecology, Management and Conservation Fish Resources of Lake Sone in Assam (India). Tiger Paper (FAO, UN), XXIII (3): 27-32

Kar, D.; Dey, S.C.; Purkayastha, M.; and, Kar, S. (1996 a). An overview of the Impediments in Conservation of Biodiversity of Lake Sone in Assam. Proc.Seminar on Conservation of Biodiversity in Indian Aquatic Ecosystems, JawaharLal Nehru University, New Delhi,1.

Kar, D.; Dey, S.C.; Roy, A.; and, Kar, S. (1996 b). Viral Disease Sy ome in Fishes of India. Proc. International Congress of Virology, 10.

Kar, D.; Purkayastha, M.; and, Kar, S. (1996 c). Biodiversity Conservation Prioritisation Project: A case study from Sone Beel in Assam. Proc. National Workshop on Biodiversity Conservation Prioritisation Project (BCPP), World Wide Fund (WWF) for Nature-India and Centre for Ecological Sciences, Indian Institute of Science (Bangalore).

Kar, D.; Saha, D.; Laskar, R.; and, Barbhuiya, M.H. (1997). Biodiversity Conservation Prioritisation Project (BCPP) in Barak valley region of Assam. Proc. National Project Evaluation Workshop on BCPP, Betla Tiger Reserve and National Park, 1 (Palamu).

Kar, D.; Saha, D.; and, Dey, S.C. (1998 a). Epizootic Ulcerative Syndrome in Barak valley of Assam: 2-4. In: Project Report submitted and presented at the National Symposium of Biodiversity Conservation Prioritisation Project (BCPP) held at WWF-India, 18-19 Jan 1998 (New Delhi).

Kar, D.; Dey, S.C.; and, Roy, A. (1998 b). Present status of Epizootic Ulcerative Syndrome (EUS) in Southern Assam. Proc.Regional Project Initiation Workshop for NATP-ICAR-NBFGR Project, 1: 9 pp.

Kar, D.; Dey, S.C.; Kar, S.; and, Roy, A. (1998 c). An account of Epizootic Ulcerative Syndrome in Assam: pp.1-3.. In: Final Project Report of Biodiversity Conservation Prioritisation Project (BCPP). Submitted and presented at the International Project Finalisation Workshop of BCPP, 1, 22-28 April 1998 (New Delhi).

Kar, D.; Dey, S.C.; and, Roy, A. (1998 d). Fish Disease Diagnosis and Fish Quarantine problems in India and South-East Asia with particular emphasis on Epizootic Ulcerative Syndrome (EUS) fish disease problems. Proc. International Workshop on Fish Disease Diagnosis and Quarantine, Indian Council of Agricultural Research (ICAR)-Ministry of Agriculture (MoA), Govt. of India (GoI)-Network of Aquaculture Centres in Asia (NACA)-Food and Agricultural Organisation (FAO) of the United Nations (UN)-OIE, France: Held at the Central Institute of Freshwater Aquaculture (CIFA), 1.

Kar, D. and Upadhyaya. T. (1998). Histopathological Studies of Fish tissues affected by Epizootic Ulcerative Syndrome in Assam. Technical Bulletin, XIII Convention and National Symposium of Indian Association of Veterinary Anatomists (IAAM), 11-13 Dec 1998, College of Veterinary Sciences, Assam Agricultural University (Guwahati).

Kar, D.; Dey, S.C.; Kar, S.; and, Ramachandra, T.V. (1999). Trawls of Lake Sone in Assam. J. Applied Zool. Res., 10 (2): 170-172.

Kar, D.; Rahaman, H.; Barnman, N.N.; Kar, S.; Dey, S.C.; and, Ramachandra, T.V. (1999 a). Bacterial Pathogens associated with Epizootic Ulcerative Syndrome in Freshwater Fishes of India. Environment and Ecology, 17 (4): 1025-1027.

Kar, D.; Mandal, M.; and Bhattacharjee, S. (1999 b). Fungal Pathogens associated with Epizootic Ulcerative Syndrome in Fishes of

Barak valley region of Assam. Proc. 1st National Conference on Fisheries Biotechnology, CIFE, 1: 34 pp.

Kar, D.; Dey, S.C.; and Roy, A. (2000 a). Fish Genetic Resources in the Principal rivers and wetlands in North-East India with special emphasis on Barak valley (Assam), in Mizoram and in Tripura, with a note on Epizootic Ulcerative Syndrome fish disease. Proc. National Project Initiation Workshop of the NATP-ICAR World Bank- aided Project on 'Germplasm Inventory, Evaluation and Gene Banking of Freshwater Fishes', National Bureau of Fish Genetic Resources (NBFGR), Lucknow, 1: 12 pp.

Kar, D.; Mandal, M.; Laskar, B.A.; Dhar, N.; and, Barbhuiya, M H. (2000 b). Ichthyofauna of some of the oxbow lake in Barak valley region of Assam. Proceedings of the National Symposium on Wetlands and Fisheries Research in the New Millennium, 1: 16 pp.

Kar, D.; Dey, S.C.; Roy, A.; and, Mandal, M. (2000 c) Epizootic Ulcerative Syndrome Fish Disease in Barak valley region of Assam, India. Proc. Nat. Symp.Current Trends in Wetlands and Fisheries Research in the New Millennium, 1: 2 pp

Kar, D.; Dey, S.C.; Mandal, M.; and, Lalsiamliana (2000 d). Epizootic Ulcerative Syndrome among the fishes of Assam. In: Proc. National Workshop of NATP-ICAR-NEC North-East Programme, Shillong, 1: 24 pp.

Kar, D.; Mandal, M.; and Lalsiamliana (2001 a). Species composition and distribution of Fishes in the rivers in Barak valley region of Assam and in the Principal rivers of Mizoram and Tripura in relation to their habitat parameters. Proc. National Workshop, NATP-ICAR Project, Mid-term Review,1: Central Marine Fisheries Research Institute, Cochin, 25 pp.

Kar, D.; Laskar, B.A.; and, Lalsiamliana (2001 b). Further studies on the Ichthyospecies composition and distribution of Freshwater Fishes in Barak drainage and in the Principal rivers in Mizoram and in Tripura with a note on their Feeding and Breeding biology. Proc. National Project Monitoring Workshop of NATP-ICAR Project, National Bureau of Fish Genetic Resources, Lucknow, 1: 22 pp.

Kar, D.; Laskar, B.A.; Mandal, M.; Lalsiamliana; and, Nath, D. (2002 a). Fish Genetic Diversity and Habitat parameters in Barak drainage, Mizoram and Tripura. Indian J. Environment and Ecoplanning, 6 (3): 473-480.

Kar, D.; Laskar, B.A; Nath, D.; Mandal, M.; and, Lalsiamliana (2002 b). *Tor progenius* (McClelland) under threat in River Jatinga, Assam. Science and Culture, 68 (7-8): 211.

Kar, D.; Laskar, B.A.; and, Nath, D. (2002 c). *Tor* sp. (Mahseer fish) in river Mat in Mizoram. Aquacult, 3 (2): 229-234.

Kar, D.; Dey, S.C.; and, Roy, A. (2002 d). Prevalence of Epizootic Ulcerative Syndrome among fishes of Assam. In: Proc. Regional Symp. on Biodiversity, S.S. College, Assam (Central) University, Hailakandi (Assam).

Kar, D.; Dey, S.C.; and, Roy, A. (2002 e). On the diseased fishes suffering from hitherto unknown Epizootic Ulcerative Syndrome in Fishes in India. Proc. All-India Congress of Zoology, Dec., 2002 (Bangalore).

Kar, D.; Dey, S.C.; and, Roy, A.(2002 f). Prevalence of Epizootic Ulcerative Syndrome (EUS) among fishes of Mizoram. In: Proc. Regional Symp. on Aquaculture, 1.

Kar, D.; Laskar, B.A.; and Lalsiamliana (2002g). Germplasm Inventory, Evaluation and Gene Banking of Freshwater Fishes. 3rd Annual Technical Report: pp.57. National Project Evaluation Workshop, 3: NATP-ICAR World Bank-aided Project: National Bureau of Fish Genetic Resources (Lucknow).

Kar, D.; Dey, S.C.; Mandal, M.; Laskar, B.A.; and, Lalsiamliana (2002h). Preliminary Survey of the Fish Genetic Resources of the Rivers in Barak Drainae, Mizoram and Tripura, pp. 73-81. In: Restoration of Lakes and Wetlands (Eds.) Ramachandra, T.R.; Rajasekahra, Murthy, C.; and, Ahalya, N., pp. xxii + 400, Allied Publishers (P) Ltd. (Bangalore).

Kar, D.; Dey, S.C.; Roy, Aurobindo; and, Mandal, M. (2002i). Epizootic Ulcerative Syndrome in Fishes of Barak valley of Assam, India: pp.303-307. In: Restoration of Lakes and Wetlands (Eds.) Ramachandra, TV; Murthy Rajasekhara, S.; and, Ahalya, N., pp.xii + 400, Allied Publishers (P) Ltd. (Bangalore).

Kar, D.; Dey, S.C.; and, Datta, N.C. (2003 a). Welfare Biology in the New Millennium, xx + 97, Allied Publishers Pvt. Ltd. (Bangalore).

Kar, D.; Dey, S.C.; and,Roy, A (2003 b). Capture Fishery in Lentic Systems with some light on Sone Beel in Assam with special reference to prevalence to EUS in the Barak valley region of Assam. Proc. Lecture Series in UGC-sponsored Vocational Course in Industrial Fish and Fisheries, Cachar College, Assam University, Silchar, Abstracts, 1.

Kar, D.; Roy, A.; and Dey, S.C. (2003 c). Epizootic Ulcerative Disease Syndrome in Freshwater Fishes with a note on its Management for Sustainable Fisheries. Proc. International Conference on Disease Management for Sustainable Fisheries, 26-29 May, 2003, Department of Aquatic Biology and Fisheries, University of Kerala, Trivandrum.

Kar, D.; Lohar, M.; Ngassepam, Z.; Sinom, S.; Tiwary, B.; and Bawari, M. (2003 d). Fish Bioresources in certain rivers of Assam and Manipur with a note on their assessment, management and conservation. Proc. Nat. Symp. Assessment and Management of Bioresources, North Bengal University and The Zoological Society, Calcutta, 28-30 May, 2003: 56 pp.

Kar, D.; Dey, S.C.; and, Roy, A. (2003 e). Epizootic Ulcerative Syndrome: A virulent disease among the freshwater fishes of India. In: Proc. Guest Lecture delivered at the Dept. of Zoology, Arunachal University, Arunachal Pradesh, 1.

Kar, D.; Dey, S.C.; and, Roy, A. (2003 f). Biodiversity of freshwater fishes of North-East India with a note on their health status. In: Proc. DBT-sponsored Guest Lecture Series delivered at the Dept. of Molecular Biology and Biotechnology, Tezpur (Central) University, Tezpur, 1.

Kar, D.; Dey, S.C.; and, Roy, A.(2003 g). Epidemiology of Epizootic Ulcerative Syndrome disease among the freshwater fishes of India. In: Proc. Regional Symp. on World Bank-aided programme entitled, 'Assam Rural Infrastructure and Agricultural Services Project (ARIASP), Silchar, 1.

Kar, D.; Dey, S.C. and Roy, A. (2003 h).Review of Epizootic Ulcerative Syndrome disease among the freshwater fishes of India. In: Proc. UGC-sponsored lecture series delivered at the Department of Environmental Engineering and Biotechnology, Guru Jambeswar University, Hisar (Haryana).

Kar, D.; Laskar, B.A,; and, Nath, D.(2003 i).Length-weight relationship of *Salmostoma phulo phulo*(Hamilton Buchanan) collected from River Karnafuli along Indo-Bangladesh border (in Mizoram). J. Applied Zoological Researches, 14 (2): 188-190.

Kar, D.; Dey, S.C.; Roy, A.; and, Mandal, M. (2003 j). Epizootic Ulcerative Syndrome Fish Disease in Barak Valley Region of Assam: pp. 73-78. In: Welfare Biology in the New Millennium (Eds.) Kar, D.; Dey, S.C.; and, Datta, N.C., pp. xx + 97, Allied Publishers Pvt.Ltd. (Bangalore).

Kar, D.; Roy, A,; and Dey, S.C. (2004 a). An overview of Fish Genetic Diversity of North-East India. In: Proc. National Workshop on Rational use of Water Resources for Aquaculture (Eds.) Garg., S.K. and Jain, K.L., 18-19 March 2004, CCS Haryana Agricultural University, 1: 164-171.

Kar, D.; Roy, A.; Mazumder, J.; and Patil, P (2004 b) Biotechnological approach for defining Epizootic Ulcerative Syndrome Disease in Freshwater Fishes. In: Proc. Mid-term Review: DBT-sponsored Project, New Delhi, 1.

Kar, D.; Dey, S.C.; Roy, A.; Mazumder, J.; Patil, P.; and, Kohlapure, R.M. (2004 c). Fish Disease Prevalence in Assam with particular reference to Epizootic Ulcerative Syndrome and its pathogens. Proc. International Conference 'Bioconvergence', 18-20 Nov 2004, Thapar Institute of Engineering and Technology, Patiala, 1: 220 pp.

Kar, D.; Mazumder, J.; Dey, S.C.; and, Roy, A. (2005). Dynamics of Interaction between environmental temperature, bacterial growth and fish in the initiation of Epizootic Ulcerative Syndrome in Freshwater Fishes. Ekologia (in press).

Khan, H. (1938). Ovulation in Fish (effect of administration of anterior lobe of pituitary gland). Curr. Sci., 7 (5): 233-234.

Khan, H. (1944). Studies in Diseases of Fish, infestation of fish leeches and fish lice. Proc. Indian Acad. Sci., 19 B (5): 171-175.

Klee, G.A. (1991). Conservation of Natural Resources, xvi + 410, Prentice Hall, Englewood Cliffs, NJ (USA).

Koops, H.; Mann, H.; Pfitzner, I.; Schmid, O.J.; and, Schubert, G. (1970). The cauliflower disease of eels: pp. 291-295. In: Symposium on Diseases of Fishes and Shellfishes (Ed.) Snieszko, S.F., American Fisheries Society, Spl. Publication No. 5 (Washington D.C.).

Kottelat, M. (1989). Zoogeography of the Fishes from Indo-Chinese Inland Waters with an annotated checklist. Bull. Zool. Mus. Univ. Amsterdam, 12: 1-54.

Kowtal, G.V. (1994). Endangered, Vulnerable and Rare foodfishes of the East Coast River System, pp. 57-61. In: Threatened Fishes of India, pp. 384, Natcon. Publication No. 4 (UP).

Kulkarni, C.V. (1950). Breeding habits, eggs and early life history of Indian shad, *Hilsa ilisha* (Ham.) in the Narbada river. Proc. Nat. Inst. Sci. India, 15: 169-176.

Kulkarni,.C.V. (1990). Mahseer: the mighty game fish of India. Farm Information Unit, pp. 23, Ministry of Agriculture (New Delhi).

Lagler, K.F. (1956). Freshwater Fishery Biology, pp. xii + 421, W_M Brown Co., Dubuque, Iowa (USA).

Lagler, K.F.; Bardach, J.E.; and, Miller, R.R. (1962). Ichthyology: The Study of Fishes, xiii + 545, John Wiley and Sons (New York).

Landolt, M.L. (1989). The relationship between diet and the immune response of Fish. Aquaculture, 79: 193-206.

LaRoe and Edward, T. (1986). Instream impacts of soil erosion on Fish and Wildlife. In: The offsite cost of Soil erosion (Ed.) Thomas E. Waddell, Washington DC (USA).

Laskar, B.A.; Nath, D.; Mandal, M.; Das, S.; and, Kar, D. (2002). Ecological Studies in Punir Haor wetland in Cachar district of Assam with special emphasis on aquatic macrophytes, ichthyofauna and wetland birds. Conservation Forum Journal, 1 (1): 15-18.

Laveran, A. and Pettit, A. (1910). *Sur une epizootic des truites. C.r. hebd. Se'anc. Acad. Sci. Paris*, 151: 421-423.

Leach, G. (1925). Artificial propagation of shad, US Bur. Fish. Washington DC, 981:459-486.

Le Cren, E.D. (1951). The Length-Weight relationship and seasonal cycle in gonad weight and condition of perch (*Perca fluviatilis*). J. Anim.Ecol., 21: 210-219.

Lehane, R. (1996). Fish with ulcers: why and what can be done? Partners in Research for Development (Australia), 9: 13-19.

Lele, S.H. (1934). On the Physical morphology of the thymus gland in Plaice of various ages wih a note on the evolution of the organ, including also notes on the other ductless glands in this species. J. Univ.Bombay, 1: 37-53.

Leung, K.Y.; Yeap, I.V.; Lam, T.J.; and, Sin, Y.M. (1994). Serum resistance as a good indicator for virulence in *Aeromonas hydrophila* strains isolated from diseased fish in South-East Asia. Journal of Fish Diseases, 18: 511-518.

Lian C.X. (1990). China Report: Pp. 8-9. In: Regional Research Programme on the relationship between Epizootic Ulcerative Syndrome in Fish and the Environment, 13-26 Aug 1990, NACA (Bangkok).

Likens, G.E. (1975). Primary Production of Inland Aquatic Ecosystems. In: Primary Productivity of the Biosphere (Eds.) Leith, H. and Whittaker, R.H., pp. 185-202, Springer-Verlag (UK).

Lindeman, R.L. (1942). The Trophic-dynamic aspect of Ecology. Ecol., 23: 339-418.

Lilley, J.H; Phillips, M.J.; and, Tonguthai, K. (1992). A Review of the Epizootic Ulcerative Syndrome (EUS) in Asia. Aquatic Animal Health Research Institute and Network of Aquacultural Centres in Asia-Pacific (Bangkok).

Lilley, J.H.; and Frerichs, G.N. (1994). Comparison of Rhabdoviruses associated with Epizootic Ulcerative Syndrome (EUS) with respect to their structural proteins, Cytopathology and serology. Journal of Fish Diseases, 17: 513-522.

Lilley, J.H.; Hart, D.; Richards, R.H.; Roberts, R.J.; Cerenius, L.; and, Söderhä ll, K. (1997 a). Pan-Asian spread of single fungal clone results in large-scale fish-kills. Veterinary Record, 140:653-654.

Lilley, J.H.; Thompson, K.D.; and, Adams, A. (1997 b). Characterisation of *Aphanomyces invadans* by electrophoretic and Western blot analysis. Diseases of Aquatic Organisms, 30: 187-197.

Lilley, J.H.; Callinan, R.B.; Chinabut, S.; Kanchanakhan, S.; Mac Rae,I.H.; and, Phillips, M.J. (1998). Epizootic Ulcerative Syndrome (EUS): Technical Handbook, pp. 88, Aquatic Animal Health Research Institute (Bangkok).

Lilley, J.H. and Roberts, R.J. (1997). Pathogenicity and culture studies comparing the *Aphanomyces* involved in Epizootic Ulcerative

Syndrome (EUS) with other similar fungi. Journal of Fish Diseases, 20: 135-144.

Llobrera, A. and Gaeutan, R.Q. (1987). *Aeromonas hydrophila* associated with ulcerative disease epizootic in Laguna de Bay, The Philippines. Aquaculture, 67: 273-278.

Lowe, J. (1874). Fauna and Flora of Norfolk, Part 4: Fishes. Trans. Norfolk Norwich Nat. Soc., 21: 56 pp.

Lowe McConnell, R.H. (1975). Fish Communities in Tropical Freshwaters, xvii + 337, Longman (London).

Mackay, W.S. (1944). Some high range hatchery and river management notes, Rajamallay, Mumar, HRAA Publication: 20 pp.

Macintosh, D.J. and Phillips, M.J. (1986). The Contribution of Environmental factors to the Ulcerative Fish Diseases condition in South-East Asia: pp.175-208. In: Field and Laboratory Investigations into Ulcerative Fish Diseases in the Asia-Pacific Region. Technical Report of the FAO Project TCP/RAS/4508 Bangkok (Thailand).

Mammen, T.A. (1962). Live fish transport using modified splashless carrier with compressed aeration. In: Training Course on Live Fish Transport, 19-21 Nov 1962; Hyderabad. Fisheries Extension Unit, Govt. of India: 5 pp.

Maitland, H.B. and Maitland, M.C. (1928). Cultivation of Vaccinia virus without tissue culture, Llancet, 215: 596.

Malhotra, J.C. (1969). Successful artificial propagation of *Hilsa ilisha* (Hamilton) near Allahabad. Curr. Sci., 38: 429-430.

Mckenzie, R.A. and Hall, W.T.K. (1976). Dermal Ulceration of mullet (*Mugil cephalus*). Australian Veterinary Journal, 52: 230-231.

Menasveta, P. (1985). Current Fish Disease Epidemic in Thailand. J. Sci Soc.Thailand, 11: 147-160.

Menon, A.G.K. (1954). Fish Geography of the Himalayas. Proc. Nat Inst. Sci. India, 20 (4): 467-493.

Menon, A.G.K. (1955). The external relations of Indian Freshwater Fishes with special reference to the countries bordering the Indian ocean. J. Asiat. Sci. Berg. (Sci.), 21 (2): 31-38.

Menon, A.G.K. (1973). Origin of the Freshwater Fish Fauna of India. Curr. Sci., 42 (16): 553-556.

Menon, A.G.K. (1974). A Checklist of the Fishes of the Himalayan and the Indo-gangetic Plains, pp. viii + 136, Inland Fish. Soc. India (Barrackpore).

Menon, A.G.K. (1988). Conservation of the Ichthyofauna of India, pp. 55-33.. In: Conservation and Management of the Inland Capture Fisheries Research Bulletin No. 57, Central Inl. Capt. Fish.Res. Inst. (Barrackpore).

Menon, A.G.K. (1994). Criteria for determining the status of threatened categories of Indian freshwater fishes: pp. 1-5. In: Threatened Fishes of India, pp. 384, Natcon Publication No. 4 (UP).

Menon A.G.K. (1999). Checklist: Freshwater Fishes of India, xviii + 366, Occasional Paper No. 175, Zoological Survey of India (Calcutta).

Michael, R.G. (1969). Seasonal trends in Physico-chemical factors and Plankton of freshwater fish pond and their role in fish culture. Hydrobiologica, 33 (1): 145-160.

Michael, R.G. and Sharma, B.K. (1988). Indian Cladocera: Fauna of India and Adjacent Countries Series, × + 261, Zoological Survey of India (Calcutta).

Mirza, M.R. (1967). *Tor zhobensis* sp.nov., a new Mahseer from the Rive Zhob. Pakistan J. Sci. Res., 19 (1): 54-56.

Misra, K.S. (1959). An aid to the Identification of Commercial Fishes of India and Pakistan. Rec. Indian Mus., 57 (1-4): 1-320.

Misra, K.S. (1976). The Fauna of India and the adjacent countries: Pisces, 2nd Ed.: 3. Teleostomi: Cypriniformes, Siluri: pp. 387.

Mitra, G.N. and Devasundaram, M.P. (1954). On the hilsa of Chilka Lake with a note on the occurrence of hilsa in Orissa. J. Asiat. Soc. Sci., 20: 33-40.

Mittermeier, R.A. and Mittemeier, C.G. (1997). Megadiversity: Earth's Biologically Wealthiest Nation. In: Global Freshwater Biodiversity (Ed.) McAllister, D.E; Hamilton, A.L.; and, Harvery, B; Sea Wind, Cemex, Mexico City, 11: 1-140.

Miyazaki, T.and Egusa, S. (1972). Studies on Mycotic Granulomatosis in Freshwater Fishes: 1, Mycotic Granulomatosis gold fish. Fish Pathology, 7: 15-25.

Mohan, C.V.; Shankar, K.M.; and, Ramesh, K.S. (1999). Is Epizootic Ulcerative Syndrome (EUS) specific fungus a primary pathogen? An opinion. Naga, The ICLARM Quarterly, Vol. 22, No. 1: 15-18.

Mohan, C.V. and Shankar, K.M. (1994). Epidemiological analysis of Epizootic Ulcerative Syndrome of fresh and brackish water fishes of Karnataka, India. Curr. Sci., 66 (9): 656-658.

Molur, S. and Walker, S. (1998). CAMP (Conservation Assessment and Management Plan) Workshop on Freshwater Fishes of Inida. Zoo Outreach Organisation/CBS, Coimbatore (India).: pp. 156.

Mookerjee, H.K. (1945). Life history of some major carps of Bengal. Sci. and Cult., 10 (9): 400-402.

Motwani, M.P.; Jhingran, V.G; and, Karamchandani, S.J. (1957). On the breeding of Indian shad, *Hilsa ilisha* (Hamilton) in freshwater. Sci and Cult., 23: 47-48.

Motwani, M.P.; Jayaram, K.C.; and, Sehgal, K.L. (1982). Fish and Fisheries of Brahmaputra River System, Assam, I. Fish Fauna with observation on their zoogeographical significance. Trop. Ecol., 3: 17-43.

Moyle, P.B. (1976). Fish Introduction in California. History and impact on Native Fishes. Biological Conservation, Vol. 9: 101-118.

Moyle, P.B.;and Herbold, B. (1987). Life history patterns and Community structure in Stream Fishes of Western Northern America and Europe: pp. 25-32. In: Matthews, W.J. and Heins, D.C. (Eds) Community and Evolutionary Ecology of North American Stream Fishes, University of Oklahoma Press, Norman, Oklahoma (USA).

Mukherjee, D.D. (1936). Yale North India Expedition, Article 18. Report on Fishes, part 2: Sisoridae and Cyprinidae. Mem. Connecticut Acad. Arts. Sci., 10: 323-359.

Murray, G. (1885). Notes on the inoculation of Fishes with *Saprolegnia ferax*. J. Bot., 23: 302-308.

Myers, G.S. (1949). Salt tolerance of freshwater Fish Groups in relation to geographical problems. *Bijdr tot de Dierk*, 28: 315-322.

Nair, K.K. (1939). On some early stages in the development of so-called Indian shad, *Hilsa ilisha* (Ham.). Rec. Indian Mus., 41: 409-418.

Nair, K.K. (1954). Dams and hilsa fisheries. J. Asiatic Soc. Sci., 20: 77-79.

Naryani, N. and Tamot, P (2002). Growth and age determination of Endangered Mahseer, *Tort or* by scales. NBFGR-NATP Pubication, 4: 1-2.

Nath, P. and Dey, S.C. (1997). Fish and Fisheries of North-East India, Vol. I: Arunachal Pradesh: 1-140.

Nath, P and Dey, S.C. (2000). Conservation of Fish Germplasm Resources of Arunachal Pradesh:pp.49-67. In: Fish Biodiversity of North-East India (Eds.) Ponniah, A.G. and Sarkar, U.K., xiii + 228, NATP Publication No. 2, NBFGR (Lucknow).

Nautiyal, P. and Lal, M.S. (1981). Recent Records of Garhwal Mahseer (*Tor putitora*) with a note on its present status. J. Bombay nat. Hist. Soc., 79: 593-595.

National Workshop on Ulcerative Disease Syndrome in Fish (1990). Proceedings of the National Workshop on Ulcerative Disease Syndrome in Fish, 6-7 March, 1990, Calcutta (India).

Naumann, E. (1917).*Undersö gnengar of Ver Fytoplankton och under den pelagiska regionen fö- rsig gaende gyttje__ Och dybildningar inom Vissa syd__ ochmellansvenska ubergsvalten. Kungla Svenska Vetenskap.* Handlingar, 56 (6): 1-165.

NBFGR (2000). Manual on Habitat Inventory, pp. 27, National Bureau of Fish Genetic Resources (ICAR) (Lucknow).

NEC (1991). Summary Report on Integrated Inland Fisheries Development Project in North-Eastern States, pp. 176, North-Eastern Council, Govt. of India (Shillong).

Needham, J.G. (1930). The Teaching of Hydrobiology and Aquaculture in American Universities. Science, 71: 265-266.

Needham, J.G. and Needham,P.R. (1962). A Guide to the Study of Freshwater Biology, × +108, Holden Day Inc. San Francisco (USA).

Needham, J.G. and Lloyd, J.T. (1930). The Life of Inland waters, 2nd.ed., Spring-field: 438 pp.

Nichols, J.T. (1928). Fishes of the White Nile (with Table of World's Freshwater Fish Faunae). American Mus. Novitates No. 319.

Nikolsky, G.V. (1963). The Ecology of Fishes, pp.352, Academic Press (London).

Nikolsky, G.V. (1978). The Ecology of Fishes, pp. xv +352, T.F.H Publications (USA).

Noble, R.L. (1980). Management of Lake, Reservoirs and Ponds: pp. 265-295. In: Fisheries Management (Eds.) Robert, T.; Larkey, and Larry A. Niellsen, Blackwell Scientific (Boston).

Noga, E.J. and Dykstra, M.J. (1986). Oomycete fungi associated with ulcerative mycosis in Menhaden, *Brevoortia tyrannus* (Latrobe). Journal of Fish Diseases, 9: 47-53.

Ogale, S.N. (1994). Endangered Deccan Mahseer, *Tor khudree*(Sykes): A case Study: pp. 213-218. In: Threatened Fishes of India, pp. 384, Natcon. Publication No. 4 (UP).

OIE (1997). Epizootic Ulcerative Syndrome: Diagnostic Manual for Aquatic Animal Diseases, pp. 127-129, *Office International Des Epizootics (Paris).*

Olsen, S.R. *et al.* (1954). Estimation of available phosphorus in soils by extraction with Sodium bicarbonate. Circ.U.S. Dep. Agric., 939.

Owen, W. (1985). Natural Resources Conservation: An Ecological Approach, 4th Ed., Macmillan (New York).

Pal, R.N. (1963). A new acanthocephalan parasite, *Acanthocentis hilsai* N. sp. from Indian shad, *Hilsa ilisha* (Hamilton). Indian J. Helminth, 15: 96-99.

Pal, R.N. (1975). Role of Fish Pathology in Aquaculture. J. Inland Fish. Soc. India, 7:131-136.

Pal, R.N. and Ghosh, A.K. (1975). An effective method of controlling tail and fin rot in Indian major carps. J. Inland Fish. Soc. India, 7: 98-99.

Palisoc, F.P. (1990).Histopathology of the Epizootic Ulcerative Syndrome (EUS) positive snakehead, *Ophicephalus striatus*, from Laaguna Lake, the Philippines: pp. 22. Abstracts from the Symposium on Diseases in Asian Aquaculture, 26-29 Nov

1990,Bali, Indonesia. Fish Health Section, Asian Fisheries Society (Manila).

Pallibare, J.O.; Catap, E.S.; and, Callinan, R.B. (1994). Fungal isolation from EUS-affected fish in the Philippines: pp. 238-243. In: Proceedings of the ODA Regional Seminar on Epizootic Ulcerative Syndrome, 25-27 Jan 1994 (Eds.) Roberts, R.J.; Campbell, B.; and, MacRae, I.H., Aquatic Animal Health Research Institute (Bangkok).

Pandey, K.D.; Khan, H.H.; and, Awarthi, S.K. (1994). Present status of *Tor putitora* in Bhimtal Lake, UP: pp. 143-147. In: Threatened Fishes of India, pp. 384, Natcon. Publication No. 4 (UP).

Pathani, S.S.(1981). Fecundity of Mahseer, *Tor putitora* (Ham.). Proc. Indian Acad. Sci., 90:255-260.

Pathani, S.S. (1994). Trends of Mahseer, Snow trout and Chittal abundance in Kumanon Himalayas: pp. 155-160. In: Threatened Fishes of India, pp. 384, Natcon. Publication No. 4 (UP).

Patil, P.R.; Kohlapure, R.M.; Mazumder, J.; Kar, D.; and, Roy, A. (2003). Isolation of a suspected viral agent from EUS-affected *Labeo calbasu* fish. Proc. International Conference on Disease Management for Sustainable Fisheries, University of Kerala (Trivandrum).

Paul, J. (1975). Cell and Tissue culture, xii + 483, Churchill Livingstone (London).

Pearce, M. (1990). Epizootic Ulcerative Syndrome Technical Report, Dec 1987 to Sep 1989, pp. 82, Northern Territory Department of Primary Industry and Fisheries. Fisheries Report No. 22, Northern Territory (Australia).

Pearsall, W.H. (1938). The Soil complex in relation to Plant communities. I. Oxidation-reduction potentials in soils. J. Ecol., 26: 180-193.

Pennak, R.W. (1953). Freshwater Invertebrates of the United States, ix + 769, The Roland Press Co. (New York).

Pfitzner, I. (1969). *Zür Atiologie der Bhemenkohler krankheit deer. Aa!e. Arch. FischWiss., 20*: 24-35.

Phillips, M.J. (1989). A Report on the NACA Workshop on the Regional Research Programme on Ulcerative Syndrome in Fish and the Environment, 20-24 March, 1989. Network of Aquaculture Centres in Asia-Pacific (Bangkok).

Phillips, M.J. and Keddie, H.G. (1990). Regional Research Programme on Relationship between Epizootic Ulcerative Syndrome in Fish and the Environment: A Report on the 2nd Technical Workshop, 13-26 Aug, 1990, 133 pp., Network of Aquaculture Centres in Asia-Pacific (Bangkok).

Pillay, T.V.R. (1959). An inexpensive tag for marking hilsa. Curr. Sci., 28:212-213.

Pillay, T.V.R.(1958). Biology of hilsa, *Hilsa ilisha* from river Hooghly. Indian J. Fish., 5 (2): 201-257.

Pillay, T.V.R. and Ghosh, A.N. (1958). A note on hilsa fisheries of Assam. J. Bombay nat. Hist. Soc., 55: 174-177.

Pillay, T.V.R.; Pillay, S.R.; and, Ghosh, K.K. (1962). A comparative study of the population of hilsa, *Hilsa ilisha* (Ham.) in Indian waters. Proc. Indo-Pacific Fish. Counc., 10 (2): 62-104.

Pisolkar, M.D. (1994). Status of Mahseer in Maharashtra, pp. 117-119. In: Threatened Fishes of India, pp. 384, Natcon. Publication No. 4 (UP).

Plumb, J.A. (1973). Neutralisation of Channel cat fish virus by serum of Channel cat fish. J. Wildl. Dis., 9: 324-330.

Plumb, J.A. and Gaines, J.L. Jun. (1975). Channel cat fish virus disease, pp. 287-302. In: The Pathology of Fishes (Ed) Ribelin, W.E. and Migaki, G., University of Wisconsin Press, Madison (USA).

Poonsuk, K.; Saitanu, K.; Navephap, O.; and, Wongsawang, S. (1983). Anti-microbial susceptibility testing of *Aeromonas hydrophila* strain isolated from freshwater fish infection: pp.436-438. In: Proc. Symposium on Freshwater Fishes Epidemic, 1982-83, 23-24 June, 1983, Chulalongkorn University (Bangkok).

Prasad, B.; Hora, S.L.; and, Nair, K.K. (1940). Observations on the seasward migration of the so-called Indian shad, *Hilsa ilisha* (Hamilton). Rec. Indian Mus., 42: 529-552.

Prasad, P.S. and Sinha, J.P. (1990). Status paper on the occurrence of Ulcerative Disease Syndrome in Fishes of Bihar. National Workshop on Ulcerative Disease Syndrome in Fish, 6-7 March, 1990 (Calcutta).

Priyadi, A. (1990). Indonesia Report: pp. 15-17. In: Regional Research Programme on Relationships between Epizootic Ulcerative Syndrome in Fish and the Environment, 13-26 Aug., 1990, NACA (Bangkok).

Purkait, M.C. (1990). Case Studies on the Epizootic Ulcerative Syndrome of Fishes in Chanditala:I of Hooghly district. National Workshop on Ulcerative Disease Syndrome in Fish, 6-7 March, 1990 (Calcutta).

Quddus, M.M.A. (1982). Two types of *Hilsa ilisha* and their population biology from Bangladesh waters. Doctoral Thesis, University of Tokyo: 180 pp.

Rahman, T.; Choudhury, B.; and, Barman, N. (1988).Role of UDS-affected fish in the health of ducks in Assam: An experimental study. In: Proceedings of the Symposium on Recent Outbreak of Fish Diseases in North-Eastern India, 30 Dec 1988, p. 10, Zoological Society of Assam (Guwahati).

Raina, H.S.; Sunder, S.; Vass, K.K.; and, Langer, R.K. (1986). Artificial breeding and nursery rearing of *S. esocinus* Heckel in Kashmir. Proc. Nat.Acad.Sci.India, 56 (B): 335-338.

Raj, B.S. (1945). Decline of Mahseer fisheries of Kumaon lakes and possible remedy. Proc. Nat. Inst. Sci. India, 11: 341-345.

Rajyalakshmi, T. (1972). The population characteristics of the Godavari hilsa over the years, 1963-1967. Indian J. Fish, 20: 78-94.

Ramakrishnaiah, M. (1972). Biology of the *Hilsa ilisha* (Hamilton) from the Chilka Lake with an account on its racial status. Indian J. Fish., 19: 35-53.

Rao, K.V. and Pathak, S.C. (1972). A note on the occurrence and spawning of *Hilsa ilisha* (Hamilton) in the River Brahmaputra (Assam). Proc.nat.Acad. Sci.India, B, 42: 231-233.

Reantaso, M.B. (1990). Philippines Report: pp. 48-53. In: Regional Research Programme on Relationship between Epizootic Ulcerative Syndrome in Fish and the Environment, 13-26 August 1990, NACA (Bangkok).

Reantaso, M.B. (1991). EUS in brackish waters of the Philippines. Fish Health Section Newsletter, 2 (1): 8-9, Asian Fisheries Society (Manila).

Reichenbach-Klinke, H. and Elkan, E. (1965). The Principal Diseases of Lower Vertebrates: Book I: Diseases of Fishes, Academic Press (New York).

Reinprayoon, S.; Lertpo-kasombat, K.; and, Chantaratchada, S. (1983). The incidence of *Aeromonas hydrophila* in asymptomatic people at Bang-Toi: pp.534-546. The Symposium on Freshwater Fishes Epidemic, 23 June 1983, Chulalongkornkorn University (Bangkok).

Richards, R.H. and Pickering, A. (1978).Frequency and Distribution patterns of *Saprolegnia* infection in wild and hatchery-reared brown trout, *Salmo trutta* L. and Char, *Salvelinus alpinus* (L). Journal of Fish Diseases, 1: 69-82.

Ricklefs, R.E. (1987). Community Diversity: Relative roles of Local and Regional processes, Science, 235: 167-171.

Roberts, R.J. (Ed.) (1978). Fish Pathology, × + 318, Bailliere Tindall (London).

Roberts, R.J. (Ed.) (1989). Fish Pathology, × + 318, Bailliere Tindall (London).

Roberts, R.J.; Frerichs, G.N.; and, Millar, S.D. (1992). Epizootic Ulcerative Syndrome: The Current Position: pp.431-436. In: Diseases in Asian Aquaculture: I. (Eds.) Shariff, I.M.; Subasinghe, R.P.; and, Arthur, J.R. Fish Health Section, Asian Fisheries Society, Manila (The Philippines).

Roberts, R.J.; Campbell, B.; and, Mac Rae, I.H. (1994 a). Proceedings of the Regional Seminar on Epizootic Ulcerative Syndrome, 25-27 Jan 1994, The Aquatic Animal Health Research Institute (Bangkok).

Roberts, R.J.; Frerichs, G.N.; Tonguthai, K.; and, Chinabut, S. (1994 b). Epizootic Ulcerative Syndrome of farmed and wild fishes: pp. 207-239. In: Recent Advances in Aquaculture V (Eds.) Muir, J.F. and Roberts, R.J., Chapter 4, Blackwell Science (UK).

Roberts, R.J.; Macintosh, D.J.; Tonguthai, K.; Boonyaratpalin, S.; Tayaputch, N.; Phillips, M.J.; and, Millar, S.D. (1986). Field and Laboratory Investigations into Ulcerative Fish Diseases in the Asia-Pacific Region, pp. 214, Technical Report of FAO Project TCP/RAS/4508 (Bangkok).

Roberts, R.J.; Willoughby, L.G.; and, Chinabut, S. (1993). Mycotic aspects of Epizootic Ulcerative Syndrome (EUS) of Asian Fishes. Journal of Fish Diseases, 16: 169-183.

Roberts, R.J.; Wootten, R.; MacRae, I.; Millar, S.; and, Struthers, W. (1989). Ulcerative Disease Survey: Bangladesh Final Report to the Govt. of Bangladesh and the Overseas Development Administration. Institute of Aquaculture, Stirling University (UK).

Rodgers, L.J. and Burke, J.B. (1981). Seasonal Variation in the prevalence of Red Spot Disease in Estuarine fish with particular reference to the sea mullet, *Mugil cephalus* L. Journal of Fish Diseases, 4: 297-307.

Rodhe, W. (1969). *Sulla productione difitoplankton in laghi transparents di alta montagna*. Mem. 1stel Idrolbiol., 15: 21-28.

Rodhe, W. (1974). The International Association of Limnology: Creation and Function. Mitt. Int. Ver. Limnol., 20: 44-70.

Ross, A.J.; Pelnar, J.; and, Rucker, R.R. (1960). A virus-like disease of Chinook salmon. Trans. Am. Fish. Soc., 89: 160-165.

Ross, A.J. and Yasutake, W.T. (1973). *Scolecobasidium humicola*, a fungal pathogen of fish. J. Fish. Res.Bd. Can., 30: 994-995.

Ross, A.J.; Yasutake, W.T.; and, Leek, S. (1975). *Phonna herbaum*, a fungal plant saprophyte as a fish pathogen. J. Fish. Res. Bd. Can., 32: 1648-1652.

Rucker, R.R.; Whipple, W.J.; Parvin, S.R.; and, Evans, C.A. (1953). A contagious disease of salmon possibly of virus origin. US Fish and Wildlife Serv.Spc.Sci.Rep.Fish. No. 138: 36.

Rukyani, A. (1994). Status of Epizootic Ulcerative Disease in Indonesia. In: Proceedings of the ODA Regional Seminar on Epizootic Ulcerative Syndrome, 25-27 January, 1994, Aquatic Animal Health Research Institute (Bangkok).

Saha, K.C. and Sen, D.P. (1955). Gammexane in the treatment of *Argulus* and fish leech infection in fish. Biochem., 15: 71-72.

Saitanu, K. and Poonsuk, K. (1983). Prophylactic and therapeutic studies of antimicrobial agents to the *Aeromonas hydrophila* infection of snakehead fish (*Ophicephalus stiratus*): pp.461-514. In: Proc. Symp. On Freshwater Fishes Epidemic, 23-24 June, 1983, Chulalongkorn University (Bangkok).

Saitanu, K.; Wongsawang, S.; Sunyasootcharee, B.; and, Sahaphong, S. (1986). Snakehead fish virus isolation and pathogenicity studies: pp. 327-330. In: Maclean, J.L.; Dizon, L.B.; and, Hosillos, L.V. (Eds.), The 1st Asian Fisheries Forum, Asian Fisheries Society, Manila (The Philippines).

Sammut, J.; White, I.; and, Melville, M.D. (1996). Acidification of an estuarine tributary in Eastern Australia due to drainage of acid sulphate soils. Marine and Freshwater Research, 47 (5): 669-684.

Sandhu, G.J.; Tandon, K.K.; and, Johal, M.S. (1994). Growth studies on an endangered *Tor putitora* (Hamilton) from Govinda Sagar (H.P., India): pp. 137-142. In: Threatened Fishes of India, pp. 384, Natcon Publication No.4 (UP).

Sarkar, H.L. (1946). On a protozoan parasite *Myxobolus mrigalae* Chakravarty, infecting the fry of *Cirrhinus mrigala* (Ham.). Curr. Sci., 15 (4): 111-112.

Saxena, S.C. and Chandy, M. (1966). Adhesive apparatus in certain Indian Hill stream Fishes. J. Zool. Soc. London, 148:315-346.

Schä perclaus, W. (1933). Text Book of Pond culture: rearing and keeping of carps, trout and allied fishes (Tr. F. Hund.). Fish Leaflet. Wash. (311): 260 pp.

Schä perclaus, W. (1954). *Fischkrankheiten* (3rd Ed.) *Akademie Verlag*, Berlin (Geermany).

Schlosser, I.J. (1982). Fish Community Structure and Function along two habitat gradients in a headwater stream. Ecol. Monogr., 52: 395-414.

Schubert, G.H. (1966). The infective agent of carp pox. Bull. off. Inst. Epizoot., 65: 1011-1022.

Schwoerbel, J. (1987). Handbook of Limnology, pp. vi + 228, Ellis Horwood Ltd.- John Wiley and Sons (USA).

Scott, M.C. and Hall, L.W. (Jr.) (1997). Fish assemblages indicators of environmental degradation in Maryland Coastal plain streams. Transactions of Amerian Fisheries Society, 1_6: 340-360.

Sehgal, K.L. (1967). Observations on seed resources of Schizothoracid fishes and their commercial collection centres in Kashmir valley. CIFRI, CWF, Report No. 3: 8 pp.

Sehgal, K.L. (1970). Report on the factors responsible for large-scale mortality of brown trout, *Salmo trutta fario* Linnaeus in Chawalagam Trout Farm in Kashmir Valley, CIFRI, CWF, Report No. 4: 10 pp.

Sehgal, K.L. (1971). Report on the Survey of Coldwater Fishery Resources and their development in Kerala, CIFRI, CWF, Report No. 5: 12 pp.

Sehgal, K.L.; Shukla, J.P.; and, Shah, K.L. (1971). Observations on the Fisheries of Kangra Valley and adjacent areas with special reference to Mahseer and other fishes. J. Inl. Fish. Soc. India, 3: 63-71.

Sehgal, K.L. (1974). Report on the Researches carried out at the Coldwater Fisheries Research Centre in Himachal Pradesh and Kashmir from 1966-73. CIFRI Spl. Report: 13 pp.

Sehgal, K.L. (1975). Report on the Survey of Coldwater Fisheries in the Hills of Uttar Pradesh along with suggestions for improvement. CIFRI, CWF, Report No. 8: 17 pp.

Sehgal, K.L. (1987). Impact of construction and completion of BSL (Beas-Sutlej Link) Project on species and habitat of coldwater fish in River Beas. In: Symposium on the Impact of Current Land use Pattern and Water Resources Development on Riverine Fisheries, Abstracts, pp. 33 (Barrackpore).

Sehgal, K.L. (1988). State-of-Art on Introduction, Transplantation, Propagation and Present status of exotic coldwater fish in India. In: Workshop on Exotic Aquatic species in India, Mangalore, Abstract, pp. 121.

Sehgal,K.L. (1991). Artificial propagation of the golden Mahseer, *Tor putitora* (Hamilton) in the Himalayas. NRCCWF Spl. Publication No. 2: 12 pp.

Sehgal, K.L. (1994). State-of-Art of Endangered, Vulnerable and Rare Coldwater Fishes of India: pp. 127-135. In: Threatened Fishes of India, pp. 384, Natcon Publication No. 4(UP).

Selye, H. (1950). Stress and the General Adaptation Syndrome. Br. Med. J., i. 1383-1392.

Sen, N. (1982). Studies on the Systematics, Distribution and Ecology of the Ichthyofauna of Meghalaya and their bearing on the Fish and Fisheries of the State, vi + 576, PhD Thesis, University of Gauhati (Assam).

Sen, T.K. (1985). The Fish Fauna of Assam and the neighbouring North-Eastern States of India. Records of Zoological Survey of India, Occasional Paper No. 64: 1-216.

Sen, N. (2000). Occurrence, Distribution and Status of Diversified Fish Fauna of North-East India: pp. 31-48. In: Fish Diversity of North-East India (Eds.)Ponniah, A.G.and Sarkar,U.K., pp. 228, National Bureau of Fish Genetic Resources, ICAR (Lucknow).

Sen, T.K. and Jayaram, K.C. (1982). The Mahseer Fishes of India: A Review. Rec. Zool. Surv. India Misc. Publ. Occasional Paper No. 39: 38 pp.

Seymour, R.L. (1970). The genus *Saprolegnia*. Nova Hedwigia, 19: 1-124.

Shafi, M.; Quddus, M.M.; and, Hussain, M. (1978). Studies on Length-Weight relationship, sex ratio, size composition, gears and abundance of *Hilsa ilisha* (Hamilton Buchanan) in the River Padma. Dacca University Stud. B, 26: 123-127.

Shannon, C.E. and Weaver, W. (1949). A Mathematical Theory of Communications, pp. 117, University of Illinois Press, Urbana (USA).

Shanor, Q.L. and Saslow, H.B. (1944). *Aphanomyces* as a fish parasite. Mycologia, 36: 413-415.

Shariff, M. and Law, A.T. (1980). An incidence of fish mortality in Bekok river, Johore, Malaysia. In: Proc. International Symp. on Conservation and Input from Life Sciences, 27-30 Oct 1980, Universiti Kebangsa a.n., Bangi, Selangor (Malaysia).

Shariff, M. and Saidin, T.H. (1994). Status of Epizootic Ulcerative Syndrome in Malaysia since 1986: pp. 48. In: Proceedings of the ODA Regional Seminar on Epizootic Ulcerative Syndrome, 25-27 Jan 1994 (Eds.) Roberts, R.J.; Campbell, B.; and, MacRae, I.H.: Aquatic Animal Health Research Institute (Bangkok).

Sharifpour, I. (1997). Histology of the Inflammatory response of the carp (*Cyprinus carpio*) to various stimuli. PhD Thesis, University of Stirling (Scotland).

Shaw, G.E. and Shebbeare, E.O. (1937). The Fishes of Northern Bengal. J. Royal Asiatic Soc. Bengal Science: 137 pp, 6 pls.

Shelford, V.E. (1911). Ecological Succession: Stream Fishes and the method of Physiographic analysis. Biological Bulletin, 21: 9-35.

Shelford, V.E. (1929). Laboratory and Field Ecology, Baltimore: 608 pp.

Shrestha, G.B. (1990). Nepal Report: pp. 39-44. In: Regional Research Programme on Relationship between Epizootic Ulcerative Syndrome in Fish and the Environment, 13-26 Aug 1990, NACA (Bangkok).

Shrestha, G.B. (1994). Status of Epizootic Ulcerative Syndrome (EUS) and its effects on Aquaculture in Nepal: pp. 49-57. In: Proceedings of the ODA Regional Seminar on Epizootic Ulcerative Syndrome, 25-27 Jan 1994 (Eds.) Roberts, R.J.; Campbell, B.; and, MacRae, I.H., Aquatic Animal Health Research Institute (Bangkok).

Silas, E.G. (1952). Further studies regarding Hora's Satpura Hypothesis. (2)Taxonomic assessment and levels of Evolutionary divergence of Fishes with the so-called Malayan affinities in Peninsular India. Proc.nat. Inst., Sci. India, 18 (5): 423-448.

Sindermann, C.J. and Rosenfield, A. (1954). Diseases of Fishes of the Western North Atlantic. 2. Ichthyosporidium disease of the sea herring (*Clupea harrengus*). Maine Dept. Sea Shore Fish Res. Bull. No. 19: 1-40.

Singh, S.B. (1969). Ecological requirements for induced breeding of Fishes. FAO/UNDP Regional Seminar of Induced Breeding of Cultivable Fishes, India. FAO: FRI/IBCF/, 10: 20 PP.

Singh, S.B.; Sukumaran, K.K.; and, Chakrabarty, P.C. (1970). Observations on the hybrid between silver carp and grass carp. SABRAO Newsletter, 1 (2): 103-108.

Singh, A.K.; Mahanta, P.C.; and, Kapoor D. (1994). Some drastically declined fishes of Uttaranchal: pp 17-30. In: The Threatened Fishes of India, pp. 384, Natcon Publication No.4 (UP).

Sinha, M. (1994). Threatened coldwater fishes of North-Eastern Region of India: pp. 173-176. In: Threatened Fishes of India, pp. 384, Natcon. Publication No. 4 (UP).

Sinha, M.and De, D.K. (1996). Impact of engineering structures on Fish and Fisheries in Hooghly estuarine system. Proc. XI Annual International Conference of Nat.Env.Sci.Acad. on Sustainable Development and Environment, 15-17 March 1996.

Sinha, M.; De, D.K.; and Jha, B.C. (1998). The Ganga Environment and Fisheries, pp. 147, CICFRI, Publication (Barrackpore).

Sinha, M. and De, D.K. (1999). Migratory behaviour of hilsa, *Tenualosa ilisha* (Hamilton) based on recovery of tagged specimens. In: Nat. Sem. On Eco-friendly Management of Resources for Doubling Fish Production strategies for 21st Century. Inland Fish. Soc. India and Central Inland Fish. Res. Inst., Barrackpore, 22-23 Dec 1999.

Smith, G.M. (1950). The Freshwater Algae of the United States, vii + 719, McGraw Hill Book Co. Inc. (New York).

Smith, R.L. (1986). Ecology and Field Biology, pp. 835, Harper and Row Publishers Inc.(USA).

Snieszko, S.F. (1974). The effects of environmental stress on outbreaks of infectious diseases of fishes. J. Fish Biol., 6: 197-208.

Soe, U.M. (1990). Myanmar Report: pp. 35-38. In: Regional Research Programme on Relationships between Epizootic Ulcerative Syndrome in Fish and the Environment, 2nd Technical Workshop, 18-26 Aug 1990 (Eds.) Phillips, M.J. and Keddie, H.G., Network of Aquaculture Centres in Asia-Pacific (Bangkok).

Soni, V. (1994). Endangered sport fishes of India: pp.205-207. In: Threatened Fishes of India, pp. 384, Natcon Publication No. 4: 10 pp.

Southwell, T and Prashad, B.(1918). On hilsa investigations in Bengal, Bihar and Orissa. Bull Dep. Fish. Bengal, Bihar and Orissa, 15.

Sprague, V.and Vermick, S.H. (1974). Fine structure of the cyst of some sporulation stages of *Ichthyosporiidum* (Microsporidia). J. Protozoology, 21: 667-677.

Sreenivasan, A. (1968). Limnology of Tropical Impoundments. V. Studies on two upland impoundments in the Nilgiris, Madras State (India). Phykos, 7: 144-160.

Steemann, Nielsen, E. (1952). The use of radio-active carbon (C_{14}) for measuring organic production in the sea. J. Cons. Int. explor. Mer., 18: 117-140.

Subasinghe, R.P. (1993). Effects of Controlled infections of *Trichodina* sp on transmission of Epizootic Ulcerative Syndrome (EUS) to naïve snakehead, *Ophicephalus striatus* Bloch. Journal of Fish Diseases, 16: 161-174.

Swarup, K. (1961). The fecundity of Indian shad, *Hilsa ilisha* (Hamilton). J. Zool. Soc. India, 13:108-112.

Swingle, H.S. (1950). Relationship and dynamics of balanced and unbalanced fish populations. Bull. Alabama Agr. Exp. Sta., 264: 73 pp.

Sharma, P.C. and Pant, M.C. (1984). Evidence for increased eutrophication of Lake Nainital as a result of human interference, Envi. Poll., 1: 149-161.

Tachushong, A. and Saitanu, K. (1983). The inactivation of *Aeromonas hydrophila* by Potassium permanganate: pp. 394-402. In: Proc. Symp. on Freshwater Fishes Epidemic, 1982-83, 23-24 June, 1983, Chulalongkurn University (Bangkok).

Talwar, P.K. (1978). Identity of the Schizothoracid fish genus *Schizothorax* Heckel 1838, with considerations of the status of *Schizothoraichthys* Misra 1962. Rec. Zool. Surv. India, Occasional Paper No. 31: 4 pp.

Talwar, P.K. and Jhingran, A.G. (1991). Inland Fishes of India and Adjacent Countries, Vol. I and II, pp. 1158, Oxford and IBH Co., Pvt. Ltd. (New Delhi).

Tangtrongpiros, J.; Wattanavijarn, W.; Thirapatsakun, T.; and, Wattanodorn, S. (1983). Fish kidney cell culture and the cytopathic effects in the culture after infected with cell-free filtrate prepared from sick snakehead (*Ophicephalus striatus*). Thai Vet. Med., 13 (2): 131-139.

Tangtrongpiros, J.; Charoennetisart, P.; Wongsatayanont, B., Tavatsin, A.; and, Chaisiri, N. (1985). Haematological change in snakehead fish during 1984 outbreak. Proceedings of the Technical Conference on Living Aquatic Resources, 7-8 March 1985, Chulalongkorn University (Bangkok).

Taylor, C.M. (2000). A large-scale comparative analysis of riffle and pool fish communities in an upland stream system. Environmental Biology of Fishes, 58: 89-95.

Temeyavanich, S. (1983). Acute toxicity of iodine, chlorine and malachite green on three freshwater fishes and its treatment on snakehead fish pond: pp. 415-427. In: Proc. Symp on Freshwater Fishes Epidemic, 1982-83, 23-24 June, 1983, Chulalongkorn University (Bangkok).

Tesprateed, T.; Rattanaphani, R.; Wongsawang, S.; Wattanawijarn, W.; Ousawaplangehai, L.; and, Tangtrongpiros, J; and, Saitanu, K. (1983). Pathology of the diseased fish with reference to the recent epidemic in Thailand in 1982: pp.255-276. In: Proc. Symp. on Freshwater Fishes Epidemic, 1982-83, 23-24 June, 1983, Chulalongkorn University (Bangkok).

Thienemann, A. (1925). *Die Binnergewä sser Mittelauropas Die Binnengewä sser, 1, Verlag Schweizerbart* (Stutgart).

Thomas, H.S. (1873). Tank Angling, pp. 435, Thacker Spink (London).

Tiffney, W.N. (1939). The host range of *Saprolegnia parasitica*. Mycologia, 31: 310-321.

Toham, A.K. and Tuegels, G.G. (1997). Patterns of Microhabitat use among fourteen abundant fish of the Lower Ntem River basin (Cameroon). Aquat. Living Resour., 10: 289-298.

Toham, A.K. and Tuegels, G.G. (1998). Diversity Patterns of Fish assemblages in the Lower Ntem River basin (Cameroon) with notes on potential effects of deforestation. Arch.Hydrobiol.,141 (4): 421-446.

Tonguthai, K. (1985). A Preliminary account of Ulcerative Fish Diseases in the Indo-Pacific Region (a Comprehensive Study based on Thai Experiences). National Inland Fisheries Institute (Bangkok): 39 pp.

Tripathi, Y.R. (1952). Studies on parasites of Indian Fishes: Protozoa: Myxosporidia together with a checklist of parasitic Protozoa described from Indian Fishes. Rec. Indian Mus., 50 (1): 63-88.

Tripathi, Y.R. (1954). Studies on parasites of Indian Fishes, 3: Protozoa, 2 (Mastigophora and Ciliophora). Rec. Indian Mus., 52 (2/4): 221-230.

Tripathi, Y.R. (1955). Experimental infection of Indian major carps with *Ichthyophthirius multifilis* Fouquet. Curr. Sci., 24 (7): 236-237.

Tripathi, Y.R. (195). Studies on paasites of Indian Fishes, 5. Acanthocephala. Rec. Indian Mus., 54 (1/2): 61-92.

Tripathy, S.D. and Chaturvedi, G.K. (1974). A new record of an isopod, *Ichthyoxenus jellinighansii* Hocklots, from Pariat Lake, Jabalpur, India. J. Inland Fish. Soc., 6: 85-87.

Ulcerative Fish Disease Committee (1983). Practical Report of the Ulcerative Fish Disease Committee, 1982-83 (Bangkok).

UNEP (1992). The Convention of Biological Diversity (Nairobi).

Vaas, K.F.and Schurman, J.J.(1949). On the Ecology and Fisheries of some Javanese Freshwaters. Common Gen.Agric.Res. Sta. Buitenzorg Java, 97: 1-60.

Vannote, R.L.; Minshall, G.W.; Cummins, K.W.; Sedell, R.; and, Cushing, C.E. (1980). The River Continuum Concept. Canadian Journal of Fisheries and Aquatic Sciences, 37: 130-137.

Vestaergaard-Jfrgensen, P.E. (1974). A study of Viral Diseases in Danish Rainbow Trout: Their Diagnosis and Control, Copenhagen: Carl Fr. Mortensen.

Virgona, J.L. (1992). Environmental factors and the prevalence of a cutaneous ulcerative disease (Red Spot) in the sea mullet (*Mugil cephalus* L) in the Clarence River, New South Wales, Australia. Journal of Fish Diseases, 15: 363-378.

Viswanath, T.S.; Mohan, C.V.; and, Shankar, K.M. (1997a). Mycotic Granulomatosis and seasonality are the consistent features of EUS of fresh and brackish water fishes of Karnataka, India. Asian Fisheries Science, 10: 155-160.

Viswanath, T.S.; Mohan, C.V.; and, Shankar, K.M. (1997 b). Chemical and Histopathological characterization of different types of lesions associated with Epizootic Ulcerative Syndrome (EUS). Journal of Aquaculture in the Tropics, 12 (1): 35-42.

Viswanath, T.S.; Mohan, C.V.; and, Shankar, K.M. (1998). Epizootic Ulcerative Syndrome (EUS) associated with a fungal pathogen in Indian Fishes: Histopathology: 'a cause for invasiveness'. Aquaculture, 165: 1-9.

Vogel, A.I. (1978). A Text Book of Quantitative Inorganic Analysis, pp.925, Longman (London).

Wada, S.; Yuasa, K.; Rha, S.; Nakamura, K; and, Hatai, K. (1994). Histopathology of *Aphanomyces* infection in dwarf gourami (*Colisa lalia*). Fish Pathology, 31 (2): 71-80.

Walker, R. (1962). Fine structure of Lymphocystis Virus in Fish. Virology, 18: 503-508.

Walkley, A.J. and Black, I.A. (1934). Estimation of soil organic carbon by chromic acid titration method. Soil Sci., 37: 29-38.

Ward, B.H. and Whipple, C.G. (1918). Freshwater Biology, xx + 1248, John Wiley and Sons Inc. (New York).

Wattanavijarn, W.; Tesprateep; Wattanodorn, S.; Sukolapong, V.; Vetchagarun, S.; and, Eongpakorn-Keaw, A. (1983). Virus-like particles in the liver of snakehead fish. Thai J. Vet. Med., 13 (1): 51-58.

Wattanavijarn, W.; Ruttanaphani, R.; Tesprateep, T.; Tangtrongpiroj, J.; and, Thirapatsakun, T. (1985). Detection of Virus in the snakehead fish during a disease epizootic by Light and Electron microscopy. Nat. Hist. Bull. Siam., Soc., 33: 47-52.

WCMC (1998). Freshwater Biodiversity: A Preliminary Gobal Assessment. A Document prepared for the 4th Meeting of the Conference of the Practices to the Convention of Biological Diversity, World Conservation Monitoring Centre.

Welcomme, R.L. (1979). Fisheries Ecology of the Floodplain rivers, viii + 317, Longmans (London).

Weissenberg, R.(1914). *Uber infectiose zellhypertrophic bei Fischen (Lymphocystiserkrankung). Sber. Preuss. Akad. Wiss.*, 30: 792-804.

Wheeler, A. (1979). Fish in an urban environment. In: Ian C Laurie (Ed.) Nature in Cities. John Wiley (New York).

Whipple, G.C. (1927). The Microscopy of Drinking Water, 4th Ed., rev. by Fair and Whipple, pp. 586 (New York).

Widago, D. (1990). Indonesia Report: pp. 18-22. In: Regional Research Programme on Relationships between Epizootic Ulcerative Syndrome in Fish and the Environment, 13-26 Aug 1990, NACA (Bangkok).

Wikramanayake, E.D. and Moylw, P.B. (1989). Ecological structure of Tropical Fish Assemblages in wet-zone streams of Sri Lanka. Journal of Zoology (London), 281: 503-526.

Williams, I.V. and Amend, D.F. (1976). A natural epizootic of Infectious Haematopoietic Necrosis in fry of sockeye salmon (*Oncorhynchus nerka*) at Chilko Lake, British Columbia. J.Fish. Res. Bd. Can., 33: 1564-1567.

Wilson, H.C. (1909). *Fischdale farm, Nilgiris.* Government Press, Madras: 10 pp.

Wilson, K.D. and LoK.S. (1992). Fish Disease in Hong Kong and the potential role of the Veterinarians. 23rd Annual Conference of the International Association for Aquatic Animal Medicine (IAAAM), 18-22 May 1992. (Hong Kong).

Willoughby, L.G. (1968). Atlantic salmon disease fungus. Nature, London, 217: 872-873.

Willoughby, L.G. (1970). Mycological aspects of a disease of young perch in Windermere. J. Fish Biol., 2: 113-116.

Willoughby, L.G.; Roberts, R.J.; and, Chinabut, S. (1995). *Aphanomyces invaderis* sp.nov., the fungal pathogen of freshwater tropical fishes affected by Epizootic Ulcerative Syndrome (EUS). Journal of Fish Diseases, 18: 273-275.

Wood, J.W. (1968). Diseases of Pacific Salmon: Their prevention and treatment, Olympia,Washington: Department of Fisheries, Hatchery Division, State of Washington (USA).

Wolf, K.; Quimby, M.C.; Pyle, E.A.; and, Dexter, R.P. (1960). Preparations of monolayer cell cultures from tissues of some lower vertebrates. Science, N.Y., 132: 1890-1891.

Wolf, K.; and Darlington, R.W. (1971). Channel cat fish virus: a new herpes virus of ictalurid fish. J. Virology, 8: 525-533.

Wolf, K.; Herman, R.L.; and, Carlson, C.P. (1972). Fish Viruses: Histopathologic changes associated with experimental channel cat fish virus disease. J. Fish Res. Bd. Can., 29: 149-150.

Xuan, T.T. (1990). Vietnam Report. In: Regional Research Programme on Relationships between Epizootic Ulcerative Syndrome in Fish and the Environment (Eds.) Phillips, M.J.; and Keddie, H.G., A Report on the 2nd Technical Workshop, 13-26 Aug 1990, Network of Aquaculture Centres in Asia-Pacific (Bangkok).

Yadava, Y.S.; Goswami, M.M.; Kar, D.; and Choudhury, M. (1989). Present status of Hilsa resources in Assam: pp.252-256. In: Fish Genetics in India (Eds.) Das, P. and Jhingran, A.G., Today and Tomorrow's Printers and Publishers (New Delhi).

Yadava, Y.S. and Chandra, R. (1994). Some threatened carps and cat fishes of Brahmaputra River System: pp. 45-55. In: Threatened Fishes of India, pp. 384, Natcon. Publication No. 4 (UP).

Yashouv, A. (1959). Studies on the Productivity of fish ponds. I. Carrying Capacity. Proc. Tech. Pap. Gen. Fish. Counc. Mediterr., 5:409-419.

Zoological Society of Assam (1988). Proceedings of the Symposium on Recent Outbreak of Fish Diseases in North-Eastern India, 30 Dec 1988, organized by Zoological Society of Assam, Guwahati, pp.23.

Zutshi, D.P.; Wanganeo, A., and, Raina, R. (1970). Limnology of a man-made lake. Geobios: 30-32.

Index

C

E